# The Framework of Plasma Physics

# The Framework
# of Plasma Physics

Richard D. Hazeltine and
François L. Waelbroeck

*Advanced Book Program*

CRC Press
Taylor & Francis Group
Boca Raton  London  New York

CRC Press is an imprint of the
Taylor & Francis Group, an **informa** business

First published 2004 by Westview Press

Published 2018 by CRC Press
Taylor & Francis Group
6000 Broken Sound Parkway NW, Suite 300
Boca Raton, FL 33487-2742

*CRC Press is an imprint of the Taylor & Francis Group, an informa business*

Visit the Taylor & Francis Web site at
http://www.taylorandfrancis.com

and the CRC Press Web site at
http://www.crcpress.com

A Cataloging-in-Publication data record for this book is available from the
Library of Congress.

ISBN 13: 978-0-8133-4213-9 (pbk)

# Frontiers in Physics
## David Pines, Editor

Volumes of the Series published from 1961 to 1973 are not officially numbered. The parenthetical numbers shown are designed to aid librarians and bibliographers to check the completeness of their holdings.

Titles published in this series prior to 1987 appear under either the W. A. Benjamin or the Benjamin/Cummings imprint; titles published since 1986 appear under the Westview Press imprint.

Volumes published from 1974 onward are being numbered as an integral part of the bibliography.

# Editor's Foreword

The problem of communicating in a coherent fashion recent developments in the most exciting and active fields of physics continues to be with us. The enormous growth in the number of physicists has tended to make the familiar channels of communication considerably less effective. It has become increasingly difficult for experts in a given field to keep up with the current literature; the novice can only be confused. What is needed is both a consistent account of a field and the presentation of a definite "point of view" concerning it. Formal monographs cannot meet such a need in a rapidly developing field, while the review article seems to have fallen into disfavor. Indeed, it would seem that the people who are most actively engaged in developing a given field are the people least likely to write at length about it.

*Frontiers in Physics* was conceived in 1961 in an effort to improve the situation in several ways. Leading physicists frequently give a series of lectures, a graduate seminar, or a graduate course in their special fields of interest. Such lectures serve to summarize the present status of a rapidly developing field and may well constitute the only coherent account available at the time. One of the principal purposes of the *Frontiers in Physics* series is to make notes on such lectures available to the wider physics community.

As *Frontiers in Physics* has evolved, a second category of book, the informal text/monograph, an intermediate step between lecture notes and formal text or monographs, has played an increasingly important role in the series. In an informal text or monograph an author has reworked his or her lecture notes to the point at which the manuscript represents a coherent summation of a newly developed field, complete with references and problems, suitable for either classroom teaching or individual study.

*The Framework of Plasma Physics* is just such a book. The authors provide an excellent introduction to the fundamental physics of plasmas, which comprise most of the matter in the universe. Their emphasis throughout on a pedagogical presentation makes their book highly suitable either as a text for a graduate or advanced graduate course in plasma physics, or for individual study. Because they have been so successful in their goal of integrating introductory and advanced material, their book should find a wide audience among members of the plasma physics research community and mature researchers in other subfields of science. It gives me great pleasure to welcome Richard Hazeltine (for the second time) and Frank Waelbroek (for the first time) to *Frontiers in Physics*.

David Pines
Urbana, Illinois
July 1998

# Contents

# Preface

Plasma physics is a necessary part of our understanding of stellar and galactic structure; it determines the magnetospheric environment of the earth and other planets; it forms the research frontier in such areas as fusion, advanced accelerators and short pulse lasers; and its applications to various industrial processes (such as computer chip manufacture) are rapidly increasing. It is a subject, in other words, with a large and growing list of scientific and technological applications.

The usefulness of plasma physics has of course enriched the subject, supporting plasma research programs and inspiring many advances in understanding the plasma state. But its very wealth of applicability has sometimes obscured the structure and intrinsic content of the field as a physics discipline. To put the matter a little too strongly, what sometimes emerges from plasma introductory literature is the impression of a collection of recipes. One may well find what is needed for the problem at hand, but a sense of the underlying intellectual framework is not always apparent. (Essentially the same issue is addressed, from a slightly different point of view, in a National Research Council Report[1].)

This book attempts to develop an understanding of plasma behavior, starting from identified physical laws and assumptions, as systematically as possible. Plasma physics is displayed as a discipline that responds to certain scientific challenges in a coherent and comprehensible way. It is a discipline with a lengthening list of scientific accomplishments, a perhaps larger list of remaining mysteries, and a characteristic point of view.

We think this perspective helps to build an appreciation, not only of a variety of methods, but of the physical circumstances in which each is pertinent: why one plasma situation is best approached with Vlasov theory, for example, while another can be understood using MHD. In particular we try to make as clear as possible the relation between the various conceptual tools and their common origin in laws of nature.

The book was written with the graduate student in mind, and is based on lectures that were used for an introductory plasma course at the graduate level. But most of the material would also fit into an upper-level undergraduate course. We have resisted the temptation to distinguish more advanced material with asterisks; however, such material is restricted to the later parts of each chapter, so that a course covering only the first part of most chapters would preserve coherence. The core of the book is its first 6 chapters.

The book and its authors have benefitted enormously from the advice and help of colleagues at the University of Texas and throughout the international plasma research community. We are particularly grateful for the help of Richard Fitzpatrick, Stewart Prager, Richard Lovelace, and Brad Shadwick.

We typeset this book in LaTeX, using Alpha™, Oztex™, teTeX™ and TeXShop™. We made the illustrations using Adobe Illustrator™, with help from Mathematica™.

# Chapter 1

# The Nature of Plasma

## 1.1 Plasma and plasma physics

### Interacting, unbound charge

What are the visible effects of electromagnetic forces? What is routinely apparent is that the electromagnetic force creates *structure*: stable atoms and molecules, crystalline solids, charge-neutral fluids. The most studied consequences of the electromagnetic force are chemistry and solid-state physics, two disciplines developed to understand stable, essentially static structure.

Structured systems in this sense have binding energies larger than the ambient thermal energy. Placed in a sufficiently hot environment, they decompose: crystals shatter or melt, molecules disassociate. At temperatures near or exceeding atomic ionization energies, the atoms similarly decompose into negatively charged electrons and positively charged ions. These charged particles are by no means "free;" they are acutely affected by each others' electromagnetic fields, as well as by external fields that may be present. But because the charges are no longer bound, their assemblage becomes capable of motions of extraordinary vigor and complexity: it becomes a plasma. The relation between a bound-charge system and a plasma is that between a cloud and a lightning bolt.

With the mention of complexity it must be noted that bound systems can display extreme complexity of structure, as in, for example, a protein molecule. Complexity in a plasma is rather different, being expressed temporally as much as spatially. It is characterized in the first place by the excitation of a enormous variety of dynamical modes. Indeed, the presence of many degrees of freedom dominates plasma phenomenology and largely sets the intellectual flavor of plasma physical research.

However, what is most characteristic of a plasma is the manner in which its multi-dimensionality is expressed. Perhaps above all else, a plasma is a system whose copious degrees of freedom are manifested through *collective*, coherent processes, involving macroscopic currents and charge densities. Thus one can find in plasma an enormous variety of waves, turbulent eddies, vortex structures and so on, with correlation lengths and times far exceeding those associated with individual particle interactions. In his science fiction novel, *The Black Cloud*[37], Fred Hoyle suggests that the wealth of collective oscillations in a plasma could store sufficient information to express super-human intelligence. Anyone who has studied plasma physics finds the suggestion entirely plausible.

Because thermal decomposition breaks interatomic bonds before ionizing, most plasmas begin as gases. Thus plasma is sometimes defined as a gas that is sufficiently ionized to show the effects of space-charge. Note that the mass of such a gas may be predominantly in neutral particles, which do not inhibit plasma behavior. That is, a fractionally ionized gas exhibits most of the exotic phenomena seen in the totally ionized state.

Plasmas resulting from ionization will usually contain equal numbers of positive and negative charge carriers: an electron for each positive charge on an ion. In this common situation the oppositely charged fluids are strongly coupled, even to the point of nearly neutralizing each other over volumes of macroscopic size. The point is that even a very small accumulations of local charge yield huge electric fields, difficult to sustain when charges of both signs are present. One speaks in this case of "quasi-neutral" plasma. Non-neutral plasmas, on the other hand, may contain only charges of one sign. Occurring primarily in laboratory experiments, their near-equilibrium depends upon strong magnetic fields, about which the charged fluid rotates.

In summary a plasma is a system of many interacting charges, under conditions that allow for collective behavior. We investigate the nature of those conditions in Section 1.2.

## Occurrence

It is sometimes said that 95% (or 99% or whatever) of the universe consists of plasma. This statement would be awkward to verify—for one thing, whether some system is a plasma depends upon the scale on which it is viewed. Yet it is important to appreciate the prevalence of the plasma state. Plasma is the stuff of stars as well as interstellar space; it is the cosmic medium. Plasma also provides the earth's local environment, in the form of the solar wind and the magnetosphere; as the medium and determinant of "space weather," the magnetosphere plasma has a important effects on, for example, communication satellites. Plasma is in some sense the natural, untamed state of matter: only in such exceptional environments as the surface of a cool planet can other forms of matter dominate.

Moreover terrestrial plasmas are not hard to find. They occur in, among other places, lightning, fluorescent lights, a variety of laboratory experiments and a growing array industrial processes. Thus the glow discharge has become a mainstay of the electronic chip industry. The campaign for fusion power has produced a large number of devices that create, heat and confine plasma—while bringing enormous gains in plasma understanding.

Liquid and even solid-state systems can display the collective electromagnetic effects that characterize plasma. Elemental mercury, as an electrically

|                     | $n\,(\mathrm{m}^{-3})$ | $T$ $(^\circ\mathrm{K})$ | $\omega_p\,(\mathrm{sec}^{-1})$ | $\lambda_D$ (m) | $\Lambda$ |
|---------------------|-----------|-----------|-----------|-----------|-----------|
| glow discharge      | $10^{19}$ | $3\cdot 10^3$ | $2\cdot 10^{11}$ | $10^{-6}$ | $3\cdot 10^2$ |
| chromosphere        | $10^{18}$ | $6\cdot 10^3$ | $6\cdot 10^{10}$ | $5\cdot 10^{-6}$ | $2\cdot 10^3$ |
| interstellar medium | $2\cdot 10^4$ | $10^4$ | $10^4$ | $50$ | $4\cdot 10^4$ |
| magnetic fusion     | $10^{20}$ | $10^8$ | $6\cdot 10^{11}$ | $7\cdot 10^{-5}$ | $5\cdot 10^8$ |

Table 1.1: Some examples of plasmas with the key parameters that characterize them. The phrase "magnetic fusion" is jargon for 'magnetically confined plasma for fusion experiments.' Glow discharges are used, among other things, to process computer chips. The listed parameters are defined and discussed later in this Chapter.

conducting fluid, is an example that is sometimes used to demonstrate and explore the behavior of gaseous plasma.

Table 1.1 lists some representative plasmas and the physical parameters that characterize them.

## Plasma physics

Like the systems it studies, the discipline of plasma physics is extraordinarily diverse. Nonetheless a few general comments are possible.

First, despite its concern with the statistics of many-body systems, plasma physics is not a subfield of equilibrium statistical mechanics. The point is that the statistical equilibria of plasma systems are rarely pertinent; the interesting features of a plasma usually concern its departure from thermal equilibrium, whether that departure is minute (as in linear plasma transport theory) or substantial (as in plasma turbulence theory).

As a result plasma physics has developed a distinctive collection of ideas and methods, similar to the development of condensed matter physics, or hydrodynamics. Theoretical understanding of plasma behavior now rests on a combination of plasma-specific *kinetic* theory—tracking the evolution of the charged-particle distribution functions in accordance Liouville's theorem and Maxwell's equations—and similarly modified *fluid* models, generalizing hydrodynamics to include the electromagnetic force. Typical advances in plasma physical understanding use both the kinetic and fluid points of view. And invariably such advances enrich the disciplines from which they draw insight: progress in general kinetic theory, for example, has been importantly stimulated by plasma research.

Quantum wavelengths are insignificantly small in common plasma environments, which are therefore usually treated by classical physics. Above all, Maxwell's equations play a pivotal role. Indeed, in an oversimplified but instructive view, plasma physics has the sole aim of closing Maxwell's equations, by providing "constitutive relations" for the sources—the current and charge densities—as functionals of the fields.

Plasma phenomena are frequently, sometimes spectacularly, nonlinear, so that the discipline has been a test-bed and incubator for numerous nonlinear concepts and methods. Quasilinear theory, chaos and soliton (or solitary wave) theory are examples of research areas whose theoretical and experimental development owes much to plasma physics.

A theme in all fundamental plasma investigations—fluid and kinetic, linear and nonlinear—is the search for *closure*: the need to find a physically realistic description that is complete, in the sense of providing as many equations as unknowns. The only obvious closures—such as combining Maxwell's equations with a Boltzmann equation for each plasma species—are in realistic situations intractable, even numerically. Simplification is essential to all plasma investigations outside the textbooks, and to most textbooks arguments as well. The challenge is to simplify in a manner that is physically convincing, in the best cases even rigorously justifiable, and that preserves the essential physics.

Plasma physics seeks coherence in complexity. It attempts to find a convincing and tractable description of an extraordinarily complex physical system. This task remains challenging despite the creative efforts, and very considerable progress, of a generation of scientists.

## 1.2   Time and distance scales

The nature of any physical system depends upon the scale on which it is observed; a flower is hardly floral when viewed on the subnuclear scale. Here we identify the most important scales for plasma physics. Three fundamental quantities emerge from the discussion: the plasma frequency, the Debye length, and the plasma parameter.

### Basic parameters

Consider an idealized plasma consisting of $N$ electrons, with mass $m_e$ and charge $-e$ (in our notation $e > 0$ is the magnitude of the electronic charge), and an equal number of ions, with mass $m_i$ and charge $e$, confined in cubic box of side $L$. We do not expect the system to have reached thermal equilibrium, but nonetheless use the symbol

$$T_s = \frac{1}{3} m_s \langle v^2 \rangle$$

to denote a kinetic temperature or average energy, measured in energy units. Here $v$ is a particle speed, and the angular brackets denote an ensemble average. Thus $T$ is the product of the Boltzmann constant and the kinetic temperature in degrees Kelvin; since we use MKS units, $T$ is measured in Joules. (We should mention that plasma physicists almost always measure energy in units of electron Volts (eV); oneeV is equal to $1.6 \times 10^{-19}$ eV.) Quasineutrality is expressed in this simple case by

$$n_i \cong n_e$$

where $n_s$ is the *density*—the number of particles per cubic meter—of species $s$. It is generally a function of position that becomes $N/L^3$ for both species in the constant density case.

Assuming that both ions and electrons are characterized by the same $T$, we can estimate typical particle speeds in the system by the so-called thermal speeds,

$$v_{ts} \equiv \sqrt{2T/m_s}.$$

An occasional nuisance of quasi-neutral plasma physics is the occurrence of (at least) two thermal speeds, differing by a substantial factor:

$$v_{ti} \sim \sqrt{m_e/m_i} \, v_{te} \ll v_{te}.$$

The difference propagates into numerous formulae, so that one must continually ask "Ions or electrons?" Our present, convenient assumption of equal temperatures, $T_i = T_e$, helps little.

The plasma fluid is characterized by the strength of the Coulomb interaction, and by two intensive variables, $T$ and the particle density (which is species-independent by our quasineutrality assumption). The Coulomb force is conveniently measured by $e^2/(\epsilon_0 m)$ where $\epsilon_0$ is the vacuum permitivity and $m$ the particle mass. Thus a useful set of basic parameters is

$$\frac{e^2}{\epsilon_0 m}, \ n, \ v_t, \tag{1.1}$$

The dimensions of $n$ and $v_t$ are obvious, while $e^2/\epsilon_0 m$ has the dimensions of length times the square of a speed.

Note that our plasma can be viewed as a system, rather than just a fluid or substance; as the former it acquires two additional parameters, the scale-size $L$ and an observation time, which we call $\tau$. The observation time corresponds generally to the time scale of the process under investigation.

## Plasma frequency

Only one frequency involving the electron charge can be constructed from powers of the fluid parameters (1.1): the *plasma frequency*,

$$\omega_p^2 = \frac{ne^2}{\epsilon_0 m}, \tag{1.2}$$

which sets the most fundamental time-scale of plasma physics. There is evidently a plasma frequency for each species. The relatively fast electron plasma frequency is especially important, and references to 'the plasma frequency' refer implicitly to the electron version.

It is easily seen that $\omega_p$ corresponds to electrostatic oscillation in response to small charge separations. Thus consider a one-dimensional quasineutral plasma as consisting of two oppositely charged, coincident slabs. If one slab is displaced from its quasineutral position by an amount $\delta x$, then charge layers appear, with surface charge density $en\delta x$. The resulting electric field $E = en\delta x/\epsilon_0$ leads to the acceleration

$$m\frac{dV}{dt} = m\omega^2 \delta x = eE$$

whence $\omega^2 = \omega_p^2$.

Of course plasma oscillation will be observed only if the plasma system is studied over time periods $\tau$ longer than than the plasma period $\tau_p \equiv 1/\omega_p$, and if external actions change the system no faster than $\omega_p$. In the opposite case one is studying something—nuclear reactions, perhaps—other than plasma physics and the system is not usefully considered to be a plasma. Similarly, investigation over length scales shorter than the distance $v_t \tau_p$ traveled by a typical particle during a plasma period will not detect plasma behavior. (Particles will exit the system before completing a plasma oscillation!) This distance, the spatial equivalent of $\tau_p$, is called the *Debye length* and denoted by

$$\lambda_D \equiv \sqrt{T/m}\, \omega_p^{-1}. \tag{1.3}$$

Note that

$$\lambda_D = \sqrt{\frac{\epsilon_0 T}{ne^2}}$$

is independent of mass, and therefore usually comparable for different species. Again, it is the electron Debye length that most often matters.

Thus our idealized system is conventionally considered to be a plasma only if it is sufficiently large and persistent in the dimensionless sense:

$$\frac{\lambda_D}{L} \ll 1, \tag{1.4}$$

and

$$\omega_p \tau \gg 1.$$

It is helpful to introduce another frequency characterizing the system: the *transit* frequency,

$$\omega_t \equiv v_t/L. \tag{1.5}$$

The transit frequency measures the rate at which particles traverse the system; it will occur frequently in later chapters. The observation time $\tau$ is often associated with the inverse transit frequency, or *transit time*, $\tau \sim \omega_t^{-1}$. Thus the small Debye-length ordering (1.4) can equivalently be expressed as

$$\frac{\omega_t}{\omega_p} \ll 1.$$

It should be noted that, despite the conventional requirement (1.4), plasma physics does consider structures on the Debye scale; the most important example is the *Langmuir sheath* found near the perimeter of a confined plasma. The Langmuir sheath is analyzed in Chapter 3. A more fundamental observation is that any local charge perturbation of a plasma near equilibrium tends to be nullified, or shielded, beyond a distance measured by $\lambda_D$. We next verify this important circumstance.

## Debye shielding

Shielding of an external electric field can be viewed as a result of plasma conductivity: plasma current flows freely enough to remove static electric fields from the plasma interior. But it is more useful, at least in the equilibrium or near equilibrium case, to consider shielding as a dielectric phenomenon: polarization of the plasma medium—the redistribution of space charge—prevents penetration by an external electrostatic field. It is not surprising that the length scale associated with such shielding is the Debye length.

For concreteness we consider the simplest example: a quasineutral plasma so close to thermal equilibrium that its particle densities are distributed according to the Maxwell-Boltzmann law,

$$n_s = n_0 e^{-e_s \Phi/T},$$

where $\Phi(\mathbf{x})$ is the electrostatic potential, and $n_0$ and $T$ are constant. From $e_i = -e_e = e$ it is clear that quasineutrality requires the equilibrium $\Phi$ to be a constant; if this equilibrium is externally perturbed by a small, localized charge $\delta\rho_{\text{ext}}$ then $\Phi$ is perturbed from its constant value by $\delta\Phi$ and the total charge density becomes

$$\begin{aligned}
\rho_c &= \delta\rho_{\text{ext}} + e(\delta n_i - \delta n_e) \\
&= \delta\rho_{\text{ext}} - 2e^2 n_0 \delta\Phi/T.
\end{aligned} \tag{1.6}$$

Now Poisson's equation,

$$\nabla^2 \delta\Phi = -\frac{1}{\epsilon_0}\left(\delta\rho_{\text{ext}} - 2e^2 n_0 \delta\Phi/T\right),$$

reduces to

$$\left(\nabla^2 - \frac{2}{\lambda_D^2}\right)\delta\Phi = -\frac{\delta\rho_{\text{ext}}}{\epsilon_0}, \tag{1.7}$$

Considering, for example, a source proportional to a Dirac $\delta$-function,

$$\delta\rho_{\text{ext}} = \epsilon_0 Q_0 \delta^{(3)}(\mathbf{x})$$

we find that

$$\delta\Phi = \frac{Q_0}{2\pi}\frac{e^{-\sqrt{2}r/\lambda_D}}{r}$$

showing decay of the potential over a "shielding cloud" of dimension $\lambda_D$.

Notice that

1. The shielding potential is not visible on length scales short compared to the Debye length—that is, when the first term on the left-hand side of (1.7) far dominates the second. On such fine scales, particles would move independently of space-charge effects, experiencing only individual Coulomb interactions with neighboring particles and "external" electromagnetic fields. Thus the collective nature of plasma behavior is observed only in systems whose characteristic dimensions exceed $\lambda_D$.

2. When viewed on macroscopic scales $L \gg \lambda_D$, the electron and ion densities are closely equal: the plasma is "quasineutral." Indeed, Poisson's equation provides the estimate

$$\frac{n_e - n_i}{\bar{n}} \sim \lambda_D^2 \nabla^2 \frac{e\Phi}{T} \sim \frac{\lambda_D^2}{L^2}$$

where $\bar{n}$ is a representative density.

3. The shielding argument, by treating $n$ as a continuous function on the $\lambda_D$-scale, implicitly assumes there to be many particles in the shielding cloud. Actually shielding remains statistically significant and physical even in the opposite case, when the cloud is barely populated: in that case shielding implies modified probabilities for observing charged particles within a Debye length of the external charge.

The question of whether the Debye shield is densely populated brings us to the last of the three most basic quantities characterizing a plasma, the *plasma parameter*.

## Plasma parameter

Returning to our basic parameter set (1.1), we observe that two independent quantities with the dimensions of length can be constructed: the average distance between particles,

$$r_d \equiv n^{-1/3}$$

and the distance of closest approach,

$$r_c \equiv \frac{e^2}{4\pi\epsilon_0 T}.$$

We remind the reader that $r_c$ is the distance at which the total energy

$$U(r, v) = \frac{1}{2}mv^2 - \frac{e^2}{4\pi\epsilon_0 r}$$

of one charged particle in the electrostatic field of another, vanishes:

$$U(r_c, v_t) = 0.$$

Thus, in particular, typical particle trajectories are strongly influenced by electrostatic interaction at separations comparable to $r_c$. Since our basic set allows only two independent lengths, the Debye length must be related to $r_d$ and $r_c$; indeed we find that

$$4\pi\lambda_D^2 = \frac{r_d^3}{r_c}$$

or, perhaps more transparently,

$$\frac{1}{r_c} = 4\pi n \lambda_D^2. \tag{1.8}$$

The significance of the ratio $r_d/r_c$ is easily understood. When it is small, charged particles are dominated by each other's electrostatic influence more or less continuously, their kinetic energies remaining small compared to the interaction potential energy. This situation is said to characterize a *strongly coupled* plasma. In the opposite case, strong electrostatic interactions between individual particles are occasional and rare events; a typical particle is electrostatically influenced by others—all those within its Debye sphere—but its motion is rarely arrested by such influence. We will find in Chapter 7 that this weakly coupled case can be described by a Fokker-Planck equation. Understanding the strongly coupled limit is more difficult.

It is important to appreciate the sense of the word "coupling" in this context. Referring to individual particle interactions, it is entirely separate from the issue of collective processes. Indeed, collective plasma dynamics—the occurrence of macroscopic plasma currents and charge densities, of plasma waves, instabilities, vortices and so on—are most commonly associated with weakly coupled plasmas. The understanding of collective effects in strongly coupled plasmas is relatively primitive, mainly because such plasmas are less accessible to laboratory study.

With these remarks, we are ready to see the point of the plasma parameter, which is defined by

$$\Lambda = \frac{\lambda_D}{r_c}. \tag{1.9}$$

The reader can verify that this ratio is indeed proportional to $(r_d/r_c)^{3/2}$. Hence the strongly and weakly coupled cases correspond to small and large $\Lambda$ respectively.

When $\Lambda$ is large, the shielding cloud of a charged particle—its electrostatic sphere of influence—can be explored perturbatively, keeping kinetic energies large compared to the potential energy of the interaction. In the opposite, strongly coupled case, particles are dominated by multiple Coulomb interactions over most of their trajectories. As we have remarked, Debye shielding continues to act in this case, but it becomes a much more complicated statistical process, involving multiple interactions and not tractable by simple perturbation theory.

When the plasma parameter is viewed in terms of the ratio $r_d/r_c$ or $\lambda_D/r_c$ as in (1.9) it is clear that large $\Lambda$ is associated with low density; indeed

$$\Lambda = \frac{4\pi\epsilon_0^{3/2}}{e^3} \frac{T^{3/2}}{n^{1/2}}. \tag{1.10}$$

Thus weakly coupled plasmas are relatively hot and not too dense. The more conventional formula,

$$\Lambda = 4\pi n\lambda_D^3$$

which follows immediately from (1.8), can confuse by seeming to suggest otherwise. In particular, the common, obviously correct observation that $\Lambda$ measures the number of particles in a "Debye sphere" (a sphere of radius $\lambda_D$) should not be taken to imply that one can increase $\Lambda$ by cramming more particles into some fixed container—that is, by increasing the density. Because $\lambda_D$ is proportional to $n^{-1/2}$, the more particles in a Debye sphere the *less* dense the plasma.

In summary, characteristic plasma behavior is observed only on time scales longer than the plasma period and on length scales larger than the Debye length; the statistical character of that behavior is controlled by the plasma parameter. The three parameters, $\lambda_D$, $\omega_p$ and $\Lambda$ are the most fundamental quantities in plasma physics. Of the many other quantities that occur, a few have sufficiently general importance to be discussed in this chapter: the collision frequency or mean free path, and two parameters that refer specifically to magnetized plasmas, the gyroradius and the plasma beta.

## Collisionality

Collisions between charged particles in a plasma differ from those in a neutral gas because of the long range of the Coulomb force. Indeed, as shown in Chapter 7, a binary collisional process can be defined only in the weak coupling case ($\Lambda \gg 1$), and even there it is modified by collective effects: the many-particle process of Debye shielding enters in a crucial way. The resulting random particle interactions yield a Fokker-Planck scattering process with little resemblance to hard–sphere collisions.

Nonetheless for large $\Lambda$ we can speak of binary collisions, and therefore of a collision frequency, denoted by $\nu_{ss'}$. Here the two species subscripts are required by the binary nature of the process: $\nu_{ss'}$ measures the rate at which particles of species $s$ are scattered by those of species $s'$. When specifying only a single subscript, one refers to the total collision rate for that species, including impacts with all other species; very roughly

$$\nu_s \cong \sum_{s'} \nu_{ss'}.$$

The more accurate version of this formula can be found in Chapter 7. Happily the species designations are not always important and can be suppressed in many contexts. However it should be noticed that the small electron mass implies, for unit ionic charge and comparable species temperatures,

$$\nu_e \sim \left(\frac{m_i}{m_e}\right)^{1/2} \nu_i.$$

The Coulomb collision frequency measures the rate at which a particle trajectory undergoes a major angular change, due to Coulomb interactions with another particle: it is sometimes called the "90° scattering rate." (In Chapter 7 we will find that such large-angle scatterings result from the accumulation of many small-angle events.) We will use the word "collisionality" to refer to a dimensionless measure, such as $\nu/\omega_t$, of the relative importance of collisions in a plasma. An equivalent but more conventional measure of collisionality is the ratio

$$\lambda_{\mathrm{mfp}}/L \sim \nu/\omega_t \tag{1.11}$$

where

$$\lambda_{\mathrm{mfp}} \equiv v_t/\nu. \tag{1.12}$$

|  | $\nu_i$ (sec$^{-1}$) | $\nu_e$ (sec$^{-1}$) | $B$ (Tesla) | $\rho_i$ (m) | $\rho_e$ (m) | $\beta$ |
|---|---|---|---|---|---|---|
| glow discharge | $6 \cdot 10^8$ | $10^9$ | $5 \cdot 10^{-2}$ | $5 \cdot 10^{-4}$ | $4 \cdot 10^{-5}$ | 0.1% |
| chromosphere | $10^6$ | $6 \cdot 10^7$ | $10^{-2}$ | $10^{-2}$ | $2 \cdot 10^{-4}$ | 0.4% |
| interstellar medium | $2 \cdot 10^{-8}$ | $10^{-6}$ | $5 \cdot 10^{-10}$ | $3 \cdot 10^5$ | $7 \cdot 10^3$ | 6% |
| magnetic fusion | $10^2$ | $6 \cdot 10^3$ | $5$ | $3 \cdot 10^{-3}$ | $7 \cdot 10^{-5}$ | 3% |

Table 1.2: Additional parameters for the representative set of plasmas listed in Table 1.1. The first two columns give the collision frequencies for ions and electrons; the other columns list various measures of plasma magnetization: the ion and electron gyroradii, $\rho_i$ and $\rho_e$, and the plasma $\beta$, which measures the ratio of plasma thermal energy to magnetic field energy.

is the *mean free path*.

The mean free path is well named: it measures the typical distance a particle travels between collisions. A collision-dominated, or simply "collisional" plasma is one in which $\lambda_{\mathrm{mfp}} \ll L$. The opposite limit of large mean free path is often said to correspond to a "collisionless" plasma.

Although the mathematics of collisions in a plasma can be awkwardly complicated, the physical effect of collisions is generally to simplify plasma behavior. The point is that collisions drive any system towards statistical equilibrium, characterized by Maxwell-Boltzmann distribution functions and nearly thermodynamic behavior. The dynamics of collision-dominated plasma is therefore dissipative, with strong damping of flow, similar to fluid dynamics at small Reynolds number. We will find that closed plasma descriptions involving a few fluid variables—the densities, temperatures and flow velocities of each plasma species—are relatively accessible in the collisional case. Furthermore, as emphasized in Chapter 8, short mean free path enforces a certain very useful *locality* on plasma transport phenomena.

Significantly, the Coulomb collision frequency decreases with temperature; we will find in Chapter 7 that

$$\nu \sim \frac{\log \Lambda}{\Lambda} \omega_p$$

whence

$$\nu \sim n T^{-3/2}.$$

As a result many interesting plasmas are collisionless; indeed, much of the excitement and novelty of plasma physics is associated with the collisionless regime. The collision frequencies of a number of plasmas are listed in Table 1.2.

Collisions involving neutral particles, including ionization events and charge exchange, can have important effects. These phenomena are discussed in Chapter 7.

## 1.3   Magnetized plasma

A magnetized plasma is one in which the ambient magnetic field $\mathbf{B}$ is strong enough to alter fluid behavior. In particular, magnetized plasmas are anisotropic, responding in different ways to forces that are parallel or perpendicular to the direction of $\mathbf{B}$. Many of the most interesting and characteristic plasma phenomena, such as Alfvén waves or turbulent dynamos, intrinsically involve the magnetic field.

Anisotropy suggests decomposing an arbitrary vector $\mathbf{K}$ into components parallel and perpendicular to the field. Thus we will frequently use the notation

$$\mathbf{K} = \mathbf{b}K_{\parallel} + \mathbf{K}_{\perp} \qquad (1.13)$$

where $\mathbf{b} \equiv \mathbf{B}/B$ is the obvious unit vector, $K_{\parallel} \equiv \mathbf{b} \cdot \mathbf{K}$ and

$$\mathbf{K}_{\perp} = \mathbf{b} \times (\mathbf{K} \times \mathbf{b})$$

One might wonder why magnetized plasmas, rather than "electrified" plasmas, have such pervasive importance. Part of the answer is that Debye shielding hinders penetration of a plasma by electric fields, while magnetic fields will eventually penetrate even highly conducting plasmas. But the more important difference lies in the peculiar properties of the magnetic force,

$$\mathbf{F}_m = q\,\mathbf{v} \times \mathbf{B}, \qquad (1.14)$$

where $q$ is the particle charge and $\mathbf{v}$ its velocity. While unable to change particle speeds, the magnetic force radically affects trajectories.

### Gyroradius

Equation (1.14) shows that charged particles freely stream in the direction of $\mathbf{B}$, while spiraling in circular Larmor or "gyro-"orbits around an axis aligned with $\mathbf{B}$. As the field strength increases, the helical orbits become more tightly wound, effectively tying particles to field lines. The resulting anisotropy between perpendicular and parallel motion is observed on many spatial scales and visible in such phenomena as the *aurora borealis*. (The nature of particle orbits in magnetic fields is reviewed in detail in Chapter 2.)

The gyration radius, Larmor radius or, in the language we will generally use, *gyroradius* of a charged particle subject to magnetic force is given by

$$\rho(\mathbf{v}) = \frac{v_{\perp}}{\Omega}$$

where $v_\perp$ is the component of the particle's velocity $\mathbf{v}$ in the direction perpendicular to $\mathbf{B}$ and

$$\Omega = eB/m$$

is the cyclotron frequency, or *gyrofrequency* associated with the gyration. For most purposes we are interested in the thermal gyroradius,

$$\rho_t \equiv \frac{v_t}{\Omega};$$

the $t$-subscript, as well as the word "thermal," will usually be omitted. As usual, there is a distinct gyroradius for each species; when species temperatures are comparable, the electron gyroradius is distinctly smaller than any ion gyroradius:

$$\rho_e \sim \left(\frac{m_e}{m_i}\right)^{1/2} \rho_i \tag{1.15}$$

In fully ionized plasmas the ionic charge has little bearing on the size of $\rho_i$, since ion mass is roughly proportional to charge; it is the difference in thermal speeds that matters.

A plasma system or process is *magnetized* if its characteristic scale length $L$ is large compared to the gyroradius; in the opposite limit $\rho \gg L$ particles will be observed to have nearly straight-line orbits. Thus the importance of the magnetic field is measured by the magnetization parameter

$$\delta \equiv \frac{\rho}{L}. \tag{1.16}$$

An equivalent measure is the ratio of transit frequency to gyrofrequency:

$$\delta = \frac{\omega_t}{\Omega}.$$

There are some cases of interest in which the electrons are magnetized and ions are not; by a 'magnetized plasma' we mean one in which both species are magnetized. For the conventional case of (1.15), this state is achieved when

$$\delta_i \equiv \frac{\rho_i}{L} \ll 1.$$

A sense of which systems and phenomena are magnetized can be gleaned from the thermal gyroradii listed in Table 1.2.

## Plasma beta

The fundamental measure of a magnetic field's effect on a plasma is the magnetization parameter $\delta$. The fundamental measure of the inverse effect of the plasma on the field—a sort of "diamagnetization parameter"—is called $\beta$, and measured by the ratio of plasma thermal energy density $(3/2)nT$ to magnetic field energy density $B^2/2\mu_0$, where $\mu_0$ is the permeability of the vacuum. It is convenient to identify the plasma energy density with the pressure,

$$p \equiv nT$$

suppressing the $(3/2)$-factor, and to define a separate $\beta_s$ for each plasma species:

$$\beta_s = \frac{2\mu_0 p_s}{B^2}.$$

The total $\beta$

$$\beta = \sum_s \beta_s$$

is often most important.

To verify that $\beta$ measures diamagnetism, we anticipate an argument that is presented in more detail in Chapter 3. First recall (or be reminded by looking at Chapter 3) that magnetic forces can be expressed in terms of the divergence of the Maxwell stress tensor. If field-line curvature is neglected, this divergence reduces to the gradient of the field energy, giving the total (magnetic plus fluid) force, $\nabla(p + B^2/2\mu_0)$. Thus a plasma near equilibrium is roughly characterized by the relation

$$\nabla\left(p + \frac{B^2}{2\mu_0}\right) \cong 0.$$

It follows that the magnetic field is reduced in regions of higher pressure; the strength of this diamagnetic effect is evidently measured by $\beta$.

Although there are no intrinsic limits on the value of $\beta$, it is often small in magnetized plasmas; see Table 1.2. In such cases the plasma evolves with only minor perturbation of the field.

## Dimensionless parameters

A magnetized plasma is characterized by the field strength $B$ as well as the parameters $e$, $m_s$, $n$ and $T$ considered in Section 1.2. Furthermore in constructing time and length scales for the magnetized case we need to allow for the occurrence of $\mu_0$ along with $\epsilon_0$. (In other words, we must allow the speed of light $c = (\mu_0 \epsilon_0)^{-1/2}$ to enter.) The resulting collection of seven basic quantities (for each species) is sufficiently large to justify systematic treatment: can we be sure that we have identified all the fundamental measuring sticks?

A simple way to address this question, due to Buckingham[14], is to consider all possible dimensionless products of powers of the quantities

$$\{p_1, p_2, \ldots, p_7\} \equiv \{\epsilon_0, \mu_0, e, m_s, n, T, B\}.$$

Thus our dimensionless parameters have the form

$$D = p_1^{\alpha_1} p_2^{\alpha_2} \cdots p_7^{\alpha_7}. \tag{1.17}$$

Now the dimensions of each $p_i$ are known in terms of the basic dimensions $(l, t, m, q)$, where $l$ stands for length, $t$ for time, $m$ for mass and $q$ for electric charge; thus

$$p_1 \sim l^{-3} t^2 m^{-1} q^2,$$
$$p_2 \sim l m q^{-2},$$
$$p_3 \sim q$$

and so on. Therefore after substitution we find that $D$ has dimensions

$$D \sim l^{\gamma_1}\, t^{\gamma_2}\, m^{\gamma_3}\, q^{\gamma_4}$$

where each $\gamma_i$ is a linear combination of the $\alpha_i$. Thus requiring $D$ to be dimensionless gives four equations

$$\gamma_i = 0, \quad i = 1, \ldots, 4$$

for the seven unknown $\alpha_i$. Suppose we solve these equations to express the seven exponents in terms of, say, $\alpha_1$, $\alpha_2$, and $\alpha_3$. It is then clear that (1.17) can be expressed in terms of just three factors, each raised to one of these three unknown powers $\alpha_i$, and each involving *known* powers of the $p_i$. In other words it is clear that the system allows for precisely three independent dimensionless ratios.

The three dimensionless quantities turn out to be

$$D_1 = \Lambda, \tag{1.18}$$
$$D_2 = \beta, \tag{1.19}$$
$$D_3 = n\rho^3, \tag{1.20}$$

where $\Lambda$ is the plasma parameter. Of course any product of powers of these three is equally fundamental; what has been shown is that *any* dimensionless parameter can be expressed in terms of $D_1$, $D_2$ and $D_3$ alone. This implies, for example, that the field strength $B$ can enter a dimensionless description only through the combination $T/B^2$.

The quantity $D_3$ is analogous to $\Lambda$ in measuring the number of particles in a "gyro-sphere." This measure of magnetization is not very illuminating, being infinite for an unmagnetized plasma and virtually always much larger than one: vast numbers of particles typically gyrate within each others' gyroradii. (Sketches of particles in magnetized plasmas universally ignore this circumstance for the sake of clarity.) However its occurrence confirms that we have correctly identified the key measure of magnetization,

$$\rho = (D_3/n)^{1/3}.$$

Note that the above technique for generating dimensionless parameters is trivially modified to produce parameters of any desired dimension. For example, to find all fundamental lengths, one simply repeats the calculation of $D$, but now letting $\gamma_1 = 1$, the other $\gamma_i$ remaining zero. In this way the gyroradius, particle spacing and Debye length re-emerge, in combination with powers of the dimensionless $D_i$ found here.

Finally note that our species-specific analysis cannot catch parameters involving more than one species mass. It turns out that the only important parameter thereby missed in the obvious one:

$$m_e/m_i \ll 1.$$

# Additional reading

Plasma physics has generated numerous good books, beginning with two gems of physics exposition: the classic, short monograph of Spitzer [67], and Alfvén's landmark treatise [2]. The latter emphasizes stellar and galactic plasmas.

The texts by Krall and Trivelpiece [43], by Sturrock [70] and by Schmidt [65] are carefully detailed, especially with regard to plasma theory. Theory is also emphasized in the short, readable books by Nicholson [53] and by Van Kampen and Felderhof [74]. Less theoretically inclined students may find the treatments by Chen [16] and by Goldston and Rutherford [30] more accessible; the latter treats fusion plasmas with particular thoroughness. A fundamental treatment of plasma kinetic theory is given by Ichimaru [38], with special attention to the strongly coupled case in [39].

Probably the best brief introduction to plasma physics, capturing the flavor and richness of the field, is the *Physics Today* article by Harold Grad [31].

# Chapter 2

# Charged Particle Motion

## 2.1 Magnetization

All descriptions of plasma behavior are based, fundamentally, on the properties of the motion of the constituent particles. This motion obeys Newton's law,

$$m_s \frac{d\mathbf{v}}{dt} = \mathbf{F}_s + e_s \mathbf{v} \times \mathbf{B}, \tag{2.1}$$

where the force $\mathbf{F}_s$ is the sum of the gravitational and electrostatic forces, $\mathbf{F}_s = m_s \mathbf{g} + e_s \mathbf{E}$, and $e_s \mathbf{v} \times \mathbf{B}$ is the magnetic force. In this chapter, we assume that the force fields are specified in order to concentrate on the motion of individual particles.

The primary consideration for the motion of a particle is the role played by the magnetic field. When the magnetic field is negligible, the motion is ballistic and its properties are trivial (unmagnetized particles). When the magnetic field is dominant, by contrast, the trajectories can be complex (magnetized particles). The motion can nevertheless be analyzed fruitfully with perturbation theory. The expansion parameter is taken to be the inverse of the number of Larmor gyrations executed by a particle during a characteristic time for the evolution of the fields. The resulting theory of magnetized motion is the principal subject of this chapter.

The central result of this theory is that particles behave like magnetic dipoles, or small magnets, aligned with $\mathbf{B}$ and with conserved magnetic dipole moments. The position of the dipole, called the guiding center, is free of oscillations at the Larmor frequency. It is approximately given by the center of the Larmor gyration. The elimination of the rapid Larmor oscillations greatly facilitates numerical integration as well as analytical description of magnetized particle motion. An important consequence of the constancy of the magnetic moment is that particles may be trapped in regions of low field.

The theory of charged particle motion in a magnetic field can be regarded as an application of classical mechanics. Plasma physics in general, however, and the theory of magnetized motion in particular, make different demands on classical mechanics than other disciplines, such as celestial mechanics. Most notably, the theory of charged particle motion must provide a suitable foundation for constructing self-consistent descriptions of *collisional* plasma evolution. It must also provide a practical means of predicting numerically the motion of a very large number of particles. Given the difficulty of satisfying these requirements, it is unsurprising that plasma physics has inspired fundamental contributions to classical mechanics.

We begin the chapter by describing the motion of charged particles in constant, homogeneous fields.

## 2.2    Motion in constant fields

The equations of motion are easily solved in the case of constant, homogeneous force fields, $\mathbf{F}_s$. The component of the equation of motion parallel to the magnetic field,

$$\frac{dv_\parallel}{dt} = \frac{F_{s\parallel}}{m_s}, \tag{2.2}$$

predicts uniform acceleration along the magnetic field. As a result, plasmas near equilibrium usually have small or vanishing $F_{s\parallel}$.

The component of the equation of motion perpendicular to the magnetic field takes a simpler form after a Galilean transformation into a frame moving with velocity $\mathbf{v}_D$,

$$\mathbf{v} = \mathbf{v}_D + \mathbf{u},$$

where

$$\mathbf{v}_D = \frac{\mathbf{F}_s \times \mathbf{B}}{e_s B^2}. \tag{2.3}$$

is called the particle drift. In the moving frame the perpendicular equation of motion is

$$\frac{d\mathbf{u}_\perp}{dt} = \frac{e_s}{m_s} \mathbf{u}_\perp \times \mathbf{B}. \tag{2.4}$$

To solve this equation, we note that the magnetic force does no work, so that the perpendicular velocity is constant:

$$m\mathbf{u}_\perp \cdot \frac{d\mathbf{u}_\perp}{dt} = \frac{1}{2}\frac{d}{dt}(mu_\perp^2) = 0,$$

where $u_\perp = |\mathbf{u}_\perp|$. The magnetic force results in a centripetal acceleration

$$\frac{u_\perp^2}{\rho_s} = \frac{e_s}{m_s} u_\perp B,$$

where the gyroradius $\rho_s$ is the radius of curvature of the trajectory in the plane perpendicular to $\mathbf{B}$. We see that

$$\rho_s = \frac{m_s u_\perp}{e_s B}$$

is constant: the particles describe a circular motion perpendicular to the magnetic field. The rate of rotation is given by the Larmor frequency, or gyrofrequency,

$$\Omega_s = \frac{u_\perp}{\rho_s} = \frac{e_s B}{m_s}.$$

The velocity in the rest frame is thus

$$\mathbf{v} = \left(v_{\|0} + \frac{F_\|}{m_s}t\right)\mathbf{b} + \mathbf{v}_D + \mathbf{u}_\perp, \tag{2.5}$$

where

$$\mathbf{u}_\perp = \rho_s \Omega_s [\mathbf{e}_1 \sin(\Omega_s t + \gamma_0) + \mathbf{e}_2 \cos(\Omega_s t + \gamma_0)]. \tag{2.6}$$

Here $\mathbf{e}_1$ and $\mathbf{e}_2$ are unit vectors such that $(\mathbf{e}_1, \mathbf{e}_2, \mathbf{b})$ form a right handed basis, and $\gamma_0$ is the initial gyrophase of the particle. The motion consists of the superposition of a steady drift and a rotation at frequency $\Omega$ (Fig. 2.1). Note

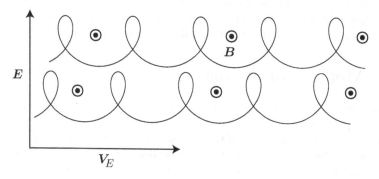

Figure 2.1: Motion of magnetized particles through a constant field of force.

that the electric part of the drift,

$$\mathbf{v}_E = \frac{\mathbf{E} \times \mathbf{B}}{B^2}, \tag{2.7}$$

is independent of the charge, so that electrons and ions will have the same electric drifts. The drifts caused by charge-independent forces (such as gravity), by contrast, are in opposite directions for electrons and ions. This has fundamental consequences for plasma dynamics.

We complete the solution by integrating the velocity to find the particle position:

$$\mathbf{r}(t) = \mathbf{R}(t) + \boldsymbol{\rho}(t),$$

where

$$\boldsymbol{\rho}(t) = \rho_s[-\mathbf{e}_1 \cos(\Omega_s t + \gamma_0) + \mathbf{e}_2 \sin(\Omega_s t + \gamma_0)] \tag{2.8}$$

$$= \frac{1}{\Omega_s}\mathbf{b} \times \mathbf{u}_\perp(t) \tag{2.9}$$

and

$$\mathbf{R}(t) = \left( v_{\|0}t + \frac{F_\|}{m_s}\frac{t^2}{2} \right) \mathbf{b} + \mathbf{v}_D \, t.$$

We find that the trajectory describes a helix, or spiral. The center $\mathbf{R}$ of the spiral, called the *guiding center*, drifts across the magnetic field at the velocity $\mathbf{v}_D$ and accelerates along the field at a rate fixed by the parallel component of the applied forces.

The guiding center concept illuminates the nature of the motion in constant fields, and provides simple explanations for a wealth of plasma phenomena. Its real power, however, is to provide a path for extending the constant field results to the important case of weakly varying fields. In the latter case, the fields acting on a particle change little in the course of a Larmor period and the motion resembles that found here for constant fields. The particles experience additional drifts, however, caused by inertial forces and the interaction of the effective magnetic moment with the equilibrium field.

The method we will use to derive the equations governing the evolution of the guiding center has broad applicability. We begin thus by presenting it in a general context, before applying it to guiding center motion in Section 2.4.

## 2.3   Method of averaging

In many problems of mechanics, the motion consists of a rapid oscillation superposed upon a slow drift. For such problems the most efficient approach is to describe the evolution in terms of the average of the dynamical variables. This is the purpose of the method presented here. Following Morozov and Solov'ev[52], we begin by presenting this method in a simple context in order to expose its structure without the geometrical complexity inherent in guiding-center motion.

Consider the differential equation

$$\frac{d\mathbf{z}}{dt} = \mathbf{f}(\mathbf{z}, t, \tau), \tag{2.10}$$

where $\mathbf{f}$ is a periodic function of its last argument with period $2\pi$ and

$$\tau = t/\epsilon. \tag{2.11}$$

Here the small parameter $\epsilon$ characterizes the separation between the short oscillation period and the time for the slow evolution of $\mathbf{f}$ through its first two arguments.

The idea of the method of averaging is to treat $t$ and $\tau$ as distinct independent variables and to look for solutions of the form $\mathbf{z}(t, \tau)$ periodic in $\tau$. Thus, we replace (2.10) by

$$\frac{\partial \mathbf{z}}{\partial t} + \frac{1}{\epsilon}\frac{\partial \mathbf{z}}{\partial \tau} = \mathbf{f}(\mathbf{z}, t, \tau) \tag{2.12}$$

and reserve (2.11) for substitution in the final result. The indeterminacy introduced by increasing the number of independent variables is lifted by the requirement of periodicity in $\tau$. All the steady drifts are thereby attributed to the $t$ variable, while the oscillations are described by the $\tau$ variation. We denote the $\tau$-average of $\mathbf{z}$ by $\mathbf{Z}$, and seek a change of variables of the form

$$\mathbf{z} = \mathbf{Z} + \epsilon\boldsymbol{\zeta}(\mathbf{Z}, t, \tau). \tag{2.13}$$

Here $\boldsymbol{\zeta}$ is a periodic function of $\tau$ with vanishing mean:

$$\langle \boldsymbol{\zeta}(\mathbf{Z}, t, \tau)\rangle \equiv \frac{1}{2\pi} \oint d\tau\, \boldsymbol{\zeta}(\mathbf{Z}, t, \tau) = 0, \tag{2.14}$$

where $\oint$ denotes the integral over a full period in $\tau$.

The evolution of $\mathbf{Z}$ is determined by substituting the expansions

$$\boldsymbol{\zeta} = \boldsymbol{\zeta}_0(\mathbf{Z}, t, \tau) + \epsilon\boldsymbol{\zeta}_1(\mathbf{Z}, t, \tau) + \epsilon^2\boldsymbol{\zeta}_2(\mathbf{Z}, t, \tau) + \ldots \tag{2.15}$$

$$\frac{d\mathbf{Z}}{dt} = \mathbf{F}_0(\mathbf{Z}, t) + \epsilon\mathbf{F}_1(\mathbf{Z}, t) + \epsilon^2\mathbf{F}_2(\mathbf{Z}, t) + \ldots \tag{2.16}$$

into the equation of motion (2.12) and solving order by order in $\epsilon$. To lowest order we find

$$\mathbf{F}_0(\mathbf{Z}, t) + \frac{\partial\boldsymbol{\zeta}_0}{\partial\tau} = \mathbf{f}(\mathbf{Z}, t, \tau). \tag{2.17}$$

The solubility condition for this equation is

$$\mathbf{F}_0(\mathbf{Z}, t) = \langle \mathbf{f}(\mathbf{Z}, t, \tau)\rangle.$$

Integrating the oscillating part of (2.17) yields

$$\boldsymbol{\zeta}_0(\mathbf{Z}, t, \tau) = \int_0^\tau (\mathbf{f} - \langle\mathbf{f}\rangle)d\hat{\tau}. \tag{2.18}$$

We carry the solution to first order so as to evaluate the slow drift of the system away from its lowest order (average) position. From the solubility condition for the first order part of the equation of motion, we find

$$\mathbf{F}_1 = \langle \boldsymbol{\zeta}_0 \cdot \nabla\mathbf{f}\rangle$$

The slow drift is thus given by the beat of the oscillation in the particle position with the oscillation of the force; this is called the quasilinear term.

Our results are summarized by the equation for the slowly varying average quantity $\mathbf{Z}$,

$$\frac{d\mathbf{Z}}{dt} = \langle\mathbf{f}\rangle + \langle\boldsymbol{\zeta}_0 \cdot \nabla\mathbf{f}\rangle.$$

This result has numerous applications, ranging from particle dynamics to turbulence theory. In the following sections, we use it to investigate the motion of charged particles in the limit where the magnetic field varies slowly compared to the Larmor frequency, and on a scale large compared to the Larmor radius. In the last section of this chapter, we show that the method of averages can also be applied to determine the drift of particles in an inhomogeneous, high frequency electromagnetic wave such as a laser beam.

## 2.4   Guiding center motion

We consider the motion of charged particles such that the field experienced by the particle changes little in a gyroperiod,

$$\rho_s|\nabla B| \ll B; \qquad \frac{\partial B}{\partial t} \ll \Omega_s B.$$

We also assume that the electric force is comparable to the magnetic force. To keep track of the order of the various quantities, we introduce the parameter $\epsilon$ as a bookkeeping device and make the substitution $\rho_s \rightarrow \epsilon\rho_s$ and $(\mathbf{E}, \mathbf{B}, \Omega_s) \rightarrow \epsilon^{-1}(\mathbf{E}, \mathbf{B}, \Omega_s)$. The parameter $\epsilon$ is set to unity in the final answer.

In order to use the results of the previous section, we write the dynamical equations in first-order form,

$$\frac{d\mathbf{r}}{dt} = \mathbf{v} \tag{2.19}$$

$$m_s \frac{d\mathbf{v}}{dt} = \frac{e_s}{\epsilon}(\mathbf{E} + \mathbf{v} \times \mathbf{B}), \tag{2.20}$$

and seek a change of variables

$$\mathbf{r} = \mathbf{R} + \epsilon\rho(\mathbf{R}, \mathbf{U}, t, \gamma) \tag{2.21}$$

$$\mathbf{v} = \mathbf{U} + \mathbf{u}(\mathbf{R}, \mathbf{U}, t, \gamma) \tag{2.22}$$

such that the new, guiding center variables $\mathbf{R}$ and $\mathbf{U}$ are free of oscillations along the trajectory. Here $\gamma$ is a new independent variable describing the phase of the gyrating particles. Note that equations (2.21)-(2.22) are analogous to (2.13) of Section 2.3. The functions $\rho$ and $\mathbf{u}$ represent the gyration radius and velocity. We require periodicity of these functions with respect to their last argument, with period $2\pi$ and with vanishing mean:

$$\langle\rho\rangle = \langle\mathbf{u}\rangle = 0. \tag{2.23}$$

Here the brackets refer to the average over a period in $\gamma$.

The equation of motion is used to determine the successive terms in the expansion of $\rho$ and $\mathbf{u}$. We adopt an expansion analogous to (2.15),

$$\rho = \rho_0(\mathbf{R}, \mathbf{U}, t, \gamma) + \epsilon\rho_1(\mathbf{R}, \mathbf{U}, t, \gamma) + \dots \tag{2.24}$$

$$\mathbf{u} = \mathbf{u}_0(\mathbf{R}, \mathbf{U}, t, \gamma) + \epsilon\mathbf{u}_1(\mathbf{R}, \mathbf{U}, t, \gamma) + \dots \tag{2.25}$$

and expand the gyrophase equation in a similar way, assuming that $d\gamma/dt \simeq \Omega_s = O(\epsilon^{-1})$

$$\frac{d\gamma}{dt} = \epsilon^{-1}\omega_{-1}(\mathbf{R}, \mathbf{U}, t) + \omega_0(\mathbf{R}, \mathbf{U}, t) + \dots \tag{2.26}$$

In the following, we will suppress the subscripts on all quantities except the guiding center velocity, $\mathbf{U}$, the only quantity for which we will evaluate the first-order corrections. We also suppress the species index $s$.

The solubility conditions on the equations of motion (2.19) and (2.20) at each order determine the evolution of the guiding center velocity and gyrophase. Substituting the expansions (2.24)-(2.26) in the equations of motion, we note first that the velocity equation, $d\mathbf{r}/dt = \mathbf{v}$, is linear. It follows that its solubility condition, to all orders in $\epsilon$, is simply

$$\frac{d\mathbf{R}}{dt} = \mathbf{U}. \tag{2.27}$$

Consider next the lowest-order, $O(\epsilon^{-1})$ momentum equation,

$$\omega \frac{\partial \mathbf{u}}{\partial \gamma} - \Omega \mathbf{u} \times \mathbf{b} = \frac{e}{m}(\mathbf{E} + \mathbf{U}_0 \times \mathbf{B}). \tag{2.28}$$

The average of this equation is

$$\mathbf{E} + \mathbf{U}_0 \times \mathbf{B} = 0.$$

This can only be satisfied if

$$\mathbf{b} \cdot \mathbf{E} \sim \epsilon|\mathbf{E}| : \tag{2.29}$$

the rapid acceleration caused by a larger parallel electric field would invalidate the drift approximation. Solving for $\mathbf{U}_0$, we find

$$\mathbf{U}_0 = \mathbf{U}_{0\|}\mathbf{b} + \mathbf{v}_E, \tag{2.30}$$

where all quantities are evaluated at the guiding-center position. The perpendicular component of the velocity, $\mathbf{v}_E$, has the same form (2.7) as for homogeneous fields. The parallel velocity is undetermined at this order.

The integral of (2.28) is

$$\mathbf{u} = \mathbf{c} + u_\perp \left( \mathbf{e}_1 \sin\left(\frac{\Omega}{\omega}\gamma\right) + \mathbf{e}_2 \cos\left(\frac{\Omega}{\omega}\gamma\right) \right), \tag{2.31}$$

where $\mathbf{e}_1$ and $\mathbf{e}_2$ are unit vectors perpendicular to $\mathbf{b}$. All quantities in (2.31) are again functions of the guiding center variables and time $(\mathbf{R}, \mathbf{U}, t)$. The periodicity constraint and (2.23), require that $\omega = \Omega(\mathbf{R}, t)$ and $\mathbf{c} = 0$. The gyration velocity is thus

$$\mathbf{u} = u_\perp \left( \mathbf{e}_1 \sin\gamma + \mathbf{e}_2 \cos\gamma \right), \tag{2.32}$$

and the gyrophase is

$$\gamma = \gamma_0 + \Omega t, \tag{2.33}$$

where $\gamma_0$ is the initial phase. The amplitude $u_\perp$ of the velocity remains undetermined at this order. We will determine $u_\perp$ below from the first order equation.

The oscillation in the particle position is determined by the first order terms of the velocity equation (2.19) taking into account the solubility condition (2.27),

$$\Omega \frac{\partial \boldsymbol{\rho}}{\partial \gamma} = \mathbf{u}. \tag{2.34}$$

This is easily integrated,

$$\boldsymbol{\rho} = \rho(-\mathbf{e}_1 \cos \gamma + \mathbf{e}_2 \sin \gamma), \tag{2.35}$$

where $\rho = u_\perp/\Omega$. We will often find it convenient to write the gyroradius in terms of the gyrovelocity as

$$\boldsymbol{\rho} = \frac{\mathbf{b}}{\Omega} \times \mathbf{u}.$$

We find that the lowest-order motion differs from that in constant fields only in the evolution of the parallel guiding center velocity and the perpendicular gyration velocity. Since the total energy is conserved, this evolution takes the form of an exchange between parallel kinetic energy and gyration energy.

## Guiding center drifts

We next consider the average of the first order of the momentum equation,

$$\frac{d\mathbf{U}_0}{dt} = \frac{e}{m}[\mathbf{U}_1 \times \mathbf{B} + E_\parallel \mathbf{b} + \langle \mathbf{u} \times \mathbf{B}(\mathbf{r}) \rangle]. \tag{2.36}$$

The averaging eliminates the first-order corrections to the gyro-velocity $\mathbf{u}$ and the oscillation in the electric field experienced by the particle, $(\boldsymbol{\rho} \cdot \nabla)\mathbf{E}$. To evaluate the last term, we consider the average over the initial gyro-phase of the virtual work $W_{\mathbf{d}}$ that must be done against the magnetic force to move a gyro-orbit from $\mathbf{R}$ to $\mathbf{R} + \mathbf{d}$ where $\mathbf{d}$ is an arbitrary displacement:

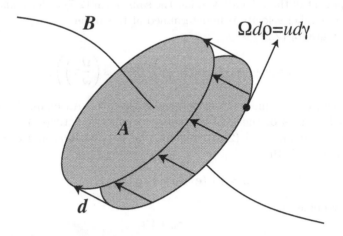

Figure 2.2: Virtual work done in moving a gyro-orbit.

$$W_{\mathbf{d}} = e\langle \mathbf{d} \cdot (\mathbf{u} \times \mathbf{B}(\mathbf{r})) \rangle = -\frac{\Omega e}{2\pi} \oint (d\boldsymbol{\rho} \times \mathbf{d}) \cdot \mathbf{B}(\mathbf{R} + \boldsymbol{\rho}), \tag{2.37}$$

where we have used (2.34) to eliminate the velocity $\mathbf{u}$. We note that the right-hand side integral represents the flux through the sides of a slanted pillbox centered on the guiding center position and with radius $\rho$ and side $d$ (see Fig. 1). Since the divergence of the magnetic field vanishes, the total flux through the box must vanish. Thus,

$$W_{\mathbf{d}} = \frac{\Omega e}{2\pi} \int dA\, B(\mathbf{R}) - \frac{\Omega e}{2\pi} \int dA\, B(\mathbf{R} + \mathbf{d}). \tag{2.38}$$

where $dA$ is the surface element on the top and bottom of the pillbox. Only the magnitude of $\mathbf{B}$ appears in the integral since the outward normal to the surfaces is parallel to $\mathbf{b}$. Taylor expansion of $B$ yields

$$W_{\mathbf{d}} = -\frac{e\Omega\rho^2}{2}\,\mathbf{d}\cdot\nabla B(\mathbf{R}) = -\frac{mu_\perp^2}{2B}\,\mathbf{d}\cdot\nabla B(\mathbf{R}), \tag{2.39}$$

Since the above result holds for any $\mathbf{d}$, it follows that

$$-e\langle\mathbf{u}\times\mathbf{B}(\mathbf{r})\rangle = \frac{mu_\perp^2}{2B}\nabla B. \tag{2.40}$$

This is the average of the gyration-induced fluctuations in the magnetic force.

The coefficient of $\nabla B$ in (2.40),

$$\mu = \frac{mu_\perp^2}{2B}, \tag{2.41}$$

plays a central role in the theory of magnetized particle motion. We can interpret this coefficient as a magnetic moment by drawing an analogy between the gyrating particle and a conducting loop. The magnetic moment for a current loop is

$$\boldsymbol{\mu} = IA\mathbf{n}, \tag{2.42}$$

where $I$ is the current, $A$ the area of the loop, and $\mathbf{n}$ the unit normal to the surface of the loop, chosen according to the right-hand rule with respect to the direction of $I$. The current associated with a gyrating particle is $I = e\Omega/2\pi$, but, because gyration is "left-handed" with respect to the field direction $\mathbf{b}$, we have $\mathbf{n} = -\mathbf{b}$. Then, for a circular loop of radius $\rho = u_\perp/\Omega$, we find

$$\boldsymbol{\mu} = -I\pi\rho^2\mathbf{b} = -\frac{mu_\perp^2}{2B}\mathbf{b}. \tag{2.43}$$

Note that the magnetic flux through the loop is proportional to its moment, $\pi\rho^2 B = 2\pi m\mu/e^2$, so that conservation of the flux requires that $\mu$ be constant. We will show below that the guiding center behaves indeed as a particle with a conserved magnetic moment $\mu$ aligned with $\mathbf{b}$.

The first-order guiding center equation is thus

$$m\frac{d\mathbf{U}_0}{dt} = e\mathbf{U}_1\times\mathbf{B} + eE_\|\mathbf{b} - \mu\nabla B. \tag{2.44}$$

The component of this equation along the magnetic field determines the evolution of the parallel guiding center velocity,

$$m\frac{dU_{0\|}}{dt} = eE_\| - \boldsymbol{\mu}\cdot\nabla B - m\mathbf{b}\cdot\frac{d\mathbf{v}_E}{dt}. \tag{2.45}$$

where we have used $\mathbf{b} \cdot d\mathbf{b}/dt = 0$. The component of (2.44) orthogonal to the magnetic field, by contrast, determines the first-order perpendicular drift velocity, $\mathbf{U}_{1\perp}$:

$$\mathbf{U}_{1\perp} = \frac{\mathbf{b}}{\Omega} \times \left[ \frac{d\mathbf{U}_0}{dt} + \frac{\mu}{m} \nabla B \right]. \tag{2.46}$$

The first term is called the inertial drift, and the second the magnetic drift. Before we examine each of these terms in more detail, we note that the first order correction to the parallel velocity, the parallel drift, is undetermined at this order. It is determined in Sec. 2.7 by deriving the guiding-center equations from the Hamilton-Lagrange variational principle.

## Magnetic drift

The magnetic drift is caused by the variation of the gyroradius as it moves across a magnetic field of varying amplitude. In the high-field part of the trajectory, the gyroradius is reduced, while in the low-field part of the trajectory it is increased. This leads to a drift perpendicular to the gradient of the magnetic field amplitude,

$$\frac{\mu \mathbf{b}}{m\Omega} \times \nabla B.$$

See Fig. 2.3. The magnetic drift can also be understood in terms of the general drift formula, (2.3), and the magnetic mirror force, $-\mu \nabla B$, as in (2.44).

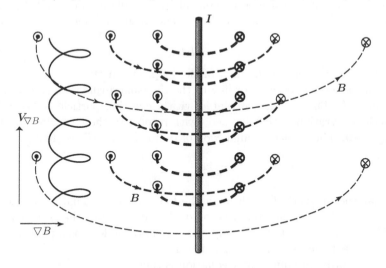

Figure 2.3: Motion of particles through an inhomogeneous magnetic field.

## Inertial drift

The inertial drift has the standard form for drifts, (2.3), where $\mathbf{F}_s$ is taken to be the "inertial force" $md\mathbf{U}_0/dt$. We can divide this drift into two parts, corresponding to changes in the parallel and perpendicular components of the guiding center velocity.

The changes in the parallel velocity give rise to the drift

$$\frac{\mathbf{b}}{\Omega} \times \frac{d}{dt}(U_{0\parallel}\mathbf{b}) = \frac{U_{0\parallel}}{\Omega}\mathbf{b} \times \frac{d\mathbf{b}}{dt}, \tag{2.47}$$

Here

$$\frac{d\mathbf{b}}{dt} = \frac{\partial\mathbf{b}}{\partial t} + \mathbf{v}_E \cdot \nabla\mathbf{b} + U_{0\parallel}\mathbf{b} \cdot \nabla\mathbf{b}. \tag{2.48}$$

The quantity

$$\boldsymbol{\kappa} = \mathbf{b} \cdot \nabla\mathbf{b} \tag{2.49}$$

in the last term represents the curvature of the magnetic field lines: it is a vector directed towards the center of the circle that most closely approximates the field-line at the given point (Fig.2.4). The magnitude of the curvature is the inverse of the radius of this circle. The last term in (2.48) represents thus the centripetal acceleration imposed by the curvature of the magnetic field on a particle following a field line. Its contribution to the drift in (2.47) is called the *curvature drift*. The first two terms in (2.48) represent the acceleration that results when changes in the direction of the magnetic field cause part of the parallel velocity to be converted to perpendicular velocity. In the important case of stationary fields and weak electric fields, the curvature drift is the dominant contribution to the inertial drift.

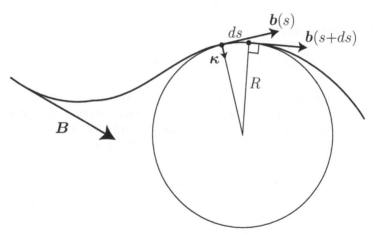

Figure 2.4: Geometric interpretation of the curvature vector $\boldsymbol{\kappa}$.

The changes in the perpendicular velocity give rise to the *polarization drift*

$$\mathbf{v}_P = \frac{\mathbf{b}}{\Omega} \times \frac{d\mathbf{v}_E}{dt}. \tag{2.50}$$

In a constant magnetic field, the polarization drift is

$$\mathbf{v}_P = \frac{1}{\Omega}\frac{d}{dt}\left(\frac{\mathbf{E}_\perp}{B}\right). \tag{2.51}$$

We can understand the polarization drift by considering what happens when we impose an electric field on a particle at rest. The particles initially accelerate in the direction of the electric field before the magnetic force can deflect them. If we hold the field to a fixed value after the initial ramp-up, the particle will undergo the normal electric drift, but the position of its guiding center is shifted from the initial particle position by $E/\Omega B$. Since ions are shifted in the opposite direction from the electrons, the drifts give rise to a *polarization current*.

## 2.5   Invariance of the magnetic moment

We have seen that the equations governing the evolution of the guiding center depend on the magnetic moment or, equivalently, the magnitude of the perpendicular velocity. The evolution of the latter is determined by the energy equation, obtained by taking the scalar product of the equation of motion with the velocity:

$$\frac{m}{2}\frac{d}{dt}(\mathbf{v}^2) = e\mathbf{v}\cdot\mathbf{E} \tag{2.52}$$

To evaluate the rate of change of the perpendicular velocity, we substitute $\mathbf{v} = \mathbf{U} + \mathbf{u}$ and average the energy equation over the gyrophase, noting that

$$\frac{d}{dt}\langle f\rangle = \langle\frac{df}{dt}\rangle$$

for any $f$. There follows

$$\frac{m}{2}\frac{d}{dt}(u_\perp^2 + \mathbf{U}_0^2) = eU_{0\|}E_\| + e\mathbf{U}_1\cdot\mathbf{E} + e\langle\mathbf{u}\cdot\mathbf{E}(\mathbf{r})\rangle. \tag{2.53}$$

This shows that the kinetic energy changes in two ways: by motion of the guiding center along the electric field, and by acceleration of the gyration due to the electromotive force around the Larmor orbit (recall that the magnetic force does no work). The last term on the right hand side of (2.53) is

$$e\langle\mathbf{u}\cdot\mathbf{E}(\mathbf{r})\rangle = \frac{e\Omega}{2\pi}\oint d\boldsymbol{\rho}\cdot\mathbf{E}(\mathbf{R}+\boldsymbol{\rho}) = -\mu\mathbf{b}\cdot\nabla\times\mathbf{E}. \tag{2.54}$$

This has a simple geometrical interpretation (Fig. 2.5): the average of $\mathbf{u}\cdot\mathbf{E}$ is the circulation of the electric field around the gyro-orbit times the angular Larmor frequency. Equation (2.54) is then seen to follow from Stoke's theorem.

Substituting (2.54) into the energy equation and applying Faraday's law, we find

$$\frac{d\mathcal{K}}{dt} = e\mathbf{U}\cdot\mathbf{E} + \mu\frac{\partial B}{\partial t}, \tag{2.55}$$

where $\mathbf{U}$ is the guiding center velocity including the first-order perpendicular drift and

$$\mathcal{K} = \frac{m}{2}(u_\perp^2 + v_E^2 + U_{0\|}^2)$$

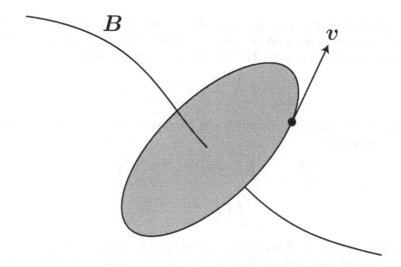

Figure 2.5: Interpretation of the change in kinetic energy as the work done by the electromotive force around the gyro-orbit in a frame such that $\mathbf{E} = 0$.

is the kinetic energy of the particle.

The kinetic energy equation clearly determines the evolution of $u_\perp$, since $\mathbf{U}_0$ is known. To find an explicit evolution equation for $u_\perp$, we eliminate the parallel velocity and the guiding center drift, $U_{0\parallel}$ and $\mathbf{U}_1$, from the equation for the evolution of the kinetic energy, (2.55), using (2.45) and (2.46) respectively. There follows

$$\frac{d}{dt}\left(\frac{mu_\perp^2}{2B}\right) = \frac{d\mu}{dt} = 0. \tag{2.56}$$

This shows that the magnetic moment is conserved to lowest order. In fact, Kruskal has shown that $mu_\perp^2/2B$ is only the first approximation to a quantity that is conserved to all orders in the perturbation expansion (note that $\exp(-1/\epsilon) > 0$, but nevertheless vanishes to all orders in $\epsilon$).[46] Such quantities are called *adiabatic invariants*. The adiabatic invariants are related to exactly conserved quantities called Poincaré invariants that we now describe.

## Poincaré invariants

The adiabatic invariants are approximations to a more general set of invariants, called the Poincaré invariants. These are given by

$$\mathcal{J} = \oint_{C(t)} \mathbf{p} \cdot d\mathbf{q}, \tag{2.57}$$

where the points on the closed curve $C(t)$ move according to the equations of motion. $\mathcal{J}$ is thus the sum of the areas of the projections of the curve $C(t)$ into the 3 planes $(q_1, p_1), (q_2, p_2)$, and $(q_3, p_3)$.

To show that $\mathcal{J}$ is exactly conserved, we introduce a periodic variable $s$ parametrizing the points on the curve $C$. The points of $C$ are thus given by $q_i = q_i(s,t)$ and $p_i = p_i(s,t)$. The rate of change of $\mathcal{J}$ is then

$$\frac{d\mathcal{J}}{dt} = \oint \left( p_i \frac{\partial^2 q_i}{\partial s \partial t} + \frac{\partial p_i}{\partial t} \frac{\partial q_i}{\partial s} \right) ds. \qquad (2.58)$$

We integrate the first term by parts, and use Hamilton's equations of motion to simplify the result:

$$\frac{d\mathcal{J}}{dt} = \oint \left( -\frac{\partial p_i}{\partial s} \frac{\partial q_i}{\partial t} + \frac{\partial p_i}{\partial t} \frac{\partial q_i}{\partial s} \right) ds \qquad (2.59)$$

$$= -\oint \left( \frac{\partial p_i}{\partial s} \frac{\partial H}{\partial p_i} + \frac{\partial H}{\partial q_i} \frac{\partial q_i}{\partial s} \right) ds, \qquad (2.60)$$

where $H(\mathbf{q}, \mathbf{p}, t)$ is the Hamiltonian for the motion. The integrand is now seen to be the total derivative of H along the curve $C$. Since the Hamiltonian is single valued, it follows that

$$\frac{d\mathcal{J}}{dt} = -\oint \frac{dH}{ds} ds = 0. \qquad (2.61)$$

The constancy of $\mathcal{J}$ is thus a direct consequence of the Hamiltonian nature of the equations motion. Note that $C$ need not bear any relationship to the the trajectories of the particles.

## Adiabatic invariants

Poincaré invariants are generally of little use since it is necessary to know the motion of the entire curve $C(t)$ before $\mathcal{J}$ can be evaluated. For the motion of magnetized particles, however, it follows from the form of the solution, (2.21) and (2.33), that all the points having the same guiding center at some time will continue to have approximately the same guiding center at a later time. Consider thus the Poincaré invariant $\mathcal{J}_{gc}$ obtained by taking the curve $C(t_0)$ to be the circle of points corresponding to a gyrophase period, $C_{gc}(X(t_0))$. At later time, $C(t) \cong C_{gc}(X(t))$, so that

$$\mathcal{J} \cong J \equiv \oint \mathbf{p} \frac{\partial \mathbf{q}}{\partial \gamma} d\gamma. \qquad (2.62)$$

Here $J$ is an *adiabatic invariant*. The adiabatic invariant is easily evaluated, but its invariance is only approximate.

To evaluate $J$ for a magnetized particle recall that the canonical momentum for charged particles is

$$\mathbf{p} = m\mathbf{v} + e\mathbf{A}, \qquad (2.63)$$

where $\mathbf{A}$ is the vector potential. The element of length along the curve $C(t)$ is

$$\frac{\partial \mathbf{q}}{\partial \gamma} d\gamma = \frac{\partial \boldsymbol{\rho}}{\partial \gamma} d\gamma = \frac{\mathbf{u}_\perp}{\Omega} d\gamma. \qquad (2.64)$$

The adiabatic invariant is thus

$$J = \frac{1}{\Omega} \oint d\gamma \, \mathbf{u}_\perp \cdot [m(\mathbf{U} + \mathbf{u}_\perp) + e\mathbf{A}(\mathbf{r})] + O(\epsilon); \tag{2.65}$$

$$= 2\pi m \frac{u_\perp^2}{\Omega} + 2\pi \frac{e}{\Omega} \langle \mathbf{u} \cdot \mathbf{A}(\mathbf{r}) \rangle \rangle + O(\epsilon). \tag{2.66}$$

The last term is evaluated as in (2.54)

$$\langle \mathbf{u} \cdot \mathbf{A}(\mathbf{r}) \rangle = -\frac{u_\perp^2}{2\Omega} \mathbf{b} \cdot \nabla \times \mathbf{A}.$$

It follows that

$$J = 2\pi \frac{m}{e} \mu + O(\epsilon). \tag{2.67}$$

The adiabatic invariant $J$ is thus proportional to the magnetic moment.

## 2.6   Case of stationary fields

The case of time-independent fields is of particular practical importance and allows several simplifications. The most important consequence of time-invariance, energy conservation, follows directly from (2.55):

$$\frac{d\mathcal{E}}{dt} = 0, \tag{2.68}$$

where

$$\mathcal{E} = \mathcal{K} + e\Phi = \frac{m}{2}(U_\parallel^2 + \mathbf{v}_E^2) + \mu B + e\Phi. \tag{2.69}$$

Here $\Phi$ is the electrostatic potential. All the field quantities ($\mathbf{v}_E$, $B$, and $\Phi$) are evaluated at the guiding center position. It is frequently advantageous to use the conserved quantity $\mathcal{E}$, instead of $U_\parallel$, as independent variable: $U_\parallel$ may then be considered to be a function of $\mathcal{E}, \mu$ and $\mathbf{R}$,

$$U_\parallel(\mathcal{E}, \mu, \mathbf{R}) = \sqrt{(2/m)[\mathcal{E} - \mu B - e\Phi] - \mathbf{v}_E^2}. \tag{2.70}$$

The parallel velocity vanishes at points where $\mathcal{E} = \mu B + e\Phi + m\mathbf{v}_E^2/2$. At such points, the particle reverses its direction along the field lines, hence they are named "bounce points" or "turning points."

### Magnetic mirror

In confined plasmas, the electric field is usually weak, $E \ll Bv$, so that the electric drift is no larger than the curvature and magnetic drifts. The lowest order guiding-center velocity is then parallel to the magnetic field: the particles behave as if they are tied to the magnetic field lines. The turning point condition, $U_\parallel = 0$, reduces to

$$\mathcal{E} = \mu B(\mathbf{R}).$$

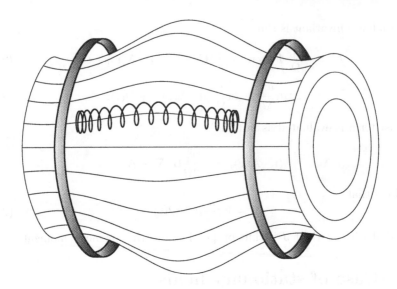

Figure 2.6: Motion of trapped particles in a magnetic mirror.

Expressing the energy in terms of the total velocity, we find that particles are trapped by the magnetic field *unless*

$$\frac{u_\perp^2}{|\mathbf{v}|^2} < \frac{B(\mathbf{R})}{B_{\max}}$$

where $B_{\max}$ is the maximum of the magnetic field amplitude on the field line the particle is tied to. This condition describes a cone in velocity-space called the *loss-cone*, since particles in this region are free to move along the field lines. The remaining, trapped particles undergo periodic motion between the bounce points. Trapping of energetic solar-wind particles by the earth's magnetic field gives rise to the Van Allen belts in the magnetosphere. The scattering of trapped particles into the loss cones causes them to stream into the upper atmosphere, creating the polar auroras.

For weak electric fields, the polarization drift (2.50) and the convective acceleration term in (2.48) are small and the expression for the guiding center drift simplifies to

$$\mathbf{U}_\perp = \frac{\mathbf{b}}{m\Omega} \times (\mu\nabla B + e\nabla\Phi + mU_\parallel^2\,\mathbf{b}\cdot\nabla\mathbf{b}). \qquad (2.71)$$

When the plasma carries no current, $\nabla \times \mathbf{B} = 0$, the inertial and gravitational drifts can be combined by using

$$\nabla \times \mathbf{B} = B\nabla \times \mathbf{b} + \nabla B \times \mathbf{b} = 0$$

to write

$$\kappa = \mathbf{b} \cdot \nabla\mathbf{b} = -\mathbf{b} \times (\nabla \times \mathbf{b}) = \frac{\nabla_\perp B}{B}.$$

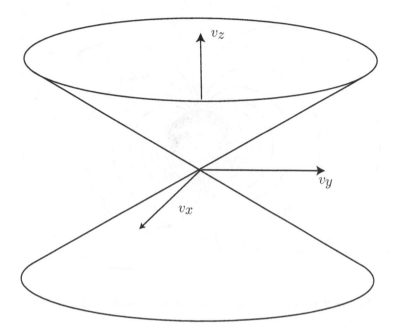

Figure 2.7: Loss cone in velocity space. The particles lying outside the cone at a given position are reflected by the magnetic mirror.

There follows

$$\mathbf{U}_\perp = \frac{\mathbf{b}}{\Omega} \times \left[ \left( U_\parallel^2 + \frac{\mathbf{u}_\perp^2}{2} \right) \frac{\nabla B}{B} + \frac{e}{m} \nabla \Phi \right] \qquad (\mathbf{J} = 0).$$

A more generally useful expression is obtained by adopting $\mathcal{E}$ as the independent variable instead of $U_\parallel$. We use

$$\frac{1}{2} m \nabla U_\parallel^2 = -\mu \nabla B - e \nabla \Phi, \tag{2.72}$$

where all the spatial derivatives are now taken at constant $\mathcal{E}$ and $\mu$. Using

$$\mathbf{b} \cdot \nabla \mathbf{b} = -\mathbf{b} \times (\nabla \times \mathbf{b}), \tag{2.73}$$

the guiding center drift takes the form

$$\mathbf{U}_\perp = \frac{U_\parallel}{\Omega} \left[ \nabla \times (U_\parallel \mathbf{b}) \right]_\perp. \tag{2.74}$$

The complete guiding center velocity may then be written

$$\mathbf{U} = \frac{U_\parallel}{B} \left[ \mathbf{B} + \frac{m}{e} \nabla \times (U_\parallel \mathbf{b}) \right]. \tag{2.75}$$

This expression is correct to lowest order, namely to order $\epsilon$ in the perpendicular direction and to order 1 in the parallel direction.

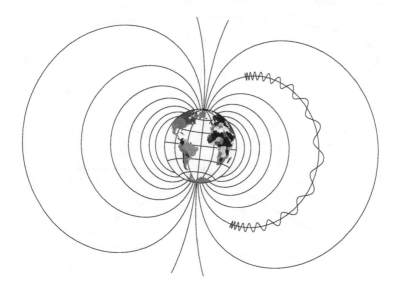

Figure 2.8: Motion of trapped particles in the earth's magnetic field.

Introducing the vector potential

$$\mathbf{A}^* = \mathbf{A} + \frac{U_\parallel}{\Omega}\mathbf{B}, \qquad (2.76)$$

we see that the guiding center velocity is

$$\mathbf{U} = \frac{U_\parallel}{B}\mathbf{B}^* \qquad (2.77)$$

where

$$\mathbf{B}^* = \nabla \times \mathbf{A}^* = \mathbf{B} + \frac{m}{e}\nabla \times (U_\parallel \mathbf{b}) \qquad (2.78)$$

is a pseudo-magnetic field. The guiding center follows the field line for the pseudo-magnetic field $\mathbf{B}^*$.

## Second adiabatic invariant

The nearly-periodic motion that occurs when a particle is trapped between two turning points has associated with it an adiabatic invariant, called the *second adiabatic invariant*. This second invariant is given by

$$J = \oint U_\parallel ds$$

where s is the element of length along the magnetic field. Conservation of the second adiabatic invariant is equivalent to the conservation of the magnetic flux through the bounce orbit of the guiding center.

For the important problem of the motion of particles in the magnetosphere, there is a third invariant associated with the drift of the particles around the earth. This third invariant accounts for the fact that particles can be confined, thus giving rise to the well-known radiation belts or Van Allen belts. In laboratory plasmas, by contrast, the occurrence of collisions severely limits the usefulness of the higher adiabatic invariants.

## 2.7 Guiding center Lagrangian

Physical systems frequently have symmetry properties that give rise to conserved quantities, such as angular momentum in azimuthally symmetric systems or energy in time-independent systems. More generally, all Hamiltonian systems conserve phase space volume (Liouville's theorem). The guiding center equations we have derived, however, can only be expected to satisfy these conservation laws approximately, within the error set by the order to which they have been derived. Fortunately, it is possible to find guiding center equations that satisfy the conservations laws to all orders in the expansion parameter. We refer to these as Hamiltonian guiding-center equations.[12, 50]

A simple method, due to Littlejohn,[51], for finding the Hamiltonian guiding-center equations is to use a variational principle. Note that in general, variational principles have the advantage of being invariant under arbitrary coordinate transformations. They have the further advantage that the invariance associated with symmetries follows transparently from the independence of the Lagrangian on the corresponding symmetry coordinates. In an azimuthally symmetric system, for example, one has immediately

$$\frac{\partial L}{\partial \zeta} = 0 \implies \frac{dp_\zeta}{dt} = 0,$$

where

$$p_\zeta = \frac{\partial L}{\partial \dot{\zeta}}$$

is the angular momentum and the dots denote derivatives with respect to time.

Littlejohn's variational principle is

$$\delta \mathcal{L} = 0, \tag{2.79}$$

where

$$\mathcal{L} = \int L(\mathbf{r}(t), \dot{\mathbf{r}}(t), \mathbf{p}(t), t) \, dt. \tag{2.80}$$

The Lagrangian $L$ is given by

$$L = \mathbf{p} \cdot \dot{\mathbf{r}} - H(\mathbf{p}, \mathbf{r}, t) \tag{2.81}$$

where $H$ is the Hamiltonian,

$$H(\mathbf{p}, \mathbf{r}, t) = e\Phi(\mathbf{r}, t) + \frac{1}{2}m\mathbf{v}^2$$

and $\mathbf{p}$ is the canonical momentum given by (2.63). This variational principle differs from Lagrange's in that the variation is taken over phase-space trajectories. That is, $\dot{\mathbf{r}}$ and $\mathbf{v}$ are to be considered independent when calculating the variation of $\mathcal{L}$; the equations of motion will specify that they should be equal. Littlejohn has shown that this variational principle guarantees the exact conservation of phase space volume.

The idea of the method is that the Lagrangian for the guiding center equations is the gyro-average of the exact Lagrangian. This follows from the observation that Lagrange's equations are invariant if the Lagrangian is replaced by

$$L' = L - \frac{dS}{dt},$$

since the total derivative of $S$ does not contribute to the integral in (2.80). The oscillating part of the exact Lagrangian can always be written approximately as a total time derivative,

$$L - \langle L \rangle = \Omega \frac{\partial S}{\partial \gamma} \sim \frac{dS}{dt} + O(\epsilon), \qquad (2.82)$$

so that this oscillating part can be eliminated term by term from the Lagrangian. The result is

$$L'(\mathbf{R}, \dot{\mathbf{R}}, U_\parallel, \mu, \gamma) = \langle L \rangle = e\mathbf{A}^* \cdot \dot{\mathbf{R}} + \frac{m}{e}\mu\dot{\gamma} - H', \qquad (2.83)$$

where

$$H' = e\Phi + \mu B + \frac{m}{2}U_\parallel^2. \qquad (2.84)$$

The variational principle (2.79), with the Lagrangian (2.83), reproduces the equations of motion to the order derived previously. To verify this, we write the Lagrange equations for the six independent variables $Z_i = (X_i, U_\parallel, \gamma, \mu)$:

$$\frac{\delta L}{\delta Z_i} = \frac{d}{dt}\left(\frac{\partial L}{\partial \dot{Z}_i}\right) - \frac{\partial L}{\partial Z_i} = 0. \qquad (2.85)$$

For the three velocity variables we find $U_\parallel = \mathbf{b} \cdot \dot{\mathbf{R}}$, $\dot{\mu} = 0$, and $\dot{\gamma} = \Omega$. These equations are identical to our previous results. For the guiding center coordinates, by contrast, we find

$$\mathbf{E}^* + \dot{\mathbf{R}} \times \mathbf{B}^* = \dot{U}_\parallel \mathbf{b} + \mu \nabla B, \qquad (2.86)$$

where the modified magnetic and electric fields are defined by (2.78) and

$$\mathbf{E}^* = \frac{\partial \mathbf{A}^*}{\partial t} - \nabla\Phi = \mathbf{E} - \frac{m}{e}U_\parallel \frac{\partial \mathbf{b}}{\partial t}. \qquad (2.87)$$

It is a simple matter to solve (2.86) for the rate of change of the parallel velocity and for the guiding center drift. Comparing (2.86) to (2.44), we note that the former contains higher order corrections. Littlejohn has shown that with these corrections, the guiding center equations *exactly* conserve the phase-space volume, energy, and (in the case of symmetry) conserved conjugate momenta.

## 2.8 Motion in oscillating fields

We have seen that particles can be confined in static magnetic fields. A more surprising fact is that particles can also be confined by rapidly oscillating, inhomogeneous electromagnetic fields. The demonstration of this fact uses again the method of averages. To lowest order, the particles oscillate in the wave-field. A weak inhomogeneity of the fields, however, causes the restoring force to be unequal on opposite parts of the trajectory. The average of the corresponding quasi-linear corrections yields a force that acts on the *center of oscillation*.

Consider a spatially inhomogeneous electromagnetic field oscillating at the frequency $\omega$,

$$\mathbf{E}(\mathbf{x}, t) = \mathbf{E}_0(\mathbf{x}) \cos \omega t.$$

The particles respond to this field through Newton's law,

$$m \frac{d\mathbf{v}}{dt} = e \left[ \mathbf{E}_0(\mathbf{x}) \cos \omega t + \mathbf{v} \times \mathbf{B}_0(\mathbf{x}) \sin \omega t \right], \qquad (2.88)$$

where

$$\mathbf{B}_0 = -\omega^{-1} \nabla \times \mathbf{E}_0. \qquad (2.89)$$

In order for the method of averages to be applicable, the amplitude $\mathbf{E}_0$ experienced by the particle must be approximately constant during a period of oscillation

$$\mathbf{v} \cdot \nabla \mathbf{E} \ll \omega \mathbf{E}. \qquad (2.90)$$

When this is satisfied, Faraday's law (2.89) implies that the magnetic force is smaller than the electric force by one order in the expansion parameter. That is, (2.90) is equivalent to the condition that the particles be unmagnetized, $\Omega \ll \omega$.

We now apply the method of averages. We make the substitution $t \to \tau$ in the oscillatory terms and seek a change of variables

$$
\begin{aligned}
\mathbf{x} &= \mathbf{X} + \boldsymbol{\xi}(\mathbf{X}, \mathbf{U}, t, \tau); \\
\mathbf{v} &= \mathbf{U} + \mathbf{u}(\mathbf{X}, \mathbf{U}, t, \tau),
\end{aligned}
$$

such that $\boldsymbol{\xi}$ and $\mathbf{u}$ are periodic functions of $\tau$ with vanishing mean. We average $d\mathbf{x}/dt = v$ and find again $d\mathbf{X}/dt = \mathbf{U}$ to all orders. The momentum equation is, to lowest order,

$$\frac{\partial \mathbf{u}(t)}{\partial \tau} = \frac{e}{m} \mathbf{E}_0(\mathbf{X}) \cos \omega \tau. \qquad (2.91)$$

The solution, taking into account (2.14), is

$$\mathbf{u} = \frac{e}{m\omega} \mathbf{E}_0 \sin \omega \tau; \qquad (2.92)$$

$$\boldsymbol{\xi} = -\frac{e}{m\omega^2} \mathbf{E}_0 \cos \omega \tau. \qquad (2.93)$$

The $\tau$-average of the force at the oscillation center vanishes for purely oscillatory fields. The acceleration of the oscillation center is thus caused solely by the quasi-linear force,

$$\frac{d\mathbf{U}}{dt} = \frac{e}{m}\langle \boldsymbol{\xi} \cdot \nabla \mathbf{E} + \mathbf{u} \times \mathbf{B}\rangle \tag{2.94}$$

$$= -\frac{e^2}{m^2\omega^2}\langle \mathbf{E}_0 \cdot \nabla \mathbf{E}_0 \cos^2 \omega\tau + \mathbf{E}_0 \times (\nabla \times \mathbf{E}_0)\sin^2 \omega\tau\rangle, \tag{2.95}$$

where we have eliminated the magnetic field with (2.89). The averages of the trigonometric functions yield a factor of $1/2$, and we find

$$m\frac{d\mathbf{U}}{dt} = -e\nabla\Phi_{\mathrm{pond}}, \tag{2.96}$$

where

$$\Phi_{\mathrm{pond}} = \frac{1}{4}\frac{e}{m\omega^2}|\mathbf{E}_0|^2. \tag{2.97}$$

The oscillation center experiences a force, called the *ponderomotive force*, proportional to the gradient of the amplitude of the wave field. The ponderomotive force is independent of the sign of the charge, so that both ions and electrons can be confined in the same well.

The total energy of the oscillation center,

$$\mathcal{E}_{oc} = \frac{m}{2}\mathbf{U}^2 + e\Phi_{\mathrm{pond}}.$$

is conserved by (2.96). Note that the ponderomotive potential-energy is proportional to the average kinetic-energy of the oscillatory motion,

$$e\Phi_{\mathrm{pond}} = \frac{m}{2}\langle \mathbf{u}^2\rangle.$$

The force on the oscillation center originates thus in a transfer of energy from the oscillatory motion to the average motion.

An important application of the ponderomotive force is to guide intense laser pulses. Laser light can propagate in a plasma if its frequency is greater than the plasma frequency: the plasma is then said to be "underdense." If the laser beam is sufficiently intense, the plasma particles are repulsed from the center of the beam by the ponderomotive force. The resulting variation in the plasma density gives rise to a cylindrical well in the index of refraction. The well acts as a waveguide for the laser light.

# Additional reading

The most complete treatment of particle motion in a magnetic field is that given by Morozov and Solov'ev in their review article.[52] This review article treats the case of strong electric field relevant to the MHD ordering, the relativistic equations, the effect of Bremsstrahlung, and even considers briefly the guiding-center hamiltonian. A more direct derivation of the guiding center drift is given by Northrop.[54] The use of averaged Lagrangians to describe evolution on a slow time scale is introduced in a paper by Whitham.[77] This paper addresses itself to the problem of wave propagation in a slowly varying medium (see Chapter 4) but is nevertheless a good general source on averaged Lagrangians. The application of the averaged Lagrangian method to the description of single particle motion is given by Littlejohn.[51]

# Exercises

1. Evaluate the Larmor frequency, bounce frequency, and drift velocity for a 10 MeV proton trapped in the Van Allen belt. Assume that this proton crosses the equatorial plane at a distance of $3R_E$ from the center of the earth (Here $R_E = 6 \cdot 10^6$m is the radius of the earth).

2. Discuss the existence of adiabatic invariants for a particle in a rapidly oscillating electromagnetic field.

3. Extend the analysis of the motion of a particle in rapidly varying electromagnetic fields to the case of a traveling wave-packet.

4. **Relativistic guiding-center equations**: Apply the method of averages to the equation for the relativistic motion of a particle in electromagnetic fields given in (10.24), using the Faraday tensor given in (10.20), in order to show that the relativistic guiding center motion obeys the equations:

$$\mathbf{U} = U_{0\|}\mathbf{b} - \frac{\mathbf{b}}{\Omega} \times \left( \frac{e}{m}\mathbf{E} - U_{0\|}^2 \boldsymbol{\kappa} - \frac{u_\perp^2}{2\Omega}\nabla\Omega \right); \qquad (2.98)$$

$$\frac{d\mathcal{E}}{dt} = e\mathbf{E} \cdot \mathbf{U} + \mu\mathbf{b} \cdot \nabla \times \mathbf{E}; \qquad (2.99)$$

$$\frac{d\mu}{dt} = 0. \qquad (2.100)$$

Here

$$\mathcal{E} = \frac{m_0 c^2}{\sqrt{1 - (U_{0\|}^2 + u_\perp^2)/c^2}}$$

and

$$\mu = \frac{mu_\perp^2}{2B} \frac{1}{1 - (U_{0\|}^2 + u_\perp^2)/c^2}.$$

5. **Speiser orbits**: Consider the magnetic field

$$\mathbf{B} = B_0\mathbf{e}_y, \qquad x > a;$$

$$\mathbf{B} = \frac{x}{a}B_0\mathbf{e}_y, \qquad |x| < a;$$

$$\mathbf{B} = -B_0\mathbf{e}_y, \qquad x < -a;$$

corresponding to a current sheet such as that in the magnetotail of the earth. Near the $x = 0$ plane of this configuration, the magnetic field vanishes so that the guiding center theory may not be applied. Use the conservation of energy and $\mathbf{e}_z$-momentum to show that the orbits are bounded in the $\mathbf{x}$ direction. More specifically, show that some orbits do not cross the neutral plane and resemble magnetized particle orbits, while others meander across the neutral plane. Is it possible to define an adiabatic invariant for the meandering orbits?

6. **Poincaré's monopolar mirror:** Show that particles moving around a magnetic monopole,

$$\mathbf{B} = B_0 \frac{\hat{\mathbf{r}}}{r^2},$$

follow geodesic lines on circular cones centered on the monopole (see Figure 2.9). Write down the Lagrangian, and find all the constants of the motion. Compare the exact solution to the predictions of guiding center theory. The motion of a charged particle around a monopole was first studied by Poincaré in the context of magnetic mirroring in the polar cusp regions.

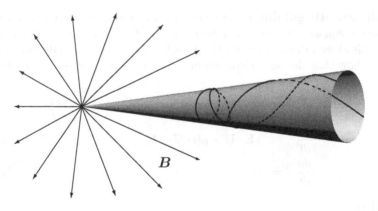

Figure 2.9: Motion of magnetized particles in a monopolar field.

7. Consider the motion of a particle in a time varying electromagnetic field described by the vector potential $\mathbf{A} = B(t)x\hat{\mathbf{y}}$, where

$$B^2(t) = \frac{1}{2}(B_i^2 + B_f^2) - \frac{1}{2}(B_i^2 - B_f^2)\tanh(t/\tau).$$

Here $B_i$ and $B_f$ are the initial and final values of the magnetic field. Solve the equation of motion,

$$\frac{d^2x}{dt^2} + \Omega^2 x = 0$$

where $\Omega = eB/m$, using the asymptotic initial condition

$$x(t) \sim \mathrm{Re}[e^{i\Omega t}], \qquad t \to -\infty.$$

Compare the asymptotic value of the magnetic moment at large times with its initial value and show that the change is proportional to $\exp(-\pi\Omega_f\tau)$, where $\Omega_f = eB_f/m$.

8. **Motion about a straight current filament** Solve the equations of motion for a particle moving in the magnetic field of an infinite, straight current filament. Find the potential energy for the radial motion, and sketch it. Compare the exact results to the predictions of guiding center theory.

# Chapter 3

# Fluid Description of a Plasma

## 3.1  Value of fluid description

In its fluid description, a plasma is characterized by a few local parameters such as the particle density, (kinetic) temperature and flow velocity. These quantities are advanced in time by means of fluid equations that are analogous to, but usually more complicated than, the equations of ordinary hydrodynamics. Thus the goal of plasma fluid theory is to construct and solve a plasma version of the Navier-Stokes equation.

We have emphasized that the key issue in any physical description of a plasma is closure: finding a set of equations that, together with boundary and initial conditions, will determine the plasma-system's evolution. Plasma fluid equations can be viewed as one strategy for achieving closure. But we will find that the derivation of a closed set of fluid equations is not easy or clean-cut, and generally requires serious approximation. Furthermore, we will note that an alternative, seemingly more straightforward closure is available at the deeper level of kinetic theory. It is therefore appropriate to begin by examining the value and role of fluid equations in plasma physics.

This and the next section will use some elementary results of kinetic theory: the Boltzmann equation and a few properties of the collision operator. Readers unfamiliar with this material might want to glance at the first section of Chapter 6 before proceeding.

We have noted that plasma physics can be viewed formally as a closure of Maxwell's equations by means of *constitutive relations*: expressions for the charge density, $\rho_c$ and current density $\mathbf{J}$ in terms of the electric and magnetic fields $\mathbf{E}$ and $\mathbf{B}$. Such constitutive relations are easily expressed in terms of the microscopic, one-particle distribution functions, $\mathcal{F}_s$, for each plasma species:

$$\rho_c = \sum_s e_s \int d^3 v \mathcal{F}_s(\mathbf{x}, \mathbf{v}, t), \tag{3.1}$$

$$\mathbf{J} = \sum_s e_s \int d^3 v \mathbf{v} \mathcal{F}_s(\mathbf{x}, \mathbf{v}, t). \tag{3.2}$$

Here $\mathcal{F}_s(\mathbf{x}, \mathbf{v}, t)$ is the exact, "microscopic" phase-space density of plasma species $s$ at the point $(\mathbf{x}, \mathbf{v})$, at time $t$. Note that we normalize the distribution such that its velocity integral is the ordinary (coordinate-space) particle density:

$$\int d^3 v \mathcal{F}_s(\mathbf{x}, \mathbf{v}, t) = n_s(\mathbf{x}, t),$$

where $n_s(\mathbf{x}, t)$ is the number of species-$s$ particles per unit volume in an infinitesimal volume surrounding the point $\mathbf{x}$ at time $t$.

If one could determine each $\mathcal{F}_s(\mathbf{x}, \mathbf{v}, t)$ in terms of the electromagnetic field, then (3.1) and (3.2) immediately give the desired constitutive relations. Furthermore it is easy to see, in principle, how each distribution evolves. Phase-space conservation requires

$$\frac{\partial \mathcal{F}_s}{\partial t} + \mathbf{v} \cdot \nabla \mathcal{F}_s + \mathbf{a}_s \cdot \frac{\partial \mathcal{F}_s}{\partial \mathbf{v}} = 0. \tag{3.3}$$

Here $\mathbf{a}_s(\mathbf{x}, \mathbf{v}, t)$ is the $s$-species particle acceleration, given by the Lorentz force:

$$m_s \mathbf{a}_s = e_s(\mathbf{E} + \mathbf{v} \times \mathbf{B}) \tag{3.4}$$

where $m_s$ is the mass.

In other words the distribution for each species, from which the constitutive relations are trivially obtained, is determined by a single first-order partial differential equation—and a rather harmless looking equation at that! (Note that its only nonlinearity is in the last term, involving acceleration.) Since plasma fluid descriptions always involve several equations and usually include second- or higher-order derivatives, it is far from obvious that fluid theory represents an advance.

This argument is misleading for several reasons, but its most egregious flaw is the view of (3.3) as a tractable equation. Note that (3.3) is easy to derive because it is exact, taking into account all scales from the quantum limit to the system size; it is also microscopic in the sense of involving no statistical average. Thus the distribution in (3.3) is essentially a collection of Dirac delta-functions, each following the detailed trajectory of a single particle. Similarly the fields in (3.3) are horrendously spiky and chaotic, reflecting as they must the instantaneous charges and currents associated with all these trajectories. Solving (3.3), in other words, amounts to nothing less than solving the classical electromagnetic many-body problem—a hopeless and hardly interesting task.

A much more useful and tractable equation can be extracted from (3.3) by ensemble-averaging. The averaged distribution

$$\bar{\mathcal{F}} = \langle \mathcal{F} \rangle_{ensemble}$$

is sensibly smooth, and closely related to experimental measurements. Similarly the ensemble-averaged fields are smoothed, although an appropriate ensemble will preserve the field (and density) fluctuations of interest. Finally the averaged equation is amenable to analysis or numerical solution—at least in limiting cases.

Examples of ensemble-averaged kinetic descriptions are studied in Chapter 6. Here we emphasize that the extraction of an ensemble-averaged equation from (3.3) is a challenging exercise in mathematical physics, always requiring severe approximation. The problem is that, since the exact electromagnetic field depends on the trajectories, it is not statistically independent of $\mathcal{F}$. In other words the nonlinear acceleration term in (3.3),

$$\langle \mathbf{a_s} \cdot \frac{\partial \mathcal{F}_s}{\partial \mathbf{v}} \rangle_{ensemble} \neq \bar{\mathbf{a}}_s \cdot \frac{\partial f_s}{\partial \mathbf{v}}$$

involves *correlations* that must be considered separately. Here we have introduced the simplified notation

$$f_s \equiv \bar{\mathcal{F}}_s. \tag{3.5}$$

The traditional goal of kinetic theory is to analyze the correlations, using approximations tailored to the parameter regime of interest, and thus to express the averaged acceleration term in terms of $f$ and the averaged fields alone. For present purposes we suppose this ambitious task has been completed, giving an expression of the form

$$\langle \mathbf{a}_s \cdot \frac{\partial \mathcal{F}_s}{\partial \mathbf{v}} \rangle_{ensemble} = \bar{\mathbf{a}}_s \cdot \frac{\partial f_s}{\partial \mathbf{v}} - C_s(f) \tag{3.6}$$

where C is a generally complicated operator that accounts for correlations. Because the most important correlations result from close encounters between particles, C is called the *collision operator*. It is not necessarily a linear operator, and usually involves the distributions of both species; the subscript on the argument of C is omitted for that reason.

We conclude that the useful version of (3.3) is both challenging to derive and difficult to solve.

There are situations in which correlations are not important and the collision operator can be neglected. Then, of course, the apparent simplicity of (3.3) is not deceptive: a useful kinetic description is obtained simply by inserting overbars on the distribution and fields. The resulting closure, called the *Vlasov* system, is tractable in sufficiently simple geometry. But even in the Vlasov limit, the fluid approach has much to contribute: it has intrinsic advantages that weigh in almost every context.

First of all, fluid equations possess the key simplicity of involving fewer dimensions: the three spatial coordinates on which each fluid variable depends, rather than the six dimensions of phase space. This advantage can be especially important in computational studies.

Second, as mentioned in Chapter 1, the fluid description is intuitively appealing. One has an immediate sense of the significance of density and temperature, for example; this familiarity can help one's thinking about plasma behavior.

Moreover the fluid variables are often the most important quantities for plasma physical applications. The density and temperature, for example, determine radiation levels, reaction rates, and a host of other processes. Being able to predict the evolution of these parameters is often the key demand made on plasma physics.

Finally, the inefficiency of the kinetic approach to closure has to be mentioned. Even when the kinetic equation can be solved, it provides vastly more information than is needed for constitutive relations. The latter, after all, involve only the two lowest moments of $\mathcal{F}_s$, as in (3.1) and (3.2). All other moments of the distribution, involving higher powers of $\mathbf{v}$, could be arbitrarily changed without affecting the evolution of the fields. While fluid theory cannot compute $\rho_c$ and $\mathbf{J}$ without reference to other moments, it can be viewed as an attempt to impose some efficiency on the task of dynamical closure.

## 3.2   Moments of the distribution function

By the *kth* moment of the (averaged) distribution function $f(\mathbf{x}, \mathbf{v}, t)$ we mean the velocity integral of the product of $f$ and $k$ powers of the velocity (sometimes including a mass or numerical factor):

$$\mathbf{M}_k(\mathbf{x}, t) = \int d^3 v f(\mathbf{x}, \mathbf{v}, t) \mathbf{v} \mathbf{v} \cdots \mathbf{v} \tag{3.7}$$

with $k$ factors of $\mathbf{v}$. Thus $\mathbf{M}_k$ is in general a tensor of rank $k$, although it is often contracted to lower rank.

For some idealized distributions, the higher-order moments may not exist. When all moments do exist, the set $\{\mathbf{M}_k, k = 1, 2, \cdots\}$ can be viewed as an alternative description of the distribution, which indeed uniquely specifies $f$ when the latter is sufficiently smooth. A (displaced) Gaussian distribution, for example, is uniquely specified by three moments: $M_0$, the vector $\mathbf{M}_1$, and the scalar formed by contracting $\mathbf{M}_2$.

The low-order moments have names and simple interpretations. First we have the (particle) density,

$$n_s(\mathbf{x}, t) = \int d^3 v f_s(\mathbf{x}, \mathbf{v}, t), \tag{3.8}$$

and the particle *flux density*,

$$n_s \mathbf{V}_s(\mathbf{x}, t) = \int d^3 v f_s(\mathbf{x}, \mathbf{v}, t) \mathbf{v} \tag{3.9}$$

The quantity $\mathbf{V}_s$ by itself is called the plasma *flow velocity*. Note that the electromagnetic sources, (3.1) and (3.2), are determined by these lowest moments:

$$\rho_c = \sum_s e_s n_s, \quad \mathbf{J} = \sum_s e_s n_s \mathbf{V}_s. \tag{3.10}$$

The second order moment, describing the flow ($\mathbf{v}$) of momentum ($m\mathbf{v}$) in the laboratory frame, is called the *stress tensor* and denoted by:

$$\mathbf{P}_s(\mathbf{x}, t) = \int d^3 v f_s(\mathbf{x}, \mathbf{v}, t) m_s \mathbf{v}\mathbf{v} \tag{3.11}$$

Finally we will need the third order moment measuring *energy flux density*,

$$\mathbf{Q}_s(\mathbf{x}, t) = \int d^3 v f_s(\mathbf{x}, \mathbf{v}, t) \frac{1}{2} m_s v^2 \mathbf{v} \tag{3.12}$$

as well as a fourth order moment, the energy-weighted stress,

$$\mathbf{R}_s = \int d^3 v f_s \frac{1}{2} m_s v^2 \mathbf{v}\mathbf{v} \tag{3.13}$$

It is often convenient to measure the second- and third-order moments in the rest-frame of the species under consideration. In this case they assume different names: the stress tensor measured in the moving frame is called the *pressure tensor*, $\mathbf{p}$, while the energy flux density becomes the *heat flux density*, $\mathbf{q}$. We introduce the relative velocity,

$$\mathbf{w}_s \equiv \mathbf{v} - \mathbf{V}_s \tag{3.14}$$

in order to write

$$\mathbf{p}_s(\mathbf{x}, t) = \int d^3 v f_s(\mathbf{x}, \mathbf{v}, t) m_s \mathbf{w}_s \mathbf{w}_s \tag{3.15}$$

and

$$\mathbf{q}_s(\mathbf{x}, t) = \int d^3 v f_s(\mathbf{x}, \mathbf{v}, t) \frac{1}{2} m_s w_s^2 \mathbf{w}_s. \tag{3.16}$$

The trace of the pressure tensor measures the ordinary pressure ("scalar pressure"), denoted by $p$,

$$p_s \equiv \frac{1}{3} Tr(\mathbf{p}_s) \tag{3.17}$$

Note that $\frac{3}{2} p_s$ is the kinetic energy density of species s:

$$\frac{3}{2} p_s = \int d^3 f_s \frac{1}{2} m_s w_s^2 \tag{3.18}$$

In thermodynamic equilibrium, the distribution becomes a Maxwellian at temperature $T$, and (3.17) gives $p = nT$. It is therefore natural to define the temperature ("kinetic temperature") by

$$T_s \equiv \frac{p_s}{n_s} \tag{3.19}$$

Of course the moments measured in the two different frames are related. By direct substitution one verifies the useful identities

$$\mathbf{P}_s = \mathbf{p}_s + m_s n_s \mathbf{V}_s \mathbf{V}_s, \tag{3.20}$$

and

$$\mathbf{Q}_s = \mathbf{q}_s + \mathbf{p}_s \cdot \mathbf{V}_s + \frac{3}{2} p_s \mathbf{V}_s + \frac{1}{2} m_s n_s V_s^2 \mathbf{V}_s. \tag{3.21}$$

## 3.3  Fluid conservation laws

In this section we compute moments of the ensemble-averaged kinetic equation—the average of (3.3),

$$\frac{\partial f_s}{\partial t} + \mathbf{v} \cdot \nabla f_s + \bar{\mathbf{a}}_s \cdot \frac{\partial f_s}{\partial \mathbf{v}} = C_s(f). \tag{3.22}$$

Here we used (3.6) to express the average of the nonlinear term in terms of the collision operator $C$. We begin by modifying (3.22) in three ways:

1. For simplicity we suppress ensemble-average overbar on $\mathbf{a}_s$.

2. For convenience we express the acceleration term as

$$\mathbf{a}_s \cdot \frac{\partial f_s}{\partial \mathbf{v}} = \frac{\partial}{\partial \mathbf{v}} \cdot (\mathbf{a}_s f_s). \tag{3.23}$$

Because flow in velocity-space under the Lorentz force is incompressible,

$$\frac{\partial}{\partial \mathbf{v}} \cdot \mathbf{a}_s = 0,$$

the two forms are equivalent. The form on the right-hand side of (3.23) expresses phase-space conservation more directly and simplifies velocity integrals.

3. For generality we include a *source* term, $I_s(\mathbf{x}, \mathbf{v}, t)$ on the right-hand side. Thus we allow new particles of species $s$ to be introduced at the rate $I_s$, as a result of atomic or nuclear processes.

A common source term is ionization, which converts a neutral atom to a charged plasma particle; the opposite process, called *recombination*, corresponds of course to negative $I_s$. Note also that $I_s$ might represent a particle-conserving process, such as charge-exchange, in which an electron is exchanged between a neutral and a charged particle, leaving the number of each species unchanged. In that case the velocity-integral of $I_s$ would vanish, while higher-order moments would not.

Thus we obtain the kinetic equation,

$$\frac{\partial f_s}{\partial t} + \nabla \cdot (\mathbf{v} f_s) + \frac{\partial}{\partial \mathbf{v}} \cdot (\mathbf{a}_s f_s) = C_s(f) + I_s \tag{3.24}$$

Its *kth* moment equation is computed simply by multiplying (3.24) by $k$ powers of $\mathbf{v}$ and integrating. The flow term is simplified by pulling the divergence outside the velocity integral; the acceleration term is treated by partial integration. Notice that these two terms couple the moment $k$ to the neighboring-order moments $(k + 1)$ and $(k - 1)$, respectively.

The only nontrivial term is that involving the right-hand side: we need some information about the collision operator.

## Moments of the collision operator

Boltzmann's famous collision operator for a neutral gas considers only binary collisions and is therefore bilinear in the distribution functions of the two colliding species:

$$C_s(f) = \sum_{s'} C_{ss'}(f_s, f_{s'}), \tag{3.25}$$

where $C_{ss'}$ is linear in each of its arguments. We will find in Chapter 6 that such bilinearity is not strictly valid for the case of plasma collisions. Because of the long-range of the Coulomb interaction, the closest analogue to ordinary two-particle interaction is mediated by Debye shielding, an intrinsically many-body effect. Fortunately the departure from bilinearity is logarithmic and usually harmless; here we use (3.25) and presume the $C_{ss'}$ to be bilinear.

We take this opportunity to emphasize a feature of $C_{ss'}$ that is elementary yet sometimes confusing: there is no simple relation between the quantity $C_{ss'}$, which describes the effect *on* species $s$ of collisions *with* species $s'$, and the quantity $C_{s's}$. The two operators can have quite different mathematical forms (for example, when the masses $m_s$ and $m_{s'}$ are disparate) and they appear in different equations. Only $C_{ss'}$ enters the kinetic equation for the evolution of $f_s$.

Neutral particle collisions are characterized by Boltzmann's collisional conservation laws: the collisional process conserves particles, momentum and energy at each point. Such *local* conservation of momentum and energy is not trivial in a medium as electrodynamically active as plasma. One could easily imagine, for example, particle momentum being transferred to waves, which might travel macroscopic distances before delivering the momentum to another particle. Indeed, wave transport of momentum and energy can have important consequences in some plasma contexts. However, we prefer to treat such processes explicitly, rather than including them in the collision operator. In other words we define $C$ to include only locally conservative physics, and therefore we use conventionally local conservation laws.

*Particle conservation* is expressed by

$$\int d^3v \, C_{ss'} = 0. \tag{3.26}$$

*Momentum conservation* requires

$$\int d^3v \, m_s \mathbf{v} C_{ss'} = - \int d^3v \, m_{s'} \mathbf{v} C_{s's}, \tag{3.27}$$

That is, the net momentum exchanged between species $s$ and $s'$ must vanish. It is useful to introduce the rate of collisional momentum exchange, called the collisional friction force, or simply friction force:

$$\mathbf{F}_{ss'} \equiv \int d^3v \, m_s \mathbf{v} C_{ss'} \tag{3.28}$$

Thus $\mathbf{F}$ is the momentum-moment of the collision operator. The total friction force experienced by species $s$ is

$$\mathbf{F}_s \equiv \sum_{s'} \mathbf{F}_{ss'}.$$

Now momentum conservation is expressed in detailed form as

$$\mathbf{F}_{ss'} = -\mathbf{F}_{s's} \tag{3.29}$$

or in non-detailed form as

$$\sum_s \mathbf{F}_s = 0. \tag{3.30}$$

Collisional *energy conservation* requires the quantity

$$W_{Lss'} \equiv \int d^3v \frac{1}{2} m_s v^2 C_{ss'} \tag{3.31}$$

to be conserved in collisions:

$$W_{Lss'} + W_{Ls's} = 0. \tag{3.32}$$

Here the $L$-subscript indicates that the kinetic energy of both species is measured in the same "Lab" frame. What is essential is that each term of (3.32) is measured in the same frame; because of Galilean invariance, the frame choice doesn't matter.

An alternative collisional energy-moment is

$$W_{ss'} \equiv \int d^3v \frac{1}{2} m_s w_s^2 C_{ss'}, \tag{3.33}$$

the kinetic energy change experienced by species $s$, due to collisions with species $s'$, measured in the rest frame of species $s$. The total energy change for species $s$ is of course

$$W_s \equiv \sum_{s'} W_{ss'}$$

The easily verified identity

$$W_{Lss'} = W_{ss'} + \mathbf{V}_s \cdot \mathbf{F}_{ss'}, \tag{3.34}$$

allows us to express collisional energy conservation as

$$W_{ss'} + W_{s's} + (\mathbf{V}_s - \mathbf{V}_{s'}) \cdot \mathbf{F}_{ss'} = 0. \tag{3.35}$$

or, in nondetailed form,

$$\sum_s (W_s + \mathbf{V}_s \cdot \mathbf{F}_s) = 0. \tag{3.36}$$

Finally we consider moments of the source term, $I_s$. There are no generally applicable conservation laws (ionization, for example, does not conserve particle number, momentum or energy of any set of *plasma* species) so we need merely summarize notation. The $kth$ moment of $I_s$ is labeled by a subscript indicating the number of powers of $\mathbf{v}$; only three are needed:

$$I_{s0} = \int d^3v I_s, \tag{3.37}$$

$$\mathbf{I}_{s1} = \int d^3v \mathbf{v} I_s \tag{3.38}$$

$$I_{s2} = \int d^3v v^2 I_s. \tag{3.39}$$

As usual, it is convenient to define analogous quantities,

$$\mathbf{i}_{sk} = \int d^3w \mathbf{w}\mathbf{w}...\mathbf{w} I_s \tag{3.40}$$

where $\mathbf{w} = \mathbf{v} - \mathbf{V}$, in the moving frame. Note that

$$
\begin{aligned}
I_{s0} &= i_{s0}, \\
\mathbf{I}_{s1} &= \mathbf{i}_{s1} + \mathbf{V} i_{s0}, \\
I_{s2} &= i_{s2} + 2\mathbf{V} \cdot \mathbf{i}_{s1} + V^2 i_{s0}.
\end{aligned}
$$

## Moments of the kinetic equation

With the collisional conservation laws in hand, it is straightforward to calculate the desired moments of (3.24). Thus the zeroth moment of (3.24) yields the particle conservation law,

$$\frac{\partial n_s}{\partial t} + \nabla \cdot (n_s \mathbf{V}_s) = I_{s0}; \tag{3.41}$$

the first moment describes the rate of change of momentum $(m\mathbf{v})$ ,

$$m_s \frac{\partial n_s \mathbf{V}_s}{\partial t} + \nabla \cdot \mathbf{P}_s - e_s n_s (\mathbf{E} + \mathbf{V}_s \times \mathbf{B}) = \mathbf{F}_s + m\mathbf{I}_{s1}; \tag{3.42}$$

and the contracted second moment expresses conservation of energy $(\tfrac{1}{2}mv^2)$:

$$\frac{\partial}{\partial t}(\frac{3}{2}p_s + \frac{1}{2}m_s n_s V_s^2) + \nabla \cdot \mathbf{Q}_s - e_s n_s \mathbf{E} \cdot \mathbf{V}_s = W_s + \mathbf{V}_s \cdot \mathbf{F}_s + \frac{1}{2}m I_{s2}. \tag{3.43}$$

We have written these laws in the form which appears most directly from the moment calculation; other versions, obtained by rearrangement using (3.20) and (3.21), are considered presently.

The interpretation of the moment equations is also straightforward, especially since they all fit the same mold. Suppose $G$ is some physical quantity (particle number, energy...) and $g(\mathbf{x}, t)$ is its density:

$$G = \int d^3x\, g.$$

Then $g$ must evolve according to

$$\frac{\partial g}{\partial t} + \nabla \cdot \mathbf{\Gamma}_g = \Delta G \qquad (3.44)$$

where $\mathbf{\Gamma}_g$ is the flux density of $G$, and $\Delta G$ is the local rate at which $G$ is created or exchanged with other entities in the fluid. Thus the density of $G$ at some point $\mathbf{x}$ changes because there is a net flow to or away from that point (measured by the divergence term), or because of local sources of G (measured by the right-hand side).

Applying this mold to (3.41), we see that that particle number,

$$N = \int d^3x n,$$

changes only is so far as the source $I$ introduces (or loses) particles: (3.41) indeed expresses particle conservation. From the momentum law (3.42), often called the "equation of motion," we see that $\mathbf{P}$ measures *momentum flux density*, and that species $s$ momentum is changed by the Lorentz force, by collisional friction or by other momentum sources. Finally, (3.43) is the statement that the energy of species $s$ can be changed by electromagnetic work, by frictional heating, by exchange with other species and so on.

A comment on language is appropriate here. The energy flux density, $\mathbf{Q}$, is often called simply the "energy flux," and similarly the word "density" is frequently omitted in describing $\mathbf{q}$ and other flux densities. The abbreviation is not necessarily misleading, but the reader should not forget what the word "flux" really means. For any vector field $\mathbf{A}(\mathbf{x})$, we define the flux of $\mathbf{A}$ through some surface S, with surface normal $\mathbf{n}(\mathbf{x})$, to be

$$\text{flux}(A) = \int_S dS \mathbf{n} \cdot \mathbf{A}.$$

Thus for example, electric current $I$ is the flux of current density $\mathbf{J}$. A true flux is always a scalar, defined at least implicitly with respect to a particular surface. A flux density, on the other hand, is a vector field.

## 3.4   Alternative versions

### Convective derivative

Certain rearrangements of (3.41) – (3.43) are perspicuous and useful in particular contexts. In deriving such expressions we can suppress species subscripts.

First we eliminate the energy flux and stress tensor in terms of the heat flux and pressure tensor respectively, using (3.20) and (3.21). Inserting (3.20) into the momentum equation, noting that

$$\nabla \cdot (mn\mathbf{V}\mathbf{V}) = m\mathbf{V}\nabla \cdot (n\mathbf{V}) + mn\mathbf{V} \cdot \nabla\mathbf{V}$$

and using (3.41), we obtain

$$mn(\frac{\partial \mathbf{V}}{\partial t} + \mathbf{V} \cdot \nabla \mathbf{V}) + \nabla \cdot \mathbf{p} - en(\mathbf{E} + \mathbf{V} \times \mathbf{B}) = \mathbf{F} + m\mathbf{i}_1. \qquad (3.45)$$

We will use a conventional abbreviation, introducing the *convective derivative*,

$$\frac{d}{dt} \equiv \frac{\partial}{\partial t} + \mathbf{V} \cdot \nabla \qquad (3.46)$$

and thus expressing the equation of motion as

$$mn\frac{d\mathbf{V}}{dt} + \nabla \cdot \mathbf{p} - en(\mathbf{E} + \mathbf{V} \times \mathbf{B}) = \mathbf{F} + m\mathbf{i}_1$$

The convective derivative evidently measures temporal change in the rest frame of the fluid: it is the time derivative seen by an observer attached to a moving fluid element. In this regard, a word of caution is in order: the argument $\mathbf{x}$ in, for example, $\mathbf{V}(\mathbf{x}, t)$, is an independent variable, describing the coordinate space of the vector field $\mathbf{V}$. It is not a function of time—as it would be in Newtonian mechanics, or in a Lagrangian description of the fluid. Therefore the convective derivative is not a total derivative with respect to time, but rather a shorthand defined by (3.46).

Particle conservation can also be expressed in terms of the convective derivative:

$$\frac{dn}{dt} + n\nabla \cdot \mathbf{V} = i_0,$$

showing that the density is constant along the fluid trajectory for a source-free, solenoidal flow. For this reason the condition

$$\nabla \cdot \mathbf{V} = 0 \qquad (3.47)$$

is said to describe an *incompressible* fluid.

## Energy conservation

Turning to energy conservation, we first note that

$$\frac{\partial}{\partial t} \frac{1}{2} mnV^2 = \frac{1}{2} mV^2 \frac{\partial n}{\partial t} + \mathbf{V} \cdot mn\frac{\partial \mathbf{V}}{\partial t}$$

Here we eliminate the two time derivatives using the appropriate conservation laws and find

$$\frac{\partial}{\partial t} \frac{1}{2} mnV^2 = -\nabla \cdot (\frac{1}{2} mnV^2 \mathbf{V}) - \mathbf{V} \cdot \nabla \cdot \mathbf{p} + \mathbf{V} \cdot (\mathbf{F} + en\mathbf{E} + m\mathbf{i}_1) + \frac{1}{2} mV^2 i_0$$

Next we consider

$$\nabla \cdot \mathbf{Q} = \nabla \cdot \mathbf{q} + \nabla \cdot (\frac{1}{2} mnV^2 \mathbf{V}) + \nabla \cdot (\mathbf{V} \cdot \mathbf{p}) + \frac{3}{2} \nabla \cdot (p\mathbf{V})$$

When these two results are substituted into (3.43), there are several cancellations. In particular there occurs the combination

$$\nabla \cdot (\mathbf{V} \cdot \mathbf{p}) - \mathbf{V} \cdot (\nabla \cdot \mathbf{p}) = \mathbf{p} : \nabla \mathbf{V}$$

where the right hand side refers to the double contraction,

$$\mathbf{p} : \nabla \mathbf{V} = p_{\alpha\beta} \frac{\partial V_\beta}{\partial x_\alpha} \tag{3.48}$$

Here the $\alpha$ and $\beta$ subscripts refer to Cartesian components, and repeated indices are summed as usual. The resulting energy conservation law is

$$\frac{3}{2} \frac{\partial p}{\partial t} + \frac{3}{2} \nabla \cdot (p\mathbf{V}) + \mathbf{p} : \nabla \mathbf{V} + \nabla \cdot \mathbf{q} = W + \frac{1}{2} m i_2$$

This version is satisfactory, but one additional refinement is worthwhile. We introduce the *generalized viscosity tensor*, $\boldsymbol{\pi}$, by writing

$$\mathbf{p} = \mathbf{I} p + \boldsymbol{\pi}, \tag{3.49}$$

where $\mathbf{I}$ is the unit (identity) tensor. We know that the first, scalar pressure term will dominate if the plasma is close to thermal equilibrium (or if collisions are sufficiently frequent). We also expect, from conventional fluid theory, that the second term will describe viscous stress—an expectation that is typically verified, although the plasma pressure tensor can include terms unrelated to conventional viscosity. In any case the special significance of scalar pressure justifies substituting (3.49) into our energy conservation law, which becomes

$$\frac{3}{2} \frac{dp}{dt} + \frac{5}{2} p \nabla \cdot \mathbf{V} + \boldsymbol{\pi} : \nabla \mathbf{V} + \nabla \cdot \mathbf{q} = W + \frac{1}{2} m i_2$$

In words: the energy density of a fluid element changes because of compressive work, viscous heating, net heat flow and energy exchange with other plasma species. Notice that the electromagnetic effect on plasma energy, apparent in (3.43), has become entirely implicit.

Let us summarize. The first three moments of the kinetic equation can be expressed as

$$\frac{dn_s}{dt} + n_s \nabla \cdot \mathbf{V}_s = i_{s0}, \tag{3.50}$$

$$m_s n_s \frac{d\mathbf{V}_s}{dt} + \nabla p_s + \nabla \cdot \boldsymbol{\pi}_s - e_s n_s (\mathbf{E} + \mathbf{V}_s \times \mathbf{B}) = \mathbf{F}_s + m_s \mathbf{i}_{s1}, \tag{3.51}$$

$$\frac{3}{2} \frac{dp_s}{dt} + \frac{5}{2} p_s \nabla \cdot \mathbf{V}_s + \boldsymbol{\pi}_s : \nabla \mathbf{V}_s + \nabla \cdot \mathbf{q}_s = W_s + \frac{1}{2} m_s i_{s2}. \tag{3.52}$$

Note that we should strictly include an $s$ subscript with each convective derivative, since, for example,

$$\frac{dn_s}{dt} = \frac{\partial n_s}{\partial t} + \mathbf{V}_s \cdot \nabla n_s.$$

## Entropy production

Another version of (3.52), displaying entropy evolution, is sometimes useful. The fluid definition of entropy density, which coincides with thermodynamic entropy density in the equilibrium limit, is

$$s_s = n_s \log \left( \frac{T_s^{3/2}}{n_s} \right). \tag{3.53}$$

The corresponding entropy flux density is

$$\mathbf{s}_s = s_s \mathbf{V}_s + \mathbf{q}_s / T_s. \tag{3.54}$$

showing that entropy is carried not only by fluid flow but also, according to the second law of thermodynamics, by the transport of heat. After using (3.50) to eliminate $\nabla \cdot \mathbf{V}$, and decomposing the time derivative $dp/dt = ndT/dt + Tdn/dt$, we find that (3.52) can be written as

$$n_s T_s \frac{d(s_s/n_s)}{dt} + \nabla \cdot \mathbf{q}_s + \boldsymbol{\pi}_s : \nabla \mathbf{V_s} = W_s + \frac{1}{2} m_s i_{s2} - \frac{5}{2} T_s i_{s0}$$

Finally we express $\nabla \cdot \mathbf{q}$ in terms of $\nabla \cdot \mathbf{s}$ to write the entropy evolution law as

$$\frac{\partial s_s}{\partial t} + \nabla \cdot \mathbf{s}_s = \Theta_s \tag{3.55}$$

where the right hand side is

$$\Theta_s \equiv \frac{W_s}{T_s} - \frac{\boldsymbol{\pi}_s : \nabla \mathbf{V}_s}{T_s} - \frac{\mathbf{q}_s}{T_s} \cdot \frac{\nabla T_s}{T_s} + \frac{m_s}{2T_s} i_{s2} - \frac{5}{2} i_{s0}. \tag{3.56}$$

Our discussion of (3.44) shows that $\Theta_s$ is the *entropy production rate* for species $s$. It is often convenient to express (3.56) in terms of the laboratory-frame energy exchange, $W_{Ls}$, rather than $W_s$. In particular, we will find that the former is small in the case of disparate species-masses. Since (3.34) implies

$$W_s = W_{Ls} - \mathbf{V}_s \cdot \mathbf{F}_s$$

we have

$$\Theta_s \equiv \frac{W_{Ls}}{T_s} - \frac{\boldsymbol{\pi}_s : \nabla \mathbf{V}_s}{T_s} - \frac{\mathbf{V}_s \cdot \mathbf{F}_s}{T_s} - \frac{\mathbf{q}_s}{T_s} \cdot \frac{\nabla T_s}{T_s} + \frac{m_s}{2T_s} i_{s2} - \frac{5}{2} i_{s0}. \tag{3.57}$$

Here the entropy production due to frictional heating is explicit.

## Center-of-mass system

For certain applications it is useful to express our three moment equations in terms of the total mass density,

$$\rho_m = \sum_s m_s n_s \tag{3.58}$$

and the center-of-mass velocity,

$$\mathbf{V}_{cm} = \frac{\sum_s m_s n_s \mathbf{V}_s}{\rho_m}. \tag{3.59}$$

Thus particle conservation is written as the mass conservation law,

$$\frac{d\rho_m}{dt} + \rho_m \nabla \cdot \mathbf{V}_{cm} = \sum_s m_s i_{s0}, \tag{3.60}$$

For higher moments we need to introduce the total pressure tensor, measured in the center-of-mass frame,

$$\mathbf{P}_{cm} \equiv \sum_s \int d^3 v f_s m_s \mathbf{w}_{cm} \mathbf{w}_{cm} \tag{3.61}$$

where

$$\mathbf{w}_{cm} \equiv \mathbf{v} - \mathbf{V}_{cm}$$

It can be decomposed in the manner of (3.49):

$$\mathbf{P}_{cm} = \mathbf{I} p_{cm} + \boldsymbol{\pi}_{cm}$$

where $p_{cm}$ and $\boldsymbol{\pi}_{cm}$ are defined in terms of $\mathbf{w}_{cm}$ in the obvious way. Then the equation of motion, (3.51), becomes

$$\rho_m \frac{d\mathbf{V}_{cm}}{dt} + \nabla p_{cm} + \nabla \cdot \boldsymbol{\pi}_{cm} - \rho_c \mathbf{E} - \mathbf{J} \times \mathbf{B} = \sum_s m_s \mathbf{i}_{cm,s1}. \tag{3.62}$$

Here

$$\frac{d}{dt} \equiv \frac{\partial}{\partial t} + \mathbf{V}_{cm} \cdot \nabla,$$

the quantity $\mathbf{i}_{cm,s1}$ is defined by (3.40) with $\mathbf{w}$ replaced by $\mathbf{w}_{cm}$, and we used (3.10).

The analogous energy conservation law has the form

$$\frac{\partial}{\partial t} \left( \frac{3}{2} p_{cm} + \frac{1}{2} \rho_m V_{cm}^2 \right) \;+\; \nabla \cdot \left[ \left( \frac{3}{2} p_{cm} + \frac{1}{2} \rho_m V_{cm}^2 \right) \mathbf{V}_{cm} + \mathbf{V}_{cm} \cdot \mathbf{P}_{cm} + \mathbf{q}_{cm} \right]$$

$$= \; \mathbf{J} \cdot \mathbf{E} + \sum_s m_s I_{cm,s2} \tag{3.63}$$

where

$$\mathbf{q}_{cm} = \sum_s \int d^3 w_{cm} f_s \frac{1}{2} m_s w_{cm}^2 \mathbf{w}_{cm}$$

is the heat flux density measured in the center-of-mass frame. Thus the only energy source for a source-free plasma is the electromagnetic work $\mathbf{J} \cdot \mathbf{E}$.

It is instructive to compare the plasma moment equations, (3.60) – (3.63) to those corresponding to an ordinary neutral gas. The two even moments, describing the evolution of $\rho_m$ and $p_{cm}$, hardly differ from the neutral case: only the $\mathbf{J} \cdot \mathbf{E}$ term is new. The equation of motion is also formally similar to that of a neutral gas, differing only in the appearance of the $\mathbf{J} \times \mathbf{B}$ force and (less importantly because of quasineutrality) the charge density term. Thus our comparison suggests a very simple characterization of a plasma: it is a *conducting fluid*—a fluid which, because it carries electric current, is strongly coupled to the electromagnetic field. However, while the conducting fluid picture of plasm behavior is revealing, it is far from a complete description. Equations (3.60) – (3.63) omit the individual species dynamics, because of the species sum, and fail to specify such higher moments as $\mathbf{Q}$. A plasma is unfortunately more complicated than a gas that conducts electricity.

An additional comment is in order concerning the equation of motion. The first two terms of (3.62) show that the plasma will accelerate in response to a pressure gradient. It is tempting to view this response in terms of collisions: each fluid element is pushed by collisional encounters between its particles and those of neighboring fluid elements, and will accelerate if the neighboring pushes fail to balance. This picture is wrong. Indeed, the form of (3.62) is independent of collisions: pressure gradients imply fluid acceleration even in an entirely collisionless plasma. Thus the pressure gradient term in (3.62) should be considered as kinematical rather than dynamical. A fluid element is accelerated, not because of any collisional push, but simply because nonuniform pressure implies more (or faster) particles are entering the element on one side than are leaving it from the other.

## Fluid closure

No manipulation of the moment equations can fix their most serious defect: lack of closure. Since each moment is coupled to the next higher one—evolution of density depends upon velocity, evolution of velocity depends upon stress, and so on—any finite set of exact moment equations will have more unknowns than equations. Therefore some additional information—a small parameter, a limiting assumption, or some additional physics input—is always needed to determine the evolution of the system. (Determining the electric and magnetic fields is not the issue here; since Maxwell's equations are always available the present discussion treats the fields as given.)

Fluid closure schemes based on rigorous exploitation of a small parameter are called *asymptotic*. Asymptotic closure has the advantage of being systematic, and of providing in principle an estimate of the error involved. It is usually possible to refine an asymptotic closure, including higher order effects when necessary. On the other hand the asymptotic approach to closure is the most demanding, often involving a mixture of fluid and kinetic investigation.

Simpler but less convincing closures can be obtained from truncation: one simply assumes that certain higher order moments vanish, or one prescribes them in terms of lower order moments. Truncation can be useful in providing quick insight, and its results are often confirmed by subsequent, more elaborate investigation. On the other hand it always involves uncontrolled approximation.

The classic example of asymptotic closure is the Chapman-Enskog theory of a gas dominated by collisions. In that case the ratio of mean-free-path to scale size provides a small parameter for systematic expansion of the kinetic equation. One quickly finds that the lowest-order distribution must be Maxwellian. In next order, one obtains simplified kinetic equations for the non-Maxwellian corrections to the distribution; once they are computed, the corrections allow such higher order moments as $\mathbf{q}$ and $\boldsymbol{\pi}$ to be expressed in terms of the density, flow and temperature in the Maxwellian. Thus the product of Chapman-Enskog theory is a set of closed, asymptotically justified fluid equations for short-mean-free-path evolution.

The application of Chapman-Enskog theory to plasmas is straightforward and studied in Chapter 8. What is not straightforward is the development of a similarly convincing closure at *long* mean-free path. It is not even clear whether any small set of fluid variables can describe a nearly collisionless system.

An approach to closure at long mean-free path is possible when the plasma is magnetized, since the ratio of gyroradius to scale size can play a role similar to the Chapman-Enskog parameter. Unfortunately this procedure works only for dynamics perpendicular to the magnetic field: the parallel degrees of freedom, unaffected by $B$, generally require kinetic treatment. Nonetheless systematic exploitation of the small gyroradius parameter leads to a sequence of extremely useful insights. We explore these in the following section; the application to explicit fluid closures is considered in Chapter 5.

## 3.5   Magnetized plasma fluid

### Magnetization

A magnetized plasma is one in which typical particle gyroradii are small compared to the scale length of the system. We recall from Chapter 1 that the thermal gyroradius is

$$\rho = v_t/\Omega,$$

where $\Omega = eB/m$ is the gyrofrequency or cyclotron frequency and

$$v_t = \sqrt{\frac{2T}{m}} \tag{3.64}$$

is the thermal speed. Denoting the scale length by $L$, we have the basic small parameter

$$\delta \equiv \rho/L.$$

Of course there is a distinct $\rho$ and $\delta$ for each species; we assume all species to be magnetized and usually suppress species subscripts.

As noted in Chapter 2, a magnetized plasma is an ensemble of magnetic moments, each moving freely along the field and drifting, according to guiding center laws, across the field. Recall that except for the possibly large $\mathbf{E} \times \mathbf{B}$ drift, the cross-field motion of the guiding center is slow, of order $\delta$ compared to parallel streaming. Thus the perpendicular and parallel motions have quite different character, and we will need to distinguish

$$\mathbf{V}_\parallel = \mathbf{b}(\mathbf{b} \cdot \mathbf{V}), \quad \mathbf{V}_\perp = \mathbf{b} \times (\mathbf{V} \times \mathbf{b}) \tag{3.65}$$

as well as the perpendicular and parallel components of other vectors, such as $\mathbf{q}$. Here, as in Chapter 2, $\mathbf{b} \equiv \mathbf{B}/B$ is a unit vector in the direction of the magnetic field.

Some care is required in translating from particle or guiding center motion to fluid motion. In the parallel direction, the translation is straightforward—the fluid moves along $\mathbf{B}$ in so far as individual guiding centers do—but perpendicular motion results as much from gyration as from any guiding center drift. The point is that flow across the magnetic field can occur even when the guiding centers are motionless; gyration itself produces mean fluid motion when the density or energy of the guiding centers is not uniform.

That a nonuniform distribution of circulating charges produces charge flow is familiar from elementary discussions of magnetization (magnetic moment per unit volume). Any magnetized material, with magnetization $\mathbf{M}(\mathbf{x})$, carries a current density given by

$$\mathbf{J}_m = \nabla \times \mathbf{M}, \tag{3.66}$$

called the *magnetization current*. Figure 3.1 shows a typical case: each magnetic moment is motionless, but because each represents circulating charge, vertical variation of $\mathbf{M}$ induces a net horizontal charge flow. As we proceed to cal-

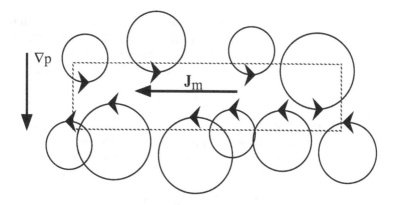

Figure 3.1: Gyrating particles with density gradient. The box sketched in dashed lines indicates a volume small on the system scale but still large enough to contain many guiding centers. It can be seen that the gradient induces a net flow in this volume, consistent with (3.66).

culate the perpendicular flow by manipulating fluid equations, this underlying mechanism should be kept in mind.

## Perpendicular dynamics

The even moment equations, (3.50) and (3.52), are independent of magnetic field, having in fact the same form as the corresponding neutral gas equations. The distinctive properties of magnetized flow are embodied in the equation of motion (3.51), in which each term has the form of a momentum density multiplied by a certain frequency (for example, the friction force on the right-hand side is measured by the momentum density times the collision frequency, $\nu$). The key observation is that the frequency associated with the $\mathbf{V} \times \mathbf{B}$ term is the gyrofrequency, which dominates all others when $\delta$ is small. In other words, the magnetic-force term in the equation of motion includes an implicit factor of $1/\delta$.

This circumstance lies at the heart of all fluid closures of a magnetized plasma. Its importance justifies displaying the ordering explicitly, in terms of dimensionless variables. Hence we introduce $\bar{n}$ as some constant measure of the density, and $\bar{v}_t$ as a constant, typical thermal speed. Then we express (3.51) in dimensionless form by introducing the following normalized fluid variables:

$$\hat{n} = n/\bar{n}, \quad \hat{\mathbf{V}} = \mathbf{V}/\bar{v}_t, \quad \hat{p} = p/(\bar{n}m\bar{v}_t^2), \quad \hat{F} = F/(\nu m \bar{n} \bar{v}_t).$$

These are defined so that the nominal size of any hatted quantity is no larger than order unity; for example $p = nT = n(\frac{1}{2}mv_t^2) \cong \bar{n}m\bar{v}_t^2$. In the same sense, it is natural to normalize the gradient operator in terms of the scale length $L$: $\hat{\nabla} = L\nabla$. Then (3.51) becomes

$$\hat{n}\frac{d\hat{\mathbf{V}}}{dt} + \frac{\bar{v}_t}{L}\hat{\nabla} \cdot \hat{\mathbf{p}} - \hat{n}(\frac{e\mathbf{E}}{m\bar{v}_t} + \Omega\hat{\mathbf{V}} \times \mathbf{b}) = \nu\hat{\mathbf{F}} \qquad (3.67)$$

This form makes visible the frequency or time scale associated with each term—with one exception. Thus the frequency of the acceleration term is the dynamical frequency, denoted by $\omega \sim d/dt$, which characterizes the process under consideration. The frequency of the second term is the *transit frequency*,

$$\omega_t \equiv \bar{v}_t/L = \delta\Omega, \qquad (3.68)$$

introduced in Chapter 1; it measures the rate at which typical particles traverse the system. The rapid gyrofrequency appears with the magnetic force term, and the collision frequency with the friction term. The exceptional term, without definite frequency, is that involving the electric field; it can be written as

$$\frac{e\mathbf{E}}{m\bar{v}_t} = \frac{e\mathbf{E}_\parallel}{m\bar{v}_t} + \Omega\mathbf{b} \times \frac{\mathbf{V}_E}{\bar{v}_t}$$

where

$$\mathbf{V}_E = \frac{\mathbf{E} \times \mathbf{B}}{B^2}$$

is the electric drift. The electric drift is exceptional because, as we recall from Chapter 2, small $\delta$ does not restrict the size of $E_\perp$. In fact Chapter 2 gives two distinctive ways to order the electric drift.

When considering magnetized particle motion in Chapter 2 we found that the parallel electric field must be relatively small. The fluid version of this requirement is visible in (3.67): the normalized parallel force is bounded by either the transit frequency or the collision frequency, both of which must be small compared to $\Omega$. Thus

$$\frac{e\mathbf{E}_\parallel}{m\bar{v}_t} \sim \delta\Omega$$

which implies

$$\frac{E_\parallel}{B} \sim \delta v_t \tag{3.69}$$

Larger electric fields would cause parallel acceleration beyond the rate permitted by the magnetized orderings. In practice such fields are associated with charged-particle "runaway," discussed in Chapter 8.

The first law of asymptotic argument is 'Put all small terms on the right-hand side.' Applying this law to (3.67), cross-multiplying by $\mathbf{b}$, and recalling (3.65) we find that

$$\hat{n}\hat{\mathbf{V}}_\perp = \hat{n}\hat{\mathbf{V}}_E + \delta\mathbf{b} \times \left( \frac{\hat{n}}{\omega_t}\frac{d\hat{\mathbf{V}}}{dt} + \hat{\nabla}\cdot\hat{\mathbf{p}} - \frac{\nu}{\omega_t}\hat{\mathbf{F}} \right) \tag{3.70}$$

Here the factor of $\delta$ mentioned previously has become explicit; note that, since $\omega$, $\omega_t$ and $\nu$ are all comparable with respect to the $\delta$–ordering, the coefficients inside the parenthesis are at most of order unity.

This Section omits source terms for simplicity, but it is not hard to see their effect. Thus $m_s \mathbf{i}_{s1}$ would add to $\mathbf{F}_s$, and introduce a new frequency, measuring the rate at which sources add momentum to the plasma. As long as this "source-frequency" is small compared to $\Omega$, our argument is essentially unchanged.

Equation (3.70) is as exact as the equation of motion from which it was derived, and it contains the same information regarding perpendicular dynamics. When $\delta$ is of order unity it brings no advantage. However, for small $\delta$, (3.70) contains the key for an asymptotic closure of the perpendicular fluid dynamics. The point is that the coupling from $\mathbf{V}_\perp$ to the higher order moment $\mathbf{p}$ appears in (3.70) multiplied by the small parameter. In the limit of vanishing $\delta$ this irksome coupling—which prevents any straightforward closure in the general case—is eliminated; for small but finite $\delta$ the closure problem becomes amenable to perturbative solution.

Suppose, for example, that a description correct through order $\delta$ is desired. Then (3.70) allows calculation of the *first* order terms in $\mathbf{V}_\perp$ from *zeroth* order calculation of the pressure tensor. (The acceleration term is evaluated iteratively.) Kinetic theory must be used for this calculation, but we will see that the kinetic equation becomes relatively simple in the zero-gyroradius limit: computing $\mathbf{p}$ for $\delta \to 0$ is a tractable task.

This procedure, with its recourse to perturbative kinetic theory, is typical of asymptotic closure schemes. Typically moment equations are used to reduce the kinetic burden, allowing one to obtain a usefully accurate fluid closure from the solution to a relatively crude (that is, lower order) kinetic equation. Chapman–Enskog theory, in which the short mean-free-path parameter plays a role analogous to our use of $\delta$, is the most famous example.

Thus (3.70) is a powerful result: much of the progress in understanding magnetized plasma dynamics can be traced to it and to analogous equations for other moments, as we discuss presently. Unfortunately, however, the outlined procedure does not provide a complete closure. Because it applies only to the perpendicular dynamics, a separate scheme for analyzing parallel motion is needed. In other words we can always add to $\mathbf{V}_\perp$ an arbitrary parallel flow, $\mathbf{V}_\parallel$—an undetermined solution to the homogeneous version of (3.51):

$$\mathbf{b} \times \mathbf{V}_\parallel = 0. \tag{3.71}$$

Until we have constructed a theory for $\mathbf{V}_\parallel$, we need to ensure that its size does not upset ordering arguments; in this regard the weak bound,

$$V_\parallel \sim V_E, \tag{3.72}$$

is sufficient.

## Diamagnetic drift

In terms of the original, unnormalized variables, (3.70) has the form

$$n\mathbf{V}_\perp = n\mathbf{V}_E + \frac{\mathbf{b}}{m\Omega} \times \left( mn\frac{d\mathbf{V}}{dt} + \nabla \cdot \mathbf{p} - \mathbf{F} \right). \tag{3.73}$$

There are two critical applications of (3.73). First, it lies at the heart of MHD closure, discussed in Chapter 5. Second, it accurately describes fluid motion in the case of moderate electric fields:

$$\frac{\mathbf{V}_E}{v_t} = O(\delta) \tag{3.74}$$

Recall from Chapter 2 that (3.74) describes the drift ordering, pertinent in particular to a plasma not too far from thermal equilibrium.

It is consistent with (3.74) to assume rather slow temporal variation,

$$\frac{d}{dt} \ll \delta\Omega. \tag{3.75}$$

In this case, under a wide range of conditions, one finds that the lowest order distribution is Maxwellian:

$$\lim_{\delta \to 0} f = f_M(\mathbf{v} - \mathbf{V}). \tag{3.76}$$

Here $f_M$ indicates a local Maxwell distribution,

$$f_{Ms}(\mathbf{x}, \mathbf{v}, t) = \frac{n_s(\mathbf{x}, t)}{\pi^{3/2} v_{ts}(\mathbf{x}, t)^3} exp\left( \frac{-[\mathbf{v} - \mathbf{V}_s(\mathbf{x}, t)]^2}{v_{ts}(\mathbf{x}, t)^2} \right), \tag{3.77}$$

Such a distribution is said to characterize *local thermal equilibrium*, or LTE. Rotational symmetry,

$$f_M(\mathbf{v}) = f_M(|\mathbf{w}|),$$

where $\mathbf{w}$ is the relative velocity defined by (3.14), implies that

$$\boldsymbol{\pi}(f_M) = 0 = \mathbf{q}(f_M); \tag{3.78}$$

an LTE plasma has neither viscosity nor heat flow. When LTE is considered as a complete plasma description, it provides a simple closure considered at the end of the Chapter. Here, however, we use only the much weaker assumption that LTE pertains in the limit of vanishing gyroradius.

At least since Boltzmann's famous argument, Maxwellian distributions have been associated with systems in which collisions play a dominant role. In fact the *lowest order* Maxwellian is pertinent even when collisions are weak, $\nu \ll \omega_t$, provided that plasma remains in quasi-static, nearly isolated state for a sufficiently long time—as generally implied by (3.75). The point is that even weak collisions will eventually Maxwellianize a isolated system.

Equation (3.78) provides the asymptotic result

$$n\mathbf{V}_\perp = n\mathbf{V}_E + \frac{1}{m\Omega}\mathbf{b} \times \nabla p + O(\delta^2) \tag{3.79}$$

Here the second term is called the *diamagnetic drift* and denoted by

$$\mathbf{V}_d = \frac{1}{eBn}\mathbf{b} \times \nabla p. \tag{3.80}$$

We conclude that the flow of a magnetized plasma across the magnetic field is given, through first order in $\delta$, by the sum of the electric and diamagnetic drifts. In the drift ordering, these two contributions are comparable, each contributing a flow that is smaller than the thermal speed by one power of $\delta$. In the MHD ordering, the electric drift is treated as comparable to the thermal speed—zeroth order in $\delta$—so it dominates all other contributions to $\mathbf{V}_\perp$. The MHD velocity is therefore

$$\mathbf{V}_{MHD} = \mathbf{V}_\| + \mathbf{V}_E. \tag{3.81}$$

with $V_{MHD} \sim v_t$.

Extraction of the first-order correction to the MHD flow is relatively complicated, because the inertial terms must be included with the pressure. That is, while the drift ordering implies

$$\mathbf{P} = \mathbf{p} + O(\delta^2), \tag{3.82}$$

MHD is characterized by

$$\mathbf{P}_{MHD} = \mathbf{p} + mn\mathbf{V}_{MHD}\mathbf{V}_{MHD} \tag{3.83}$$

where both terms on the right hand side are formally comparable. Hence the convective inertial term in $d/dt$,

$$mn\mathbf{V}_{MHD} \cdot \nabla \mathbf{V}_{MHD} \sim \nabla p,$$

in MHD is too large to allow (3.75). This fact would seriously complicate the derivation of MHD equations, were it not the case that MHD *omits* all $O(\delta)$ corrections.

Returning to (3.79), we stress that the diamagnetic drift is a macroscopic effect, not visible at the guiding center level. No single guiding center is affected by the pressure gradient. In fact diamagnetic motion comes from the curl of the plasma magnetization, as in (3.66). To see the relation, recall that the magnetic moment of a guiding center is

$$\mathbf{m} = -\mathbf{b}\mu = -\mathbf{b}\frac{mw_\perp^2}{2B}. \tag{3.84}$$

Hence the single-species magnetization is

$$\mathbf{M}_s = -\mathbf{b}\int d^3v f_s \frac{m_s w_\perp^2}{2B}, \tag{3.85}$$

The integral on the right hand side here is related to the perpendicular pressure,

$$p_\perp \equiv \int d^3v f \frac{1}{2}mw_\perp^2$$

whose significance we consider at the end of this Section. Here we note that LTE implies $p_\perp = p$, whence $\mathbf{M} = -\mathbf{b}p/B$. Therefore

$$\nabla \times \mathbf{M}_s = \frac{1}{B}\mathbf{b} \times \nabla p + p_s\left(\mathbf{b} \times \nabla\frac{1}{B} - \frac{\nabla \times \mathbf{b}}{B}\right). \tag{3.86}$$

Because

$$\nabla \times \mathbf{b} = (\nabla \times \mathbf{b})_\parallel + \mathbf{b} \times \boldsymbol{\kappa},$$

where $\boldsymbol{\kappa}$ is the curvature, we see that the two parenthesized terms in (3.86) correspond to the gradient-$B$ and curvature drifts respectivly. Thus (3.80) is simply the statement

$$e_s n_s \mathbf{V}_{ds} = e_s n_s \mathbf{V}_{gcs} + \nabla \times \mathbf{M}_s, \tag{3.87}$$

where

$$n_s \mathbf{V}_{gcs} \equiv \int d^3v f_s \mathbf{v}_{gc}$$

is the flow corresponding to guiding-center motion. (The term involving $(\nabla \times \mathbf{b})_\parallel$ corresponds to the parallel guiding-center drift.) We conclude that the diamagnetic drift is an artifact of gyration.

The origin of the name "diamagnetic" is also clear. The electric current corresponding to diamagnetic flow,

$$\mathbf{J}_\perp = \sum_s \mathbf{V}_{ds}$$

is given by

$$\mathbf{J}_\perp = \frac{1}{B}\mathbf{b} \times \nabla p_{cm}. \tag{3.88}$$

The point is that $\mathbf{J}_\perp$ generates a correction $\delta\mathbf{B}$ that attenuates the original magnetic field. Indeed, from Ampere's law,

$$\nabla \times \mathbf{M} = \frac{1}{\mu_0}\nabla \times \delta\mathbf{B},$$

whence

$$\delta\mathbf{B} = \mu_0\mathbf{M} = -\mu_0\mathbf{b}\frac{p_{cm}}{B}.$$

Note that (3.88) approximates the exact species sum, with each pressure measured in the rest frame of its species, by the center-of-mass pressure. Such approximation is common, and harmless unless there are several ion species with large and distinct flow velocities.

## Other magnetized moments

The manipulation we have used to unravel the perpendicular particle flow is equally effective when applied to other perpendicular moments. Here we consider two examples with especially important applications.

The perpendicular energy flux is obtained from the $\frac{1}{2}mv^2\mathbf{v}$-moment of the kinetic equation:

$$\frac{\partial\mathbf{Q}}{\partial t} + \nabla \cdot \mathbf{R} - \frac{e}{2m}[\mathbf{E}Tr(\mathbf{P}) + 2\mathbf{E} \cdot \mathbf{P}] + \Omega\mathbf{b} \times \mathbf{Q} = \mathbf{C}_3 \tag{3.89}$$

where $\mathbf{R}$ is defined by (3.13) and

$$\mathbf{C}_3 \equiv \int d^3v \frac{1}{2}mv^2\mathbf{v}C_{ss'}.$$

Evidently,

$$\mathbf{Q}_\perp = \frac{\mathbf{b}}{\Omega} \times \left\{ \frac{\partial\mathbf{Q}}{\partial t} + \nabla \cdot \mathbf{R} - \frac{e}{2m}[\mathbf{E}Tr(\mathbf{P}) + 2\mathbf{E} \cdot \mathbf{P}] - \mathbf{C}_3 \right\} \tag{3.90}$$

As before, we recognize the $1/\Omega$ factor as equivalent to $\delta$, and use (3.90) to compute $\mathbf{Q}_\perp$ through first $\delta$-order from the zeroth order distribution. Also as before, the MHD-ordered and drift-ordered cases should be treated separately, although they give equivalent results; we consider the drift ordering for simplicity. A simple calculation from (3.13) shows that

$$\mathbf{R}_M \equiv R(f_M) = \mathbf{I}\frac{5}{2}\frac{pT}{m} \tag{3.91}$$

whence

$$\mathbf{Q}_\perp = \frac{5}{2}\frac{\mathbf{b}}{m\Omega} \times [\nabla(pT) - ep\mathbf{E}] + O(\delta^2) \tag{3.92}$$

The heat flux has a simpler expression. Recalling (3.21) and using (3.74) to neglect some terms involving $\mathbf{V}$, we see that through first order

$$\mathbf{Q} = \mathbf{q} + \frac{5}{2}p\mathbf{V}. \tag{3.93}$$

Thus (3.79) and (3.92) combine to give $\mathbf{q}_\perp = \mathbf{q}_d + O(\delta^2)$, where

$$\mathbf{q}_d \equiv \frac{5}{2}\frac{p}{m\Omega}\mathbf{b} \times \nabla T. \tag{3.94}$$

Here the $d$-subscript emphasizes the analogy with the drift velocity of (3.80).

The "free-lunch" character of (3.92) is worth emphasizing. The only kinetic theory needed is the LTE statement of (3.76). Note, however, that a straight-forward calculation of $\mathbf{q}$ from (3.76) would yield *vanishing* heat flux, as noted in (3.78). Thus (3.92), which is rigorously valid within the indicated error, knows something about the departure from LTE, given by the first order correction to $f_M$. Our manipulation of the moment equations has spared us the labor of computing that correction.

It is also worth emphasizing that because $\mathbf{b} \times \mathbf{Q}_\parallel = 0$, our theory for heat transport leaves undetermined a parallel term,

$$\mathbf{q}_\parallel = \mathbf{Q}_\parallel - \frac{5}{2}p\mathbf{V}_\parallel. \tag{3.95}$$

Finally we consider the stress tensor. From the $m\mathbf{vv}$-moment of the kinetic equation we obtain a relation analogous to (3.90),

$$\mathbf{b} \times \mathbf{P} + (\mathbf{b} \times \mathbf{P})^\dagger = -\frac{1}{\Omega}\left[\frac{\partial \mathbf{P}}{\partial t} + m\nabla \cdot \mathbf{M}_3 - en(\mathbf{EV} + \mathbf{VE}) - C_2\right]. \tag{3.96}$$

Here the $\dagger$ on the left indicates the transposed tensor, $M_3$ is the $\mathbf{vvv}$-moment of the distribution and $C_2$ is the second moment of the collision operator. Each of these terms has some interest; for example, the $M_3$ terms yield *gyroviscosity*, a dissipationless momentum transport mechanism that is studied in Chapter 5. But for the present we consider only the limit of vanishing $\delta$.

As $\delta \to 0$, we must as usual distinguish the drift-ordered case, in which the right hand side of (3.96) vanishes, from the MHD-ordered case, in which the electric field terms on the right hand side survive. Assuming first the drift ordering, we study the homogeneous equation

$$\mathbf{b} \times \mathbf{p}^0 + (\mathbf{b} \times \mathbf{p}^0)^\dagger = 0. \tag{3.97}$$

Thus $p^0$ is the second-rank analog to $\mathbf{V}_\parallel$ and $\mathbf{q}_\parallel$; we use a lower-case $p$ in view of (3.82). After making the components of (3.97) explicit, one finds that $p^0$ must have the form of a *gyrotropic* stress tensor, $\mathbf{p}_{gt}$, where

$$\mathbf{p}_{gt} = \mathbf{bb}p_\parallel + (\mathbf{I} - \mathbf{bb})p_\perp. \tag{3.98}$$

In a coordinate system $(\mathbf{b}, \mathbf{e}_2, \mathbf{e}_3)$ where the first coordinate is aligned along $\mathbf{B}$, $\mathbf{p}_{gt}$ is diagonal:

$$\mathbf{p}_{gt} = \begin{pmatrix} p_\parallel & 0 & 0 \\ 0 & p_\perp & 0 \\ 0 & 0 & p_\perp \end{pmatrix} \tag{3.99}$$

The MHD case differs only slightly.  After expressing $\mathbf{b} \times \mathbf{P}_{MHD} + (\mathbf{b} \times \mathbf{P}_{MHD})^\dagger$ in terms of $p$, using (3.83), we find that the inertial terms precisely cancel the electric field terms on the right hand side.  Since the remaining terms on the right vanish with $\delta$, (3.97) is reproduced, along with its gyrotropic solution.

The importance of stress tensors having the form of (3.99) was emphasized in a famous analysis by Chew, Goldberger and Low; thus the gyrotropic stress is sometimes called the CGL stress, and denoted by $p_{CGL} = p_{gt}$. We prefer the term "gyrotropic" because it evokes the isotropic stress of a collisional plasma: $p_{gt}$ is isotropic in the two directions transverse to the magnetic field, corresponding to the rapid gyration of magnetized particles—for any collisionality. The only anisotropy ("gyro-anisotropy") results from the distinctive character of particle streaming along the field. The kinetic definitions are

$$p_\parallel = \int d^3 w f m w_\parallel^2, \ p_\perp = \int d^3 w f \frac{1}{2} m w_\perp^2. \tag{3.100}$$

According to (3.76) the gyro-anisotropy $p_\parallel - p_\perp$ must vanish in lowest order LTE:

$$p_\parallel - p_\perp = O(\delta)$$

Yet it can be as important as other first-order effects, such as diamagnetic drifts, in the drift-ordering. The most popular version of MHD closure also uses (3.76); however, a more rigorous formulation allows gyro-anisotropy even in lowest order.

We next comment on two features of magnetized plasma dynamics that do not depend upon specific closure schemes or approximations. The first and perhaps most important concerns the Maxwell stress tensor.

## Maxwell stress

Maxwell observed that the magnetic force could be expressed in terms of the stress tensor

$$\mathbf{T}_m = \mathbf{I} \frac{B^2}{2\mu_0} - \frac{\mathbf{B}\mathbf{B}}{\mu_0} \tag{3.101}$$

in the form

$$\mathbf{J} \times \mathbf{B} = -\nabla \cdot \mathbf{T}_m.$$

The electric force term, $\rho_c \mathbf{E}$, can be similarly expressed, but here we consider a quasineutral plasma and neglect $\rho_c$. Notice that $\mathbf{T}_m$ is expressed in terms of the same tensor quantities, $\mathbf{I}$ and $\mathbf{bb}$, as the gyrotropic stress, (3.99); indeed this form is forced by symmetry.

Now the equation of motion (3.62) takes a form,

$$\rho_m \frac{d\mathbf{V}_{cm}}{dt} + \nabla \cdot (\mathbf{p}_{cm} + \mathbf{T}_m) = \sum_s m_s \mathbf{i}_{s1}, \tag{3.102}$$

that is nearly tautological: in the absence of momentum input, the fluid acceleration is given by the divergence of the total (plasma plus magnetic) stress.

The form of this total stress tensor,

$$\mathbf{T} = \mathbf{p}_{cm} + \mathbf{T}_m$$

is especially instructive in the gyrotropic case. One finds that

$$\mathbf{T} = \mathbf{bb}T_{\parallel} + (\mathbf{I} - \mathbf{bb})\, T_{\perp},$$

with

$$T_{\parallel} = p_{cm\parallel} - \frac{B^2}{2\mu_0}, \quad T_{\perp} = p_{cm\perp} + \frac{B^2}{2\mu_0} \tag{3.103}$$

Thus the "magnetic pressure," $B^2/2\mu_0$, adds to the perpendicular stress while subtracting from the parallel stress.

The physical interpretation is clear. In the perpendicular direction, the magnetic field exerts pressure in proportion to its energy density, much like an ordinary fluid. The *negative* magnetic pressure in the parallel direction, however, amounts to *field line tension*: the field lines try to shorten and straighten. They behave, in other words, like guitar strings.

Field line tension underlies several plasma physical phenomena, including Alfvén waves. It provides a crucial stabilizing tendency for numerous plasma perturbations, and thus, as we will see in Chapter 4, strongly affects the structure of linear modes. The guitar string analogy aids intuition in thinking about such phenomena; (3.103) shows that the analogy is reliable.

## Virial theorem

Consider a plasma without external current or non-electromagnetic force: it is affected only by $\mathbf{J} \times \mathbf{B}$, in which the magnetic field mujst be generated by plasma currents alone. According to (3.102), an equilibrium state of such a plasma is characterized by

$$\rho_m \mathbf{V}_{cm} \cdot \nabla \mathbf{V}_{cm} + \nabla \cdot (\mathbf{p}_{cm} + \mathbf{T}_m) = 0.$$

Noting that $\nabla \cdot (\rho_m \mathbf{V}_{cm}) = 0$ in the steady state, we can equivalently write

$$\nabla \cdot (\mathbf{P}_{cm} + \mathbf{T}_m) = 0, \tag{3.104}$$

where

$$\mathbf{P}_{cm} \equiv \mathbf{p}_{cm} + \rho_m \mathbf{V}_{cm} \mathbf{V}_{cm}$$

Next we compute the plasma *virial* by multiplication with $\mathbf{x}$, the vector displacement from an origin inside the plasma, and integration. Beginning with

$$\mathbf{x} \cdot \nabla \cdot \mathbf{T} = x_\beta \frac{\partial}{\partial x_\alpha} T_{\alpha\beta} = \frac{\partial}{\partial x_\alpha} (x_\beta T_{\alpha\beta}) - \delta_{\alpha\beta} T_{\alpha\beta} = \nabla \cdot (\mathbf{x} \cdot \mathbf{T}) - Tr(\mathbf{T}),$$

we integrate over some volume $V$ with bounding surface $S$. After applying Gauss' law to the divergence term we find

$$\int_S d\mathbf{S} \cdot [\mathbf{x} \cdot (\mathbf{P}_{cm} + \mathbf{T}_m)] = \int_V d^3x \left( 3p_{cm} + \rho_m V_{cm}^2 + \frac{B^2}{2\mu_0} \right). \qquad (3.105)$$

The right hand side of this "virial theorem" is the trace computed from (3.15) and from $Tr(\mathbf{T}_m) = 3B^2/2\mu_0 - B^2/\mu_0 = B^2/2\mu_0$; the left hand side is the fluid virial.

The most important application of the virial theorem concerns a *confined* plasma—one that has finite extent. In that case we can choose the integration volume $V$ to be so large that its surface $S$ lies far outside the plasma, where $\mathbf{p}_{cm}$ and $\mathbf{T}_m$ vanish. By assumption the magnetic field is due entirely to plasma currents and must also vanish on a sufficiently distant surface. Thus, for a bounded plasma, the virial becomes arbitrarily small. But in the same large-$V$ limit the right hand side of (3.105) becomes independent of $V$ and remains finite: its integrand is positive definite.

We conclude that a plasma cannot be confined by its self-generated electromagnetic field alone: external forces are required. This observation bears on attempts to understand such phenomena as ball lightning, whose equilibrium evidently requires some agency, presumably chemical, beyond macroscopic electrodynamics.

## 3.6 Exact fluid closures

Chapter 5 treats the most important fluid closure schemes applicable to a hot, magnetized plasma: magnetohydrodynamics, or MHD, and the drift approximation. Both of these schemes have rigorously asymptotic versions, but both are almost always used in truncated form. We conclude the present Chapter by briefly considering two limiting cases in which *exact* closure of the moments is possible: the cold plasma system and the system in local thermal equilibrium.

### Cold plasma

The cold plasma equations are obtained from (3.50)–(3.52) simply by allowing $p$ and $\pi$ to vanish. The corresponding physical situation is a system in which all particles of a single species move in lock step, without thermal motion. Such equations will describe approximately a plasma in which velocities of interest far exceed the thermal velocity,

$$V_s \gg v_{ts}. \qquad (3.106)$$

Cold plasma equations are of interest mainly because of the large variety of linear waves whose phase velocities satisfy (3.106); such perturbations are studied in Chapter 4.

The cold plasma dynamics are fixed by density and velocity alone; (3.50) and (3.51) become

$$\frac{dn_s}{dt} + n_s \nabla \cdot \mathbf{V}_s = i_{s0}, \tag{3.107}$$

$$m_s n_s \frac{d\mathbf{V}_s}{dt} - e_s n_s (\mathbf{E} + \mathbf{V}_s \times \mathbf{B}) = \mathbf{F}_s + m\mathbf{i}_{s1}, \tag{3.108}$$

Here the source terms are presumed to be known; a separate calculation is needed to express the friction force in terms of $n$ and $\mathbf{V}$, but otherwise we have the necessary two equations for our two unknowns. (In practice, friction is often omitted.)

## Application: the cold ion sheath

We apply this closure to a special case of the Langmuir sheath: the narrow region where a plasma abuts some material wall. As noted in Chapter 1, the defining property of the wall is that it absorbs most incident electrons; since electrons move more quickly than ions, this effect by itself would rapidly accumulate electric charge in the plasma. Such charge accumulation is avoided by the formation of a Langmuir sheath, an electrostatic "cliff" at the plasma edge which reflects all but the most energetic electrons, keeping the ion and electron loss rates nearly equal. The sheath width is of course measured by the Debye length, $\lambda_D$. The sheath is accompanied by an ion flow toward the wall, accounting for the unreflected electrons.

Calculation of the loss rate requires kinetic theory, since electron loss is local in velocity space. Indeed, when the ion and electron temperatures are comparable, any deep understanding of the sheath structure depends upon kinetic insights. However in many cases, such as plasmas used for materials processing, the ions in the vicinity of the wall are relatively cold

$$T_i \ll T_e. \tag{3.109}$$

If, in addition, the reflected electrons are sufficiently collisional to be approximately Maxwellian, then cold-ion fluid equations yield useful information about sheath structure.

We consider here the simplest case of a one dimensional sheath, with coordinate $x$ increasing away from the wall at $x = 0$. Any magnetic field that may be present is supposed to be oriented along the $x$-axis (normal to the wall), so it does not enter the dynamics. We also ignore sources and collisional effects for simplicity.

The electrons in the sheath are assumed to have reached full thermodynamic equilibrium, with constant temperature,

$$T_e = constant$$

and a Maxwell-Boltzmann distribution in density:

$$n_e(x) = n_0 e^{e\Phi(x)/T_e}, \tag{3.110}$$

Here $\Phi$ is the electrostatic potential; the constant $n_0$ measures electron density in the quasineutral region, far from the wall, where $\Phi$ is taken to vanish. This description obviously ill serves the unreflected electrons, energetic enough to reach the wall, and therefore requires that the untrapped electron fraction be very small. In other words we assume

$$|e\Phi_s/T_e| \gg 1 \tag{3.111}$$

where $\Phi_s \equiv \Phi(0)$ is the so-called sheath potential.

Consider next the sheath ions. The steady-state form of the particle conservation law, (3.107), requires $\nabla \cdot (n\mathbf{V}) = 0$ ($i$-subscripts are suppressed). Hence, in the one-dimensional case,

$$n(x)V(x) = n_0 V_0.$$

Here $V_0$ is evidently the ion flow in the asymptotic region, far from the wall, where the fluid parameters have become constant on the $\lambda_D$-scale. As first noted by Bohm, sheath equilibrium demands such an asymptotic ion flow. (The interior layer in which the ion flow accelerates from zero to $V_0$ is called the *presheath*.)

The same simplifications applied to the cold-ion momentum law, (3.108), yield the obvious energy conservation law

$$\frac{1}{2}mV^2(x) + e\Phi(x) = \frac{1}{2}mV_0^2.$$

Finally we use Poisson's equation,

$$\epsilon_0 \frac{d^2\Phi}{dx^2} = e(n_e - n)$$

to obtain four equations for the four functions $n, n_e, \Phi$ and $V$. These are straightforwardly combined into a single nonlinear equation for the potential, called the *sheath equation*:

$$\epsilon_0 \frac{d^2\Phi}{dx^2} = en_0 \left[ e^{e\Phi/T} - \left(1 - \frac{2e\Phi}{mV_0^2}\right)^{-1/2} \right] \tag{3.112}$$

Notice that each term in the sheath equation becomes an exact derivative after multiplication by $d\Phi/dx$. With this observation the equation can be integrated once to obtain

$$\frac{\epsilon_0}{2} \left(\frac{d\Phi}{dx}\right)^2 + C = n_0 T \left( e^{e\Phi/T} + \frac{mV_0^2}{T}\sqrt{1 - \frac{2e\Phi}{mV_0^2}} \right)$$

Here the integration constant, $C$, is fixed by noting that $d\Phi/dx$ must vanish in the asymptotic region. Thus we find

$$\frac{\epsilon_0}{2n_0 T}\left(\frac{d\Phi}{dx}\right)^2 = e^{e\Phi/T} - 1 + \frac{mV_0^2}{T}\left(\sqrt{1 - \frac{2e\Phi}{mV_0^2}} - 1\right) \qquad (3.113)$$

It can be seen that the scale length in (3.113) is

$$\frac{\epsilon_0 T}{e^2 n_0} = \lambda_D^2,$$

the (squared) Debye length. Indeed, numerical solution to (3.113) displays the decay of $|\Phi|$ from $|\Phi_s|$ to zero, over the several Debye lengths; see Figure 3.2. More interesting is the solubility condition for the sheath equation. Since

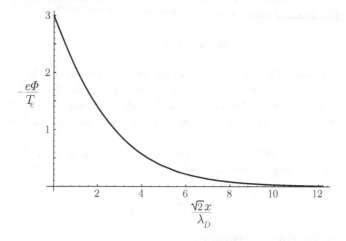

Figure 3.2: Cold ion sheath. The normalized electrostatic potential, as determined by (3.113), is plotted as function of the normalized distance from the wall, which is placed at $x = 0$.

$(d\Phi/dx)^2$ is positive we must require

$$F(\phi) \equiv e^{-\phi} - 1 + 2K\left(\sqrt{1 + \phi/K} - 1\right) \geq 0$$

where

$$\phi \equiv -e\Phi/T$$

is a positive, normalized measure of the potential and

$$K \equiv \frac{mV_0^2}{2T}$$

is the normalized kinetic energy. It is clear that the lower bound for $K$ is nonzero, since $e^{-\phi} - 1 \leq 0$. To find the lower bound, we consider $F$ as a function of $x$ and $K$, and choose an initial $K$-value that makes $F$ positive for all $x$. Now suppose we vary $K$, causing the curve $F(x)$ to deform until, at some value $K = K_c$, the minimum of $F$ reaches zero. If the corresponding value of $\phi$ is $\phi_0$, then we have the simultaneous equations

$$
\begin{aligned}
F(\phi_0, K_c) &= 0, \\
\frac{dF(\phi_0, K_c)}{d\phi} &= 0,
\end{aligned}
$$

which can only be satisfied for $\phi_0 \to 0$. Thus we expand for small $\phi$ to find

$$
F = \left( \frac{1}{2} - \frac{1}{4K} \right) \phi^2 + O(\phi^3)
$$

and conclude that $K_c = 1/2$. Restoring the ion subscript we can write the condition $K \geq 1/2$ as

$$
V_0 \geq \sqrt{T_e/m_i} \tag{3.114}
$$

This result, first derived by Bohm, is called the Bohm sheath condition.

Notice that if the ion and electron temperatures were comparable, the lower bound on $V_0$ would be close to the ion thermal speed, making our neglect of ion thermal effects inconsistent. Hence the cold-ion assumption, (3.109), is indeed essential.

Because $F$ is quadratic for small $\phi$, the asymptotic form of $\phi$ is easily found. We introduce the scaled coordinate

$$
y \equiv \sqrt{2} \frac{x}{\lambda_D}
$$

to express (3.113) as

$$
\left( \frac{d\phi}{dy} \right)^2 = F(\phi)
$$

Hence for small $\phi$ we have

$$
\frac{d\phi}{dy} = \pm \alpha \phi
$$

where

$$
\alpha = \sqrt{\frac{1}{2} - \frac{1}{4K}}
$$

and the minus sign corresponds to decay. Thus $\phi$ asymptotes to zero exponentially,

$$
\phi = \phi_s e^{-\alpha y}.
$$

The decay distance becomes large as $K$ approaches the Bohm limit; at the Bohm limit, $K = 1/2$, a separate analysis shows that the decay of $\phi$ becomes algebraic.

## Local thermal equilibrium

The second limiting case that permits an exact closure is local thermal equilibrium, or LTE, defined by

$$f = f_M;$$

recall (3.76). Since a Maxwellian is determined by quantities $n_s(\mathbf{x}, t)$, $T_s(\mathbf{x}, t)$ and $\mathbf{V}_s(\mathbf{x}, t)$, the evolution of an LTE system is determined by three moment equations.

The applicability of LTE is somewhat harder to characterize than (3.106). Its conventional context is the collision-dominated regime, where the collisional mean-free path,

$$\lambda_{\mathrm{mfp}} \equiv v_t/\nu \tag{3.115}$$

is short compared to other scale lengths of interest:

$$\lambda_{\mathrm{mfp}} \ll L.$$

In fact the issue of LTE validity is more complicated. On the one hand, certain processes that LTE omits are proportional to $\nu$ and therefore largest in the collision-dominated regime. Examples are the transport processes associated with collisional relaxation. Unless the system is maintained extremely close to its equilibrium state—a severe demand—one expects relaxation, described by non-Maxwellian corrections to $f_M$, to enter the dynamics.

On the other hand, LTE can provide a good first approximation even when collisions are weak. We have already noted, in (3.76), that the lowest order distributions in strongly magnetized plasmas are typically Maxwellian. The distribution of a *confined* plasma will become nearly Maxwellian after a few collision times, even when the mean-free path is longer than the system size; all that is required is that the collision time be short compared to the confinement time. However in this case the non-Maxwellian parts of $f$, while relatively small, can be especially important.

In summary, the Maxwellian *ansatz* of LTE often approximates the particle distributions well, but it rarely specifies them exactly. LTE closure is deficient in leaving out the effects of non-Maxwellian corrections.

We have already noted that the LTE plasma has neither viscosity nor heat flow. Hence the LTE moment equations have the form

$$\frac{dn_s}{dt} + n_s \nabla \cdot \mathbf{V}_s = i_{s0}, \tag{3.116}$$

$$m_s n_s \frac{d\mathbf{V}_s}{dt} + \nabla p_s - e_s n_s (\mathbf{E} + \mathbf{V}_s \times \mathbf{B}) = \mathbf{F}_s + m_s \mathbf{i}_{s1}, \tag{3.117}$$

$$\frac{3}{2} \frac{dp_s}{dt} + \frac{5}{2} p_s \nabla \cdot \mathbf{V}_s = W_s + \frac{1}{2} m_s i_{s2}. \tag{3.118}$$

Here of course $p_s$ represents the product of the density and the temperature appearing in the Maxwellian. In Chapter 7 we will compute the collisional moments $F$ and $W$ for the Maxwellian case. Presuming they are known, we see that (3.116)–(3.118) provide a closed description of the LTE fluid.

The neglect of heat flow and viscous heating corresponds to adiabatic evolution, characterized by the law,

$$\frac{p_s}{n_s^{5/3}} = constant, \tag{3.119}$$

familiar from elementary thermodynamics. Indeed it is easily seen that (3.119) applies to (3.116) – (3.118) when sources and collisional energy exchange are neglected.

# Additional reading

An outstandingly clear and systematic development of plasma fluid theory may be found in the famous review by Braginskii [13]. Most of the texts mentioned at the end of Chapter 1 also derive fluid equations. An approach to fluid equations that does not use moments can be found in [74].

A closed fluid description that allows for anisotropic pressure is found in [17]. The Chapman-Enskog procedure is systematically exposed by Chapman and Cowling [15]. More detailed discussions of the virial theorem are provided by Shafranov [66] and Schmidt [65]. Bohm's treatment of his sheath criterion may be found in [11].

# Problems

1. Verify the relation (3.21) between energy flux and heat flux by explicit integration. In the special case of weak flow, $\mathbf{V} = O(\Delta)$, with $\Delta \ll 1$, show that

$$\mathbf{Q} = \mathbf{q} + \frac{5}{2}p\mathbf{V} + O(\Delta^2).$$

2. Show that the collisional kinetic-energy moment, $W_{ss'}$, also measures the collisional change in *total* energy, $(1/2)mv^2 + e\Phi$.

3. Compare the statistical definition of entropy density,

$$s_s \equiv - \int d^3v f_s \log f_s$$

to the fluid expression (3.53), in the case of a Maxwellian distribution. Show that differences between the two expressions do not enter the entropy evolution law.

4. Use Figure 3.1 to infer the estimate

$$nV_d \sim \rho\nabla(nv_t),$$

and verify its agreement with (3.80).

5. To confirm the significance of gyrotropic stress, demonstrate that any second-rank tensor $\mathbf{T}$ satisfying

$$\mathbf{b} \times \mathbf{T} + (\mathbf{b} \times \mathbf{T})^\dagger = 0$$

must have the form

$$\mathbf{T} = \mathbf{bb}T_\| + (\mathbf{I} - \mathbf{bb})T_\perp$$

where the components $T_\|$ and $T_\perp$ are arbitrary.

6. Derive the adiabatic law (3.119) from the LTE fluid equations, neglecting energy exchange and source terms.

7. General thermodynamic principles require the thermal equilibrium of any fluid system to have a uniform temperature. Does the LTE fluid closure of (3.116)–(3.118) yield this conclusion? Discuss in this context the difference between a *closed* description and a *complete* one.

# Chapter 4

# Waves in a Cold Plasma

## 4.1  A speedy-wave closure

The "cold plasma" model describes waves and other perturbations that propagate faster than the thermal speed of the plasma. This model enjoys a unique role as the only closure of the moment equations that describes an unrestricted range of frequencies. It is also the simplest closure. Its simplicity allows us to completely characterize and classify a rich variety of waves. The resulting picture provides a unifying framework for theories that take thermal motion into account.

It is instructive to consider the relationship between the collective oscillations described by the cold plasma model and the oscillations of individual particles that we studied in Chapter 2. The key observation is that in the cold plasma model, all the particles at a given position have the same velocity. That is, the fluid velocity is identical to the particle velocity and is thus governed by the same equations. The cold plasma model goes beyond the single-particle description, however: it determines the fields *self-consistently* in terms of the charge and current densities generated by the motion of the plasma.

A difficulty that we must address when considering plasma waves is the role that the geometry of the equilibrium plays. Modes with wavelengths comparable to the characteristic dimension of the plasma, L, clearly depend on the shape of the plasma. We will show that modes with wavelengths much smaller than L, however, are in a local sense independent of the geometry. The local properties of small-wavelength oscillations are thus universal. To investigate these properties, we may represent the plasma by a homogeneous equilibrium (corresponding to the limit $kL \to \infty$, where $k$ is the magnitude of the wave-vector). Small wavelength modes are often preferentially excited, so they are of particular practical importance.

We begin the chapter by considering the propagation of small-amplitude plane waves in a homogeneous plasma. The principal result of the analysis is a dispersion relation that relates the frequency to the wave-vector. We examine this dispersion relation and the waves it describes. We then extend the theory to inhomogeneous plasma, and show how to determine the amplitude and phase of a wave from the dispersion relation. We conclude with an overview of the most important instabilities described by the cold plasma closure.

## 4.2   Plane waves in a homogeneous plasma

The propagation of small amplitude waves is described by *linearized equations*. These are obtained by expanding the equations of motion in powers of the wave amplitude and neglecting terms of order higher than one. We use the subscript 0 to distinguish equilibrium quantities from the perturbed quantities, for which we retain the same notation.

We consider a homogeneous plasma in which all species are at rest, so that $\mathbf{E}_0 = 0$ and $\mathbf{J}_0 = \nabla \times \mathbf{B}_0/\mu_0 = 0$. In a homogeneous medium, the general solution of a system of linear equations can be constructed as a superposition of plane wave solutions,

$$\mathbf{E}(\mathbf{x}, t) = \mathbf{E}_{\mathbf{k},\omega} \exp(i\mathbf{k} \cdot \mathbf{x} - i\omega t), \tag{4.1}$$

with similar expressions for $\mathbf{B}$ and $\mathbf{V}$. In order for the total field to be real the coefficients $\mathbf{E}_{\mathbf{k},\omega}$ must be in pairs such that

$$\mathbf{E}_{\mathbf{k},\omega}^* = \mathbf{E}_{-\mathbf{k},-\omega}.$$

The surfaces of constant phase, $\zeta = \mathbf{k} \cdot \mathbf{x} - \omega t = $ constant, are planes perpendicular to $\mathbf{k}$ traveling at velocity

$$\mathbf{v}_{\mathrm{ph}} = \frac{\omega}{k}\hat{\mathbf{k}}, \tag{4.2}$$

where $k = |\mathbf{k}|$ and $\hat{\mathbf{k}}$ is the unit vector in the direction of $\mathbf{k}$. The velocity $\mathbf{v}_{\mathrm{ph}}$ is called the *phase velocity*. In this Section, we examine the nature of the plane wave solutions. We will describe the general solution and its generalization to weakly inhomogeneous plasma in the following section. Henceforth, we omit the subscripts $\mathbf{k}$, $\omega$ from the field variables.

We substitute the plane wave solution, (4.1), in Maxwell's law of induction and in Faraday's law,

$$\mathbf{k} \times \mathbf{B} = -i\mu_0 \mathbf{J} - \frac{\omega}{c^2}\mathbf{E}, \tag{4.3}$$

$$\mathbf{k} \times \mathbf{E} = \omega \mathbf{B}. \tag{4.4}$$

Consistent with the linear approximation, the current is related to the electric field by

$$\mathbf{J} = \boldsymbol{\sigma} \cdot \mathbf{E}, \tag{4.5}$$

where the conductivity tensor $\boldsymbol{\sigma}$, a function of $\mathbf{k}$ and $\omega$, is anisotropic due to the presence of the equilibrium magnetic field. This tensor contains all the necessary information concerning the plasma response.

Substituting (4.5) in the induction law yields

$$\mathbf{k} \times \mathbf{B} = -\frac{\omega}{c^2}\mathbf{K} \cdot \mathbf{E}, \tag{4.6}$$

where we have introduced the dielectric permittivity tensor $\mathbf{K}$,

$$\mathbf{K} = \mathbf{I} + \frac{i\sigma}{\epsilon_0 \omega}. \tag{4.7}$$

Here $\mathbf{I}$ is the unit tensor. To obtain a closed set of equations, we eliminate the magnetic induction from Maxwell's equation by using Faraday's law (4.4). There follows

$$\mathbf{M} \cdot \mathbf{E} = 0, \tag{4.8}$$

where

$$\mathbf{M} = (\hat{\mathbf{k}}\hat{\mathbf{k}} - \mathbf{I})n^2 + \mathbf{K}. \tag{4.9}$$

Here $n$ is the index of refraction,

$$n = \frac{kc}{\omega} = \frac{c}{v_{\mathrm{ph}}}. \tag{4.10}$$

Equations (4.8)-(4.9) state that $\mathbf{K} \cdot \mathbf{E}$ is the projection of the electric field in the plane perpendicular to $\mathbf{k}$ multiplied by the index of refraction squared. The vectors $\mathbf{k}$, $\mathbf{K} \cdot \mathbf{E}$, and $\mathbf{B}$ are thus mutually orthogonal.

The solubility condition for (4.8),

$$\mathcal{M}(\omega, \mathbf{k}) \equiv \det(\mathbf{M}) = 0, \tag{4.11}$$

is called the *dispersion relation*. The dispersion relation relates the frequency to the wave-vector. Its name indicates that it describes the rate at which different Fourier components in a wave-train disperse due to the variation of their phase velocity with wavelength. We now consider the specific form of the permittivity tensor $\mathbf{K}$ and the dispersion relation for the cold plasma model.

## Cold plasma model

The current is determined by the velocity of the various species through

$$\mathbf{J} = \sum_s n_s e_s \mathbf{V}_s. \tag{4.12}$$

In order to determine the conductivity we must express the fluid velocities $\mathbf{V}_s$ in terms of the electric field. We accomplish this by solving the linearized equations of motion,

$$m_s \frac{d\mathbf{V}_s}{dt} = e_s(\mathbf{E} + \mathbf{V}_s \times \mathbf{B}_0), \tag{4.13}$$

where $\mathbf{V}_s$ is the fluid velocity for species $s$ and $\mathbf{B}_0$ is the equilibrium magnetic field.

We specialize to a two component plasma and rewrite the moment equations in terms of the current and the velocity of the center of mass. This yields useful physical insight, and allows important simplifications related to the smallness of the ratio of electron and ion masses.

The electron and ion densities are related by charge neutrality, $n_e = Z n_i$, where $Z = e_i/e$ measures the charge state of the ions. We denote the mass ratio by $\mu$,

$$\mu = Z m_e/m_i.$$

Summing the momentum equations over both species and neglecting terms of relative order $\mu$ yields

$$m_i n_i \frac{d\mathbf{V}}{dt} = \mathbf{J} \times \mathbf{B}_0, \tag{4.14}$$

where

$$\mathbf{V} = \frac{m_i n_i \mathbf{V}_i + m_e n_e \mathbf{V}_e}{m_i n_i + m_e n_e} \simeq \mathbf{V}_i + \mu \mathbf{V}_e$$

is the center of mass velocity. We eliminate the electron velocity in favor of the current and center of mass velocities,

$$\mathbf{V}_e = \mathbf{V} - \frac{\mathbf{J}}{n_e e}. \tag{4.15}$$

Substituting this in the electron momentum equation yields a generalized Ohm's law for alternating currents,

$$\mathbf{E} = -\mathbf{V} \times \mathbf{B}_0 + \frac{1}{n_e e} \mathbf{J} \times \mathbf{B}_0 + \frac{m_e}{n_e e^2} \frac{d\mathbf{J}}{dt}, \cdot \tag{4.16}$$

where terms of order $\mu$ have been neglected. The first term in this equation represents the electromotive force induced when a conductor moves through a magnetic field, the second term represents the Hall effect, and the last term is caused by electron inertia. We eliminate the velocity from this equation by means of the momentum equation (4.14), and write the result in terms of the dimensionless parameters

$$\mathcal{N} = \frac{\omega_{pe}^2}{\omega^2}; \qquad \mathcal{B} = \frac{|\Omega_e|}{\omega}. \tag{4.17}$$

$\mathcal{N}$ is proportional to the density and $\mathcal{B}$ to the magnetic field. We find

$$i\omega\epsilon_0 \mathbf{E} = (\mathbf{J} - \mu\mathcal{B}^2 \mathbf{J}_\perp + i\mathcal{B}\, \mathbf{J} \times \mathbf{b})/\mathcal{N}. \tag{4.18}$$

The parallel component of this equation is readily solved,

$$J_\parallel = \mathcal{N} i\omega\epsilon_0 E_\parallel. \tag{4.19}$$

In order to solve for $\mathbf{J}_\perp$, we note that $\mathbf{b} \times \mathbf{J}$ is the vector obtained by rotating $\mathbf{J}_\perp$ around $\mathbf{b}$ by $90^o$. The right-hand side of (4.18) may thus be cast in diagonal form by expressing $\mathbf{J}$ in terms of the basis formed by the eigenvectors of the rotation operator,

$$\mathbf{e}_+ = (\mathbf{e}_x + i\mathbf{e}_y)/\sqrt{2}; \tag{4.20}$$

$$\mathbf{e}_- = (\mathbf{e}_x - i\mathbf{e}_y)/\sqrt{2}, \tag{4.21}$$

where $\mathbf{e}_- = \mathbf{e}_+^*$. The action of $\mathbf{b}\times$ on these vectors is

$$\mathbf{b} \times \mathbf{e}_\pm = \mp i\mathbf{e}_\pm. \tag{4.22}$$

The perpendicular response in the rotation basis is now easily evaluated:

$$J_\pm = \frac{\mathcal{N}}{1 \mp \mathcal{B} - \mu\mathcal{B}^2} i\omega\epsilon_0 E_\pm \tag{4.23}$$

$$\simeq \frac{\mathcal{N}}{(1 \mp \mathcal{B})(1 \pm \mu\mathcal{B})} i\omega\epsilon_0 E_\pm, \tag{4.24}$$

where the $J_\pm$ are the components of the current along the vectors $\mathbf{e}_\pm$. Note that the denominator in (4.24) vanishes at the cyclotron frequencies of the electrons and ions. This expresses the fact that at these frequencies, a current oscillation can take place in the absence of electric field by virtue of the gyration of particles about the magnetic field.

## Dielectric permittivity

In the basis $(\mathbf{e}_+, \mathbf{e}_-, \mathbf{b})$, the conductivity tensor is diagonal. Its elements are given by the coefficients of $E_\parallel$ and $E_\pm$ in (4.19) and (4.24). The dielectric permittivity (4.7) is then

$$\mathbf{K}_{\text{circ}} = \begin{pmatrix} R & 0 & 0 \\ 0 & L & 0 \\ 0 & 0 & P \end{pmatrix}, \tag{4.25}$$

where

$$R = 1 - \frac{\mathcal{N}}{(1 - \mathcal{B})(1 + \mu\mathcal{B})}; \tag{4.26}$$

$$L = 1 - \frac{\mathcal{N}}{(1 + \mathcal{B})(1 - \mu\mathcal{B})}; \tag{4.27}$$

$$P = 1 - \mathcal{N}. \tag{4.28}$$

Here R and L represent the permittivity for right and left circularly polarized perturbations. The permittivity along the magnetic field, $P$, is identical to that of an unmagnetized plasma.

In order to obtain the dispersion relation it is necessary to transform back into the cartesian basis $(\mathbf{e}_1, \mathbf{e}_2, \mathbf{b})$. The components of an arbitrary vector $\mathbf{W}_\perp$ in the cartesian basis are related to those in the rotational eigen-basis by

$$\begin{pmatrix} W_x \\ W_y \\ W_z \end{pmatrix} = \mathbf{U} \begin{pmatrix} W_+ \\ W_- \\ W_z \end{pmatrix}, \tag{4.29}$$

where the unitary matrix $\mathbf{U}$ is given by

$$\mathbf{U} = \frac{1}{\sqrt{2}} \begin{pmatrix} 1 & i & 0 \\ 1 & -i & 0 \\ 0 & 0 & \sqrt{2} \end{pmatrix}. \tag{4.30}$$

The dielectric permittivity in the cartesian basis is thus

$$\mathbf{K} = \mathbf{U}\mathbf{K}_{\mathrm{circ}}\mathbf{U}^{\dagger}, \tag{4.31}$$

where $\mathbf{U}^{\dagger} = \mathbf{U}^{-1}$ is the hermitean conjugate (the transpose of the complex conjugate) of U. We find

$$\mathbf{K} = \begin{pmatrix} S & -iD & 0 \\ iD & S & 0 \\ 0 & 0 & P \end{pmatrix}, \tag{4.32}$$

where

$$S = \frac{1}{2}(R + L), \qquad D = \frac{1}{2}(R - L) \tag{4.33}$$

represent the sum and differences of the right and left dielectric permittivities.

A characteristic of the dielectric permittivity (4.32) is its independence on the wavevector $\mathbf{k}$. This is a consequence of the neglect of thermal motion in the cold plasma closure: since the particles are immobile in the reference state, they respond only to the local electric field. We will see in Chapter 6 that in the Vlasov description of the plasma, by contrast, the dielectric does depend on $\mathbf{k}$. Other significant properties of $\mathbf{K}$ are

$$\mathbf{K}(-\omega) = \mathbf{K}^{*}(\omega),$$

reflecting invariance under time reversal, and

$$\mathbf{K}(-\mathbf{B}_0) = \mathbf{K}^{t}(\mathbf{B}_0),$$

corresponding to the Onsager symmetry relations. Lastly, we note the hermitean nature of the permittivity tensor, $\mathbf{K} = \mathbf{K}^{\dagger}$. This is a consequence of the energy-conserving property of the equations of motion.

The effect of friction with a background neutral gas can be considered by adding a Langevin force $-m_s\nu\mathbf{V}_s$ to the equations of motion. It is easily seen that this is equivalent to replacing the frequency $\omega$ by $\omega + i\nu$ in the dielectric permittivity. The hermiticity property of $\mathbf{K}$ is clearly lost when friction is included.

We next consider the properties of the dispersion relation (4.11) for the dielectric permittivity (4.32).

## Dispersion relation

We may assume without loss of generality that the wave vector lies in the $x$-$z$ plane (Fig. 4.1). We denote the angle between the wave-vector and the magnetic field by $\theta$. The eigenmode equation (4.8) is then

$$\begin{pmatrix} S - n^2 \cos^2 \theta & -iD & n^2 \sin \theta \cos \theta \\ iD & S - n^2 & 0 \\ n^2 \sin \theta \cos \theta & 0 & P - n^2 \sin^2 \theta \end{pmatrix} \begin{pmatrix} E_x \\ E_y \\ E_z \end{pmatrix} = 0, \qquad (4.34)$$

where $n$ is the index of refraction (4.10).

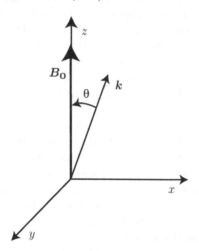

Figure 4.1: Position of the wave-vector and axes for the eigenmode equation.

The dispersion relation is obtained by setting the determinant of the above system equal to zero, as in (4.11). With the help of the identity

$$S^2 - D^2 = RL, \qquad (4.35)$$

we find

$$\mathcal{M}(\omega, \mathbf{k}) = An^4 - Bn^2 + C = 0, \qquad (4.36)$$

where

$$\begin{aligned} A &= S \sin^2 \theta + P \cos^2 \theta; & (4.37) \\ B &= RL \sin^2 \theta + PS(1 + \cos^2 \theta); & (4.38) \\ C &= PRL. & (4.39) \end{aligned}$$

The dispersion relation is thus quadratic in $n^2$. Its discriminant is non-negative,

$$B^2 - 4AC = 4P^2 D^2 \cos^2 \theta + (RL - SP)^2 \sin^4 \theta \geq 0, \qquad (4.40)$$

so that there are always two real roots for $n^2$. It can be shown from (4.40) that these two roots never cross as the propagation angle varies, although they sometimes merge for $\theta = 0$ or $\pi/2$. That is, one of these roots has a phase velocity less than or equal to that of the other, for all $\theta$. The corresponding two waves are called the slow and fast waves. Unfortunately, the slow and fast waves can exchange identities across regions where one of the roots of the dispersion relation is complex, $n^2 < 0$. This undermines the usefulness of the phase velocity as a classification scheme.

The simplicity of having only two waves to consider at any given frequency should be contrasted with the converse problem: solving the dispersion relation for $\omega$ at a given $\mathbf{k}$, as required for solving an initial value problem. Inspection of (4.36)-(4.39) shows that the dispersion relation is a tenth order polynomial in $\omega$! The order corresponds to the twelve field variables ($E_j, B_j, V_j$ and $J_j$ for $j = 1, 2, 3$) minus the two constraints $\nabla \cdot \mathbf{B} = 0$ and $\epsilon_0 \nabla \cdot \mathbf{E} = \rho$. We consider the initial value problem in the next section.

It is frequently useful to characterize waves as being either *longitudinal* or *transverse* according to the orientation of the electric field with respect to the wave-vector. By virtue of Faraday's law, longitudinal waves leave the magnetic field unperturbed, and are thus equivalently referred to as electrostatic waves. They can be described by a single field, namely the electrostatic potential. Transverse waves, by contrast, are always electromagnetic.

The eigenmode equation (4.8)-(4.9) shows that waves become electrostatic when the *resonance condition*, $n^2 = \infty$, is satisfied. Resonant waves are of special interest, in part because they are comparatively easy to excite. Since their phase velocity is much less than the speed of light in vacuum they can be excited, for example, by Cerenkov emission from particle beams. Recall, however, that when the phase velocity of the wave is comparable to the thermal velocity the cold plasma model becomes inadequate.

## 4.3    Wave propagation

We now consider the propagation of an arbitrary initial wavetrain in a uniform plasma. To simplify our investigation, we begin by assuming that the dispersion relation has been solved for the eigenfrequencies $\omega = \Omega_\ell(\mathbf{k})$, where $\ell = 1, \ldots, 10$ labels the various roots. Corresponding to each eigenfrequency there is an eigenvector $\mathbf{\Psi}_{\mathbf{k},\ell} = (\mathbf{E}_{\mathbf{k},\ell}, \mathbf{B}_{\mathbf{k},\ell}, \mathbf{V}_{\mathbf{k},\ell}, \mathbf{J}_{\mathbf{k},\ell})$ describing, in particular, the polarization of the mode $\ell$. The general solution of the initial value problem for the propagation of a wave in a homogeneous plasma is then

$$\mathbf{\Psi}(\mathbf{x}, t) = \sum_\ell \int d\mathbf{k} \, a_{\mathbf{k},\ell} \mathbf{\Psi}_{\mathbf{k},\ell} e^{i\mathbf{k}\cdot\mathbf{x} - i\Omega_\ell(\mathbf{k})t}. \tag{4.41}$$

The amplitude $a_{\mathbf{k},\ell}$ may be calculated from the initial conditions by inverting the Fourier integral and separating the initial field into its components in the eigenbasis,

$$a_{\mathbf{k},\ell} = \frac{1}{2\pi} \int d\mathbf{k} \, \langle \mathbf{\Psi}_{\mathbf{k},\ell} | \mathbf{\Psi}(\mathbf{x}, 0) \rangle e^{-i\mathbf{k}\cdot\mathbf{x}}, \tag{4.42}$$

where $\mathbf{\Psi}(\mathbf{x}, 0)$ represents the initial perturbation and $\langle \mathbf{\Psi}_{\mathbf{k},\ell} | \mathbf{\Psi}(\mathbf{x}, 0) \rangle$ represents its component along the $\ell$-th eigenvector of the eigenbasis.

The Fourier representation of the solution, although exact, conveys little insight into the nature of wave propagation in a dispersive plasma. In order to gain understanding, we consider the asymptotic properties of this solution at large distances and long times.

## Asymptotic behavior of the general solution

We are interested in the asymptotic behavior of the wave viewed by an observer moving with constant, finite velocity. The time and place of observation are thus related by $\mathbf{x} = \mathbf{v}_{\mathrm{obs}}t$. We henceforth suppress the subscript $\ell$. For large $\mathbf{x}$ and $t$, the integral is dominated by the contribution from the vicinity of points of *stationary phase* (Fig. 4.2) defined by

$$x_j - \frac{\partial \Omega}{\partial k_j} t = 0. \tag{4.43}$$

Away from these points the integrand oscillates rapidly and the positive and negative contributions approximately cancel each other. We can calculate the integral by expanding the argument of the exponential around the points of stationary phase. There follows

$$\boldsymbol{\Psi}(\mathbf{x}, t) = a_{\mathbf{k}} \boldsymbol{\Psi}_{\mathbf{k}} \left( \frac{2\pi}{t} \right)^{3/2} \left( \det \left| \frac{\partial \Omega}{\partial k_i \partial k_j} \right| \right)^{-1/2} \exp(i\mathbf{k}\cdot\mathbf{x} - i\Omega(\mathbf{k})t + i\varphi), \tag{4.44}$$

where $\varphi$ is a phase that depends on the structure of $\Omega$ near the point of stationary phase.

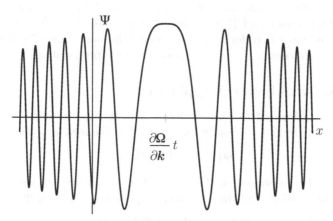

Figure 4.2: Integrand of the Fourier integral at large $\mathbf{x}, t$ illustrating the method of stationary phase.

The resemblance of the asymptotic solution to the plane wave solution is deceptive: the wave-vector $\mathbf{k}$ in (4.44) is a function of position and time through the stationary phase condition, (4.43). Likewise, the frequency may be viewed as a function of position and time through the dispersion relation, $\omega(\mathbf{x}, t) = \Omega(\mathbf{k}(\mathbf{x}, t))$. This raises the following question: what wave-vector and frequency will be observed at $(\mathbf{x}, t)$?

To answer this question, consider the phase of the wave $\zeta$,

$$\zeta(\mathbf{x}, t) = \mathbf{k}(\mathbf{x}, t) \cdot \mathbf{x} - \omega(\mathbf{x}, t)t. \tag{4.45}$$

The wavevector and frequency observed at $(\mathbf{x}, t)$ are most naturally defined as $\nabla\zeta$ and $\partial\zeta/\partial t$. Adopting the convention of summation over repeated indices, we calculate

$$\nabla\zeta = \mathbf{k} + \left[x_i - \frac{\partial\Omega}{\partial k_i}t\right]\nabla k_i; \tag{4.46}$$

$$\frac{\partial\zeta}{\partial t} = -\Omega(\mathbf{k}) + \left[x_i - \frac{\partial\Omega}{\partial k_i}t\right]\frac{\partial k_i}{\partial t}. \tag{4.47}$$

By virtue of the stationary phase condition, the terms in brackets vanish and we have

$$\nabla\zeta = \mathbf{k}; \tag{4.48}$$

$$\frac{\partial\zeta}{\partial t} = -\Omega(\mathbf{k}). \tag{4.49}$$

This shows that the $\mathbf{k}$ and $\Omega$ appearing in 4.44 are indeed the locally observed wavevector and frequency. It further shows that the local wavenumber and frequency satisfy the dispersion relation *even in a nonuniform wavetrain*. We may now propose an interpretation for the velocity $\mathbf{v}_g = \nabla_{\mathbf{k}}\Omega(\mathbf{k})$ that appears in the stationary phase condition: $\mathbf{v}_g$, called the group velocity, is the velocity at which an observer must move in order to always observe the same wave-vector $\mathbf{k}$. The phase velocity (4.2), by contrast, represents the velocity at which an observer must move in order to always observe the same phase.[78] Both of these velocities play a central role in the theory of wave propagation.

## Waves in weakly varying plasma

The asymptotic form of the phase function in homogeneous media, (4.48)-(4.49), suggests generalization to weakly varying media. We seek a solution of the form

$$\mathbf{\Psi}(\mathbf{x}, t) = \hat{\mathbf{\Psi}}(\mathbf{x}, t)e^{i\zeta(\mathbf{x}, t)}, \tag{4.50}$$

where $\hat{\mathbf{\Psi}}$ is the amplitude of the wave. The exponential factor is called the *eikonal*. The gradient and time derivative operators acting on $\mathbf{\Psi}$ yield

$$\nabla\mathbf{\Psi} = i\mathbf{k}\mathbf{\Psi} + \nabla\hat{\mathbf{\Psi}}e^{i\zeta(\mathbf{x}, t)}; \tag{4.51}$$

$$\frac{\partial\mathbf{\Psi}}{\partial t} = -i\omega\mathbf{\Psi} + \frac{\partial\hat{\mathbf{\Psi}}}{\partial t}e^{i\zeta(\mathbf{x}, t)}, \tag{4.52}$$

where $\mathbf{k}$ and $\omega$ are defined as in (4.48) and (4.49). Consider waves such that

$$\mathbf{k}\mathbf{\Psi} \gg \nabla\hat{\mathbf{\Psi}} \qquad \text{and} \qquad \omega\mathbf{\Psi} \gg \frac{\partial\hat{\mathbf{\Psi}}}{\partial t}. \tag{4.53}$$

If we neglect the small terms in (4.53), the wave equations will clearly have the same form as in a homogeneous plasma. Equation (4.50) is thus an approximate solution provided that $\omega$ and $\mathbf{k}$ satisfy the local dispersion relation for the homogeneous system,

$$\mathcal{M}(\omega, \mathbf{k}, \mathbf{x}, t) = 0, \tag{4.54}$$

and that $\hat{\boldsymbol{\Psi}}$ is directed along the corresponding local eigenvector,

$$\hat{\boldsymbol{\Psi}}(\mathbf{x}) = a_{\mathbf{k},j}(\mathbf{x})\boldsymbol{\Psi}_{\mathbf{k},j}(\mathbf{x}). \tag{4.55}$$

Note that the dispersion relation and eigenvectors depend now on $\mathbf{x}$ through the slow variation of the equilibrium parameters.

The solution (4.50), (4.55) is called the WKB approximation, after Wenzel, Kramers and Brillouin. The WKB approximation leads to a greatly simplified description of the evolution of short wavelength disturbances, and provides rules for constructing *global* eigenmodes in weakly varying plasmas. It is closely related to the method of averages described in Chapter 2, the phase of the wave playing a role analogous to the gyrophase in single-particle motion.

We can find $\zeta$ if we know its gradient $\mathbf{k}$ at each $\mathbf{x}$ and $\omega$. The dispersion relation, however, only gives the magnitude of $\mathbf{k}$ as a function of its direction and the other two variables. The indeterminacy is removed by using the fact that since $\mathbf{k}$ is a gradient, it must satisfy

$$\nabla \times \mathbf{k} = 0. \tag{4.56}$$

We next show how to use (4.54) and (4.56) to solve for the eikonal $\zeta$.

## Ray tracing

We follow the common usage of allowing the symbols $\omega$ and $\mathbf{k}$ to denote both the functions of $\mathbf{x}$ and $t$ obtained by differentiating $\zeta$, and the independent variables that appear as arguments of the dispersion relation. In order to avoid confusion, we adopt the thermodynamics convention of noting the independent variables (held constant) in the differentiations as subscripts outside parentheses. To obtain an equation for $\zeta$, we start from the cross-differentiation rules:

$$\left(\frac{\partial k_i}{\partial t}\right)_{\mathbf{x}} + \left(\frac{\partial \omega}{\partial x_i}\right)_{x_m, t} = 0; \tag{4.57}$$

$$\left(\frac{\partial k_j}{\partial x_i}\right)_{x_m, t} - \left(\frac{\partial k_i}{\partial x_j}\right)_{x_n, t} = 0, \tag{4.58}$$

where $x_m$ represents all the spatial variables other than $x_i$, and $x_n$ all the spatial variables other than $x_j$. If we consider that $\omega = \Omega(\mathbf{x}, \mathbf{k}, t)$ is defined implicitly by the dispersion relation, we may eliminate it from the first of the cross-differentiation rules, (4.57), so as to obtain an equation for $\mathbf{k}$. Substituting $\Omega$ in (4.57) yields

$$\left(\frac{\partial k_i}{\partial t}\right)_{\mathbf{x}} + \left(\frac{\partial \Omega}{\partial k_j}\right)_{x_j, t} \left(\frac{\partial k_j}{\partial x_i}\right)_{x_j, t} = -\left(\frac{\partial \Omega}{\partial x_i}\right)_{\mathbf{k}, x_j}, \tag{4.59}$$

where we recognize the group velocity $\mathbf{v}_g$ multiplying the gradient of $k_j$. The term involving $k_j$ may be eliminated by means of the second cross-differentiation rule (4.58). There follows

$$\frac{\partial k_i}{\partial t} + \mathbf{v}_g \cdot \nabla k_i = -\left(\frac{\partial \Omega}{\partial x_i}\right)_{\mathbf{k}, x_j}. \tag{4.60}$$

Equation (4.60) may be viewed as defining the evolution of $k_i$ as seen by an observer moving at the group velocity. That is, $k_i$ may be found by solving the *ray equations*

$$\frac{dk_i}{dt} = -\frac{\partial \Omega}{\partial x_i}; \qquad \frac{dx_i}{dt} = \frac{\partial \Omega}{\partial k_i}, \tag{4.61}$$

with initial conditions on an antenna where the waves are launched. Differentiation of $\omega = \Omega(\mathbf{x}, \mathbf{k}, t)$ along the ray yields

$$\frac{d\omega}{dt} = \frac{\partial \Omega}{\partial t}. \tag{4.62}$$

Equations (4.61)-(4.62) are clearly in canonical form, with $\Omega$ playing the role of the Hamiltonian.

The derivatives of $\Omega$ appearing in (4.61)-(4.62) are obtained from the dispersion relation by differentiation with respect to $k_j$,

$$\left(\frac{\partial \mathcal{M}}{\partial k_i}\right)_{\mathbf{x}, k_m, t} + \left(\frac{\partial \mathcal{M}}{\partial \omega}\right)_{\mathbf{x}, \mathbf{k}, t} \left(\frac{\partial \Omega}{\partial k_i}\right)_{\mathbf{x}, k_m, t} = 0, \tag{4.63}$$

whence

$$(v_g)_i \equiv \left(\frac{\partial \Omega}{\partial k_i}\right)_{\mathbf{x}, k_m, t} = -\frac{(\partial \mathcal{M}/\partial k_i)_{\mathbf{x}, k_m, t}}{(\partial \mathcal{M}/\partial \omega)_{\mathbf{x}, \mathbf{k}, t}}. \tag{4.64}$$

Analogous expressions for the remaining derivatives of $\Omega$ are found in similar fashion.

## Wave amplitude

The most important objective of the wave propagation theory is to predict the evolution of the wave amplitude, or equivalently the transport of the energy in the wave. This objective may be met by considering the first-order corrections to the WKB solution (4.50). The solubility condition for the first order equation yields an equation for the evolution of the amplitude of the lowest-order solution. Here we follow a different procedure, pioneered by Whitham,[77] based on averaging the Lagrangian. The Lagrangian variational procedure is strikingly simpler, and has the additional advantage of bringing out the similarities between the theory of adiabatic invariants and the eikonal theory.

The Lagrangian for the cold plasma system is obtained by converting the single-particle Lagrangian of Chapter 2 into a Lagrangian density (describing a fluid, as opposed to a single particle), and adding the vacuum Lagrangian density for the electromagnetic field. Thus,

$$L = \int dt \int d\mathbf{x}\, \mathcal{L}, \tag{4.65}$$

where $\mathcal{L}$ is the complete Lagrangian density:

$$\mathcal{L} = \mathcal{L}_M(\mathbf{A}, \Phi) + \sum_s \mathcal{L}_s(\boldsymbol{\xi}_s, \mathbf{A}, \Phi). \tag{4.66}$$

Here $\mathcal{L}_s$ is the Lagrangian density for the particles,

$$\mathcal{L}_s = n_s \left( \frac{m_s}{2} \dot{\boldsymbol{\xi}}_s^{\,2}(\mathbf{x}, t) + e_s \dot{\boldsymbol{\xi}}_s(\mathbf{x}, t) \cdot \mathbf{A}(\mathbf{x} + \boldsymbol{\xi}_s, t) - e_s \Phi(\mathbf{x} + \boldsymbol{\xi}_s, t) \right), \tag{4.67}$$

and $\mathcal{L}_M$ is that for the electromagnetic fields,

$$\mathcal{L}_M = \epsilon_0 \left( \frac{\partial \mathbf{A}(\mathbf{x}, t)}{\partial t} + \nabla\Phi(\mathbf{x}, t) \right)^2 - \frac{1}{\mu_0} (\nabla \times \mathbf{A}(\mathbf{x}, t))^2. \tag{4.68}$$

To investigate linear waves we expand the Lagrangian in powers of the perturbation amplitude, keeping only the lowest order, quadratic terms. The electromagnetic Lagrangian is already quadratic, and the particle Lagrangian density becomes

$$(\mathcal{L}_s)_{\text{lin}} = n_s \left( m_s \dot{\boldsymbol{\xi}}_s^{\,2} + e_s \boldsymbol{\xi} \cdot (\dot{\boldsymbol{\xi}}_s \times \mathbf{B}_0) + e_s \mathbf{A} \cdot \dot{\boldsymbol{\xi}}_s - e_s \boldsymbol{\xi}_s \cdot \nabla\Phi \right), \tag{4.69}$$

where $A$ and $\Phi$ now stand for the *perturbed* field amplitudes.

We next substitute the eikonal solution (4.50) in (4.72) and average the Lagrangian with respect to the rapid variation of the phase of the eikonal. The real quantities entering the Lagrangian are expressed in terms of the eikonal solutions by

$$\mathbf{A}(x, t) = \frac{1}{2} \left( \hat{\mathbf{A}} e^{i\zeta(\mathbf{x}, t)} + \hat{\mathbf{A}}^* e^{-i\zeta(\mathbf{x}, t)} \right) \tag{4.70}$$

The averaging eliminates the harmonic contributions proportional to $\exp(\pm 2i\zeta)$.

The resulting averaged Lagrangian is a functional of the perturbation amplitudes $\hat{\boldsymbol{\xi}}_s$, $\hat{\mathbf{A}}$, $\hat{\Phi}$, and of $\zeta$ through $\mathbf{k} = \nabla\zeta$ and $\omega = \partial\zeta/\partial t$. We may reduce the number of independent variables by using the equations of motion to eliminate the displacements in favor of the electric field. We also eliminate the vector potential with

$$\mathbf{E} = -\frac{\partial \mathbf{A}}{\partial t} - \nabla\Phi \tag{4.71}$$

There follows the Lagrangian density

$$(\mathcal{L})_{\text{lin}} = \epsilon_0 \hat{\mathbf{E}}^* \cdot \mathbf{M} \cdot \hat{\mathbf{E}}. \tag{4.72}$$

where $\mathbf{M}$ is given by (4.9).

We now show that this Lagrangian reproduces the mode equations (4.8). Taking the variation with respect to $\hat{\mathbf{E}}$ yields

$$\mathbf{M} \cdot \hat{\mathbf{E}} = 0, \tag{4.73}$$

showing that $\hat{\mathbf{E}}$ must be an eigenvector of the mode equation and that $\omega, \mathbf{k}$ must satisfy the dispersion relation.

Variation with respect to $\zeta$ yields

$$\frac{\partial}{\partial t}\left(\hat{\mathbf{E}} \cdot \frac{\partial \mathbf{M}}{\partial \omega} \cdot \hat{\mathbf{E}}\right) - \frac{\partial}{\partial x_j}\left(\hat{\mathbf{E}} \cdot \frac{\partial \mathbf{M}}{\partial k_j} \cdot \hat{\mathbf{E}}\right) = 0. \tag{4.74}$$

This is the desired equation for the amplitude $|\hat{\mathbf{E}}|$ of the wave. The quantity

$$\mathcal{J} = \epsilon_0 \hat{\mathbf{E}} \cdot \frac{\partial \mathbf{M}}{\partial \omega} \cdot \hat{\mathbf{E}} = \frac{\partial \mathcal{L}}{\partial \dot{\zeta}}$$

is the momentum conjugate to the phase $\zeta$. It is thus a generalization of the adiabatic invariant studied in Chapter 2. Equation (4.74) is a conservation equation for $\mathcal{J}$. It shows that $\mathcal{J}$ is carried at the velocity

$$-\left(\hat{\mathbf{E}}\frac{\partial \mathbf{M}}{\partial k_j}\hat{\mathbf{E}}\right) \bigg/ \left(\hat{\mathbf{E}}\frac{\partial \mathbf{M}}{\partial \omega}\hat{\mathbf{E}}\right) = (v_g)_j. \tag{4.75}$$

The conservation equation may thus be written in the form

$$\frac{\partial \mathcal{J}}{\partial t} + \nabla(\mathbf{v}_g \mathcal{J}) = 0, \tag{4.76}$$

showing that the characteristics for the propagation of the adiabatic invariant are the same as for the wave-vector. Thus, the adiabatic invariant increases when the rays converge, and decreases when they diverge.

A significant shortcoming of the Lagrangian method is its limitation to conservative (ideal) systems. We next show that for time-independent plasmas, Poynting's theorem leads to an evolution equation for the energy that is applicable even in the presence of dissipation.

## Wave energy

The general electromagnetic energy conservation law, known as Poynting's theorem, is obtained by multiplying Maxwell's law of induction by $\mathbf{E}$ and rearranging terms. It has the form

$$\frac{1}{2}\frac{\partial}{\partial t}\left(\frac{\mathbf{B}^2}{\mu_0} + \epsilon_0 \mathbf{E}^2\right) = -\mathbf{J} \cdot \mathbf{E} - \nabla \cdot \mathbf{S}. \tag{4.77}$$

Here,

$$\mathbf{S} = \frac{1}{\mu_0}\mathbf{E} \times \mathbf{B} \tag{4.78}$$

is the Poynting flux. Poynting's theorem states that the rate of change of the energy is equal to the sum of the work done by the field on the plasma and the radiative energy losses.

We use Poynting's theorem to obtain an evolution law for the amplitude of a wave in a dissipative system. The first steps are similar to those used in the evaluation of the averaged Lagrangian: we substitute the eikonal representation of the fields (4.70) in (4.77) and average over the phase of the eikonal, keeping the slow time variable constant. To lowest order in $kL$, we find

$$\hat{\mathbf{E}}^* \cdot \boldsymbol{\sigma}_h \cdot \hat{\mathbf{E}} = 0, \tag{4.79}$$

where

$$\boldsymbol{\sigma}_h = \frac{1}{2}(\boldsymbol{\sigma} + \boldsymbol{\sigma}^\dagger)$$

is the hermitean part of the conductivity. Equation (4.79) shows that the dissipative losses, represented by $\boldsymbol{\sigma}_h$, must vanish to lowest order in the WKB expansion:

$$\boldsymbol{\sigma}_h = O(kL). \tag{4.80}$$

We next consider the first-order terms. For simplicity, we restrict ourselves to the cold plasma model and assume that the reference state does not change with time, so that the frequency is constant. The current is then given, through first order, by

$$\begin{aligned} \hat{\mathbf{J}} + \delta\hat{\mathbf{J}} &= \boldsymbol{\sigma}(\omega + i\frac{\partial}{\partial t})(\hat{\mathbf{E}} + \delta\hat{\mathbf{E}}) \\ &= \boldsymbol{\sigma}(\omega)\hat{\mathbf{E}} + \boldsymbol{\sigma}(\omega)\delta\hat{\mathbf{E}} + i\frac{\partial\boldsymbol{\sigma}}{\partial\omega}\frac{\partial\hat{\mathbf{E}}}{\partial t}, \end{aligned} \tag{4.81}$$

where $\delta\hat{\mathbf{E}}$ is the first-order correction to the electric field amplitude. Faraday's law yields the first-order correction to the magnetic field, $\delta\hat{\mathbf{B}}$:

$$i\omega\delta\hat{\mathbf{B}} = i\mathbf{k} \times \delta\hat{\mathbf{E}} + \frac{\partial\hat{\mathbf{B}}}{\partial t} + \nabla \times \hat{\mathbf{E}}, \tag{4.82}$$

where

$$\hat{\mathbf{B}} = \frac{\mathbf{k} \times \hat{\mathbf{E}}}{\omega}.$$

Substituting these expansions in the averaged Poynting theorem (4.77) yields

$$\frac{\partial\mathcal{E}}{\partial t} + \nabla \cdot \hat{\mathbf{S}} = \mathcal{Q}. \tag{4.83}$$

Here

$$\mathcal{E} = \frac{1}{2}\left(\epsilon_0\hat{\mathbf{E}}^* \cdot \frac{\partial(\omega\mathbf{K})}{\partial\omega} \cdot \hat{\mathbf{E}} + \frac{\hat{\mathbf{B}}^* \cdot \hat{\mathbf{B}}}{\mu_0}\right)$$

is the energy density,

$$\hat{\mathbf{S}} = \mathrm{Re}(\hat{\mathbf{E}}^* \times \hat{\mathbf{B}})/\mu_0$$

is the flux of energy carried by the wave and

$$\mathcal{Q} = \hat{\mathbf{E}}^* \cdot \sigma_h \cdot \hat{\mathbf{E}}$$

is the power dissipated through nonideal processes.

Substituting $\mathbf{K}$ by its expression in terms of $\mathbf{M}$ yields a more compact form of the energy density,

$$\mathcal{E} = \frac{1}{2}\,\epsilon_0\,\hat{\mathbf{E}}^* \cdot \frac{\partial(\omega \mathbf{M})}{\partial \omega} \cdot \hat{\mathbf{E}}. \tag{4.84}$$

Recalling that $\mathbf{M} \cdot \hat{\mathbf{E}} = 0$, we see that

$$\mathcal{E} = \omega \mathcal{J}$$

as for a simple oscillator. The dielectric energy, or wave energy, is equal to the product of the frequency with the adiabatic invariant for the wave.

Comparison of (4.83) with the conservation law obtained by the average Lagrangian method(4.74) suggests that we evaluate

$$\frac{1}{2}\,\epsilon_0\,\frac{\partial}{\partial k_j}\left(\omega \hat{\mathbf{E}}^* \cdot \mathbf{M} \cdot \hat{\mathbf{E}}\right) \;=\; -\mathrm{Re}\left(\hat{\mathbf{E}}^* \times \left(\frac{\mathbf{k} \times \hat{\mathbf{E}}}{\omega \mu_0}\right)\right)_j$$

$$= -\hat{S}_j. \tag{4.85}$$

This implies that

$$\hat{\mathbf{S}} = \mathbf{v}_g \mathcal{E}.$$

The Poynting flux amplitude is the product of the group velocity with the energy density. The energy equation now takes a form similar to the equation for the conservation of the adiabatic invariant,

$$\frac{\partial \mathcal{E}}{\partial t} + \nabla \cdot (\mathbf{v}_g \mathcal{E}) = \mathcal{Q}. \tag{4.86}$$

As expected, the energy is conserved for time-independent systems in the absence of dissipation.

## 4.4  Representation of the dispersion relation

The properties of the waves described by the mode equation (4.8) depend on the frequency, the direction of propagation, and the plasma parameters. Since waves are generally the only means of probing the plasma, it is extremely important to understand their properties. We begin by reviewing the various graphical representations of the dielectric permittivity and dispersion relation.

In crystal optics, the dielectric tensor is represented by the *Fresnel indicatrix* surface. This surface is defined by

$$\epsilon_0 \mathbf{E} \cdot \mathbf{K} \cdot \mathbf{E} = 1. \tag{4.87}$$

For real, symmetric $\mathbf{K}$, the Fresnel surface provides a complete description of the dielectric; its principal axes (eigenvectors) and corresponding eigenvalues. In a crystal, the Fresnel indicatrix is an ellipsoid. When this indicatrix has axial symmetry, it intersects the plane perpendicular to $\mathbf{k}$ along an ellipse such that one axis has length $K_\perp^{-1/2}$ regardless of the direction of $\mathbf{k}$. The corresponding wave is called the ordinary wave by crystallographers. Its phase velocity does not depend on the direction of propagation, and it thus obeys Snell's law. The second root of the dispersion relation is called the extraordinary wave.

Plasma physicists use a different convention for naming the waves. The plasma physics convention is based on the observation that one of the waves propagating perpendicular to the magnetic field has a phase velocity that is independent of the magnetic field. This wave is called the ordinary wave, and the wave whose propagation does depend on the magnetic field is called the extraordinary wave. The plasma-physical convention is opposite to that used by crystallographers.

The Fresnel indicatrix is unsuited for describing magnetized plasma, because it discards the antisymmetric part of the permittivity. The most useful representations of the dispersion relation are as follows:

1. The *surface of normal slowness*, or *index surface* is the polar plot of the index of refraction $n = kc/\omega$. Its principal virtue is that the Poynting flux is normal to it.

2. The phase-velocity surface or *wave-normal* surface is given by a polar plot of the phase velocity. This is the most commonly used representation, and the one we will use in the remainder of this chapter. It is sometimes mistakenly stated that the wave-normal surface describes the shape of the ripples that result from "throwing a rock in a plasma pond." These ripples are in fact described by the ray surface.

3. The *ray surface* is the surface of constant phase resulting from an impulse disturbance. It is the envelope of the wave planes that originated from the point disturbance at time $t = 0$ and propagated during a time $\delta t$. It is thus described by

$$\mathbf{v}_{\text{ray}} = \frac{\omega \nabla_\mathbf{k} \mathcal{M}}{\mathbf{k} \cdot \nabla_\mathbf{k} \mathcal{M}}, \qquad (4.88)$$

where $\mathcal{M}$ is given by (4.36). This is the appropriate surface to use in Huygens' construction. The geometrical construction of the ray surface from the wave-normal surface is illustrated in Fig. 4.7.

The ray surface depends on the *gradients* of the equilibrium quantities, unlike the wave-normal and index surfaces that depend only on the local values of these quantities. The index surface can be unbounded, a consequence of the fact that $A$ in (4.37) may vanish as $\theta$ varies. The wave-normal surface, by contrast, is bounded whenever it exists. For this reason the wave-normal surfaces are the most convenient to study and represent.

As a wave propagates across a plasma, the topology of the wave-normal surface changes only across definite lines called bounding lines. The qualitative features of the dispersion relation may thus be conveniently summarized by a diagram of the $(\mathcal{N}, \mathcal{B})$ parameter-space showing the bounding lines and the wave-normal surfaces for the fast and slow waves in each region (Fig. 4.3). Increasing the magnetic field or the density corresponds respectively to moving upwards or to the right in this diagram. Changing the frequency at constant plasma parameters corresponds to moving towards the origin along a parabola such as that shown in dashed lines in Fig. 4.3. Note that the diagram and all dispersion relations in this chapter are drawn for an artificial mass-ratio $\mu = 1/4$. This is necessary in order to allow the different frequency regimes to be shown in the same graphs.

The above representation of the dispersion relation is known as the CMA diagram after its originators Clemmow, Mullaly and Allis. It provides an indispensable overview of the properties of waves in plasma.

A very helpful feature of the CMA diagram is that it is easily reconstructed from a few basic facts about plasma waves. Most important among those facts is the behavior of the wave-normal surfaces across the bounding lines. The bounding lines are related to the conditions for cutoff and resonance, which we now describe.

## Cutoff and resonance

The cutoff surfaces separate regions of space where the wave propagates ($k^2 > 0$) from those where it is evanescent and decays exponentially ($k^2 < 0$). Cutoff corresponds thus to $n^2 = 0$, or

$$PRL = 0. \tag{4.89}$$

This is clearly independent of the angle of propagation. The ordering assumed by the WKB theory, (4.53), breaks down near cutoff surfaces. Analysis of the wave equation close to these surfaces show that they reflect incident waves. The reflection coefficients are found by matching the WKB solution to an approximate analytic solution of the wave equation valid in the cutoff region.

The WKB ordering also breaks down near points such that $k \to \infty$. These points are called resonances: they occur at frequencies that correspond to natural modes of oscillation, such as the plasma oscillation. It is generally the case that some fraction of the wave energy is absorbed at the resonance. The behavior of the rays near cutoff and resonance is shown in Fig. 4.4.

In the cold plasma model, resonance takes place for $A = 0$, or

$$\tan\theta = -\frac{P}{S}. \tag{4.90}$$

The resonance condition, unlike the cutoff condition, depends on the angle of propagation. The resonant wave-vectors describe a cone the axis of which lies along the magnetic field. The locus of frequencies such that the angle of resonance is either $0°$ or $90°$ has a special importance, since it separates regions of

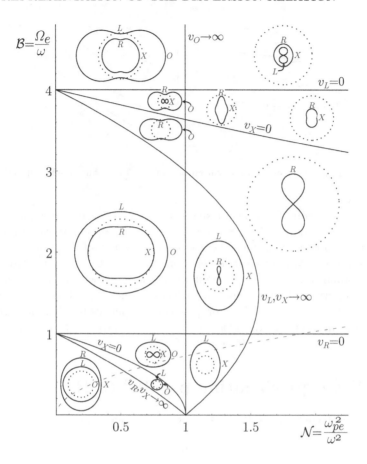

Figure 4.3: Clemmow Mullaly Allis (CMA) diagram showing the cross-sections of the wave-normal surfaces in the various regions of parameter space. The cross-sections are oriented so that the magnetic field is in the vertical direction. The dotted circles correspond to propagation at the speed of light in vacuum. The scale used to draw each surface can be inferred by comparison with these circles. The dashed parabola is an example of the sequence of points sampled by varying the frequency while maintaining the plasma parameters constant.

the plasma where a resonance occurs from regions where (4.90) has no real roots and no resonance occurs. When no resonance occurs the wave-normal surface is spheroidal.

For $\theta = 90^o$ ($S = 0$), the resonance cone flattens into the plane perpendicular to the magnetic field. The frequencies corresponding to $S = 0$ are known as the hybrid frequencies: we will describe them in greater detail below. For $\theta = 0^o$ the resonance cone collapses into a line. This occurs when the quotient $P/S = 0$. The first solution, $P = 0$, shows that the plasma frequency is at the same time a resonance and a cutoff. An unfortunate consequence of this is that the behavior of the wave-normal surface as the plasma resonance is crossed

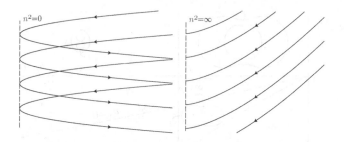

Figure 4.4: Behavior of the rays near cutoff (left) and resonance (right) surfaces.

depends on the region of parameter space where the crossing takes place. The remaining two solutions, $R = \infty$ and $L = \infty$, correspond to the cyclotron resonances. As pointed out when the cyclotron resonances first appeared in the current response (4.24), the electric field vanishes for $RL = \infty$. Away from the cyclotron resonances, however, resonant waves are electrostatic in character.

The continuity of the wave-normal surfaces implies that their shape can be inferred from the knowledge of the propagation properties parallel and perpendicular to the magnetic field. This motivates us to investigate the dispersion relation in more detail for $\theta = 0^\circ$ and $\theta = 90^\circ$.

## 4.5   Waves propagating parallel to $\mathbf{B}_0$

For parallel propagation, $\theta = 0$, the roots of the dispersion relation are

$$P = 0; \qquad n^2 = R; \qquad n^2 = L. \tag{4.91}$$

In a two-component plasma, the dispersion relation can be factored. Making use of the smallness of the mass ratio, we find

$$\omega^2 = \omega_{pe}^2; \tag{4.92}$$

$$\frac{k^2 c^2}{\omega^2} = \frac{\omega^2 \mp \omega \Omega_e - \Omega_e \Omega_i - \omega_{pe}^2}{(\omega \pm \Omega_i)(\omega \mp \Omega_e)}, \tag{4.93}$$

where the upper and lower choices of signs correspond to the R and L waves respectively.

The polarization of these waves may be found from the middle line of (4.34),

$$\frac{iE_x}{E_y} = \frac{n^2 - S}{D}. \tag{4.94}$$

For $n^2 = R$ we find $iE_x/E_y = 1$, while $n^2 = L$ we find $iE_x/E_y = -1$. This shows that the waves corresponding to R and L are circularly polarized in the right-hand and left-hand sense respectively.

The dispersion curve for parallel propagation is shown in Fig. 4.5.

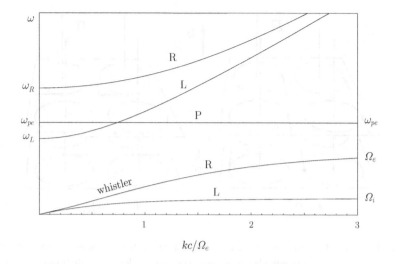

Figure 4.5: Dispersion relation for waves propagating parallel to the magnetic field. The parameters correspond to the dashed parabola shown in the CMA diagram, Fig. 4.3

The cutoffs correspond to the points where the dispersion curve intersects the ordinate ($k = 0$). The horizontal asymptotes correspond to the electron and ion cyclotron resonances. We consider the various features of the dispersion relation in greater detail.

## The cyclotron resonances

The cyclotron resonances are easily understood: They correspond to oscillations in which all the particle of a given species lying in a slab perpendicular to the magnetic field are given an impulse in the same direction, perpendicular to the magnetic field (Fig. 4.6). These particles subsequently follow their Larmor orbits, thus generating a current that oscillates at the corresponding cyclotron frequency. In the limit of vanishing wavelength, where the current in nearby slabs flows in opposite directions, the electromagnetic fields produced by the oscillating current are small and do not affect the gyration of the particles.

The polarization of the cyclotron oscillations matches the rotation of the corresponding resonant particle. Thus, the electron cyclotron oscillation is circularly polarized in the right-hand sense and the ion oscillation in the left-hand sense.

## Whistler waves

The distinguishing feature of the dispersion curve for the R waves is the increase of the group velocity (the slope of the dispersion curve) for the low-frequency branch of the R wave near the origin in frequency. As a result of this feature,

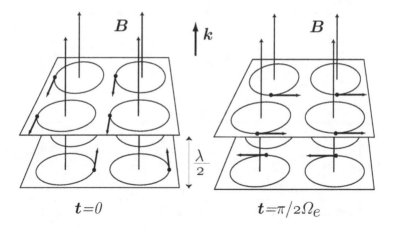

Figure 4.6: Sketch of the cyclotron resonance at two different times. Note that all the particles in the plane perpendicular to **k** have the same velocity at all times. The points represent the particles and the arrows their velocity vectors. The circles represent the particles' orbits.

the higher-frequency components of disturbances travel faster than the low-frequency components.

A remarkable demonstration of the properties of the R wave is provided when wave-pulses excited by lightning in one hemisphere are received in the other after traveling along the magnetic field lines. The high-frequency components arrive before the low-frequency components, causing a distinctive whistling-down tone. The characteristics of this phenomenom were explained by Storey[69] on the basis of the simplified dispersion relation

$$n^2 = \frac{\omega_{pe}^2}{|\omega \Omega_e \cos \theta|}$$

obtained from the cold-plasma dispersion relation (4.36) in the limit $\Omega_i \ll \omega \ll \Omega_e$, $\omega \ll \omega_{pe}$. The corresponding wave-normal and ray surfaces are shown in Fig. 4.7. The group velocity is

$$v_g = 2c \frac{\sqrt{\omega \Omega_e}}{\omega_{pe}}.$$

The frequency at the receiver varies thus according to $\omega \propto t^{-2}$ in agreement with the observations.

It was originally thought that the extraordinarily small attenuation observed as whistlers propagate back and forth between the hemispheres could be explained by the smallness of the angle between the ray and the magnetic field,

$$\alpha < \arctan(1/\sqrt{8}) \approx 19°28'.$$

It is now believed, however, that field-aligned density inhomogeneities act to create waveguides for Whistler propagation.

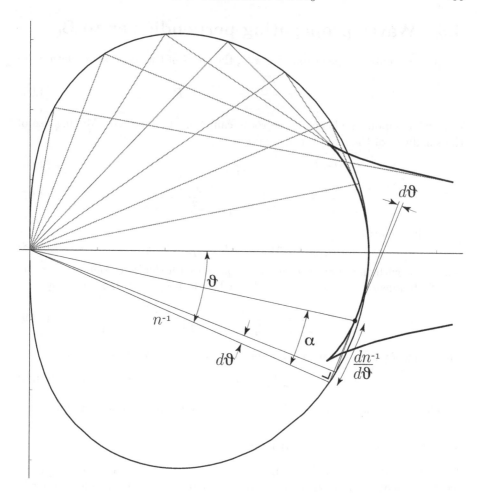

Figure 4.7: Wave-normal and ray surfaces for the whistler wave. Note the narrow propagation angle subtended by the ray surface.

## Faraday rotation

As we discussed in Section 4.2, the two roots of the dispersion relation are non-intersecting. In the case of parallel propagation, in fact, the $R$ wave always has a greater phase velocity than the $L$ wave. This can be used to measure the magnetic field in a plasma if the density is known, by measuring the phase difference between the $R$ and $L$ waves after they have traversed the plasma. This phase difference determines the polarization plane when the two waves recombine outside the plasma. The density can usually be measured by comparing the phase shift between a beam that has traveled through the plasma and a reference beam that travels around the plasma. In astrophysical applications, however, a reference beam is unavailable and all information must be deduced from the variation of the Faraday rotation with frequency.

## 4.6    Waves propagating perpendicular to $B_0$

For perpendicular propagation, $\theta = \pi/2$, the roots of the dispersion relation are

$$n^2 = \frac{RL}{S}; \qquad n^2 = P. \tag{4.95}$$

In a two-component plasma these roots can again be factored. Making use of the smallness of the mass ratio, we find

$$k^2 c^2 = \omega^2 - \omega_{pe}^2; \tag{4.96}$$

$$\frac{k^2 c^2}{\omega^2} = \frac{(\omega^2 - \omega_R^2)(\omega^2 - \omega_L^2)}{(\omega^2 - \omega_{LH}^2)(\omega^2 - \omega_{UH}^2)}, \tag{4.97}$$

where the equation

$$\omega^2 \mp \omega \Omega_e - \Omega_e \Omega_i - \omega_{pe}^2 = 0 \tag{4.98}$$

is satisfied with the upper sign for $\omega = \omega_R$ and with the lower sign for $\omega = \omega_L$.

The frequency $\omega_{LH}$ is called the lower hybrid frequency. It is given by

$$\frac{1}{\omega_{LH}^2} = \frac{1}{\Omega_i^2 + \omega_{pi}^2} + \frac{1}{\Omega_i \Omega_e}. \tag{4.99}$$

The frequency $\omega_{UH}$ is called the upper hybrid frequency. It is given by

$$\omega_{UH}^2 = \Omega_e^2 + \omega_{pe}^2. \tag{4.100}$$

The upper and lower hybrid frequencies may be understood as follows:

### The lower hybrid resonance

To visualize the forces at work near the lower hybrid resonance, consider the short wavelength limit where the inductive part of the electric field is negligible. At low frequency, an ion that has received an impulse in the direction perpendicular to the magnetic field will be subjected to two forces: the magnetic force, that would cause oscillation at the frequency $\Omega_i$ if it were acting by itself, and the electrostatic force caused by the charge perturbation associated with the ion's displacement. The velocity of the electrons is primarily due to the $\mathbf{E} \times \mathbf{B}$ drift perpendicular to the electric field (Fig. 4.8). Due to electron inertia, however, there is a small polarization drift in the direction of $\mathbf{E}$,

$$\mathbf{V}_{Pe} = -\frac{1}{|\Omega_e|} \frac{d}{dt} \frac{\mathbf{E}}{B}$$

The ion velocity, by contrast, is given by

$$\mathbf{V}_i = \frac{\omega}{\omega^2 - \Omega_i^2} \frac{e\mathbf{E}}{m_i}$$

When the two velocities are equal there results a quasi-neutral oscillation. This occurs when $\omega = (\Omega_e \Omega_i)^{1/2}$, the hybrid frequency.

More generally, the electron and ion velocities differ and the conduction current must be balanced by a displacement current corresponding to a charge oscillation. Substituting the ion velocity and the electron polarization drift into Poisson's law then gives (4.99).

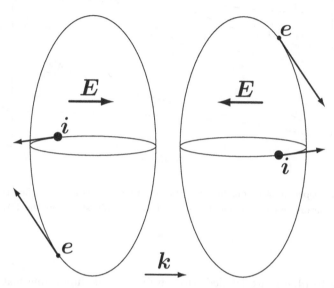

Figure 4.8: Lower hybrid resonance. The magnetic field is directed perpendicularly to the plane of the figure. The points represent the particles and the arrows their velocity vectors. The ellipses represent the particles' orbits. The ions are oscillating primarily along $\mathbf{E}$, and the electrons in the $\mathbf{E} \times \mathbf{B}$ direction.

## The upper hybrid resonance

At high frequency, the ions become essentially immobile but an electron resonance occurs: this is the upper hybrid resonance. This resonance results from the combination of the electrostatic restoring force with the magnetic restoring force. The mechanical analogue of this resonance is an oscillator with two springs in parallel: the square of the oscillation frequency is then proportional to the sum of the spring constants for each spring, or equivalently to the sum of the squares of the oscillation frequencies for each spring,

$$\omega_{UH}^2 = \Omega_e^2 + \omega_{pe}^2.$$

This is the upper hybrid frequency.

## Extraordinary and ordinary waves

The dispersion relations for waves propagating perpendicular to the magnetic field are shown in Fig. 4.9. The dispersion curves for the ordinary wave, cor-

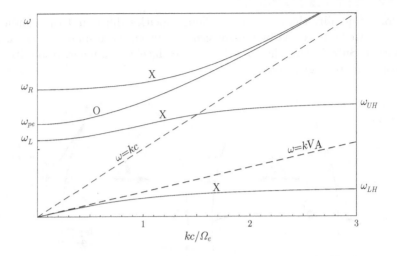

Figure 4.9: Dispersion relations for waves propagating perpendicular to the magnetic field. The plasma parameters correspond to the dashed parabola shown in the CMA diagram, Fig. 4.3

responding to (4.96), are labeled with an $O$. The distinguishing feature of the ordinary wave is that it is independent of the magnetic field. The electric field for the $O$ wave is directed along the equilibrium magnetic field.

The extraordinary wave, by contrast, does depend on the magnetic field. Its dispersion relation, corresponding to (4.97), are labeled with an $X$ in Fig. 4.9. The electric field of the X wave is directed perpendicular to $\mathbf{B}_0$. Note that it has two bands of propagation, separated by a "stop band" extending from $\omega_{UH}$ to $\omega_R$. Since the density of laboratory plasmas increases towards the center of the plasma, exciting the upper hybrid resonance with an external antenna requires ingenuity.

## 4.7   Propagation at arbitrary angle

The dispersion relations for propagation at arbitrary angle are deduced by making use of the property that the index of refraction is a monotone function of the propagation angle. The qualitative features of the dispersion curves can thus be sketched by interpolating between the dispersion curves for parallel and perpendicular propagation as shown in Fig. 4.10. Note that there are five branches for any given wavevector, corresponding to the $2 \times 5 = 10$ degrees of freedom of the cold plasma (the factor of two accounts for the two directions of propagation)

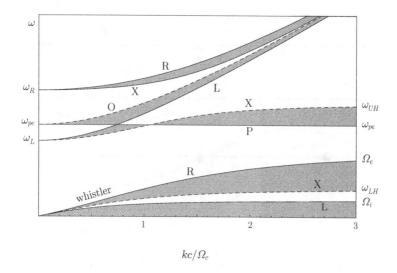

Figure 4.10: Combined sketch of dispersion relation for propagation parallel (solid lines) and perpendicular (dashed lines) to the magnetic field. The dispersion curves for intermediate angles lie in the shaded areas.

## 4.8 Alfvén waves

The low frequency case is of such importance as to deserve special consideration. We consider the range of frequencies such that $\omega \ll \Omega_i$ and $\omega \ll \omega_{pe}$. From the first inequality, we deduce that inertia is small compared to the magnetic force for both electrons and ions. The velocity across the magnetic field is thus given, to lowest order, by the electric drift. Inertia, however, gives rise to small polarization drifts. The corresponding current is

$$\mathbf{J}_\perp = (m_i n_i + m_e n_e)\frac{\mathbf{B}_0}{B_0^2} \times \frac{d\mathbf{V}_E}{dt}. \tag{4.101}$$

It follows that the dielectric permittivity tensor is diagonal. It is also isotropic in the perpendicular direction, where its diagonal component is given by

$$K_\perp = 1 + \frac{\mu_0 m_i n c^2}{B_0^2}. \tag{4.102}$$

From the second inequality, $\omega \ll \omega_{pe}$, we deduce that the parallel electric field is negligible. The two eigenmodes have the following characteristics.

The first eigenmode corresponds to an electric field lying along $\mathbf{e}_x$ (perpendicular to $\mathbf{B}_0$ in the plane formed by $\mathbf{B}_0$ and $\mathbf{k}$). The velocity is thus perpendicular to the wave-vector, $\mathbf{k} \cdot \mathbf{V} = 0$, so that no plasma compression takes place. The eigenfrequency for this mode is

$$\omega^2 = \frac{k_\parallel^2 V_A^2}{1 + \frac{V_A^2}{c^2}}, \tag{4.103}$$

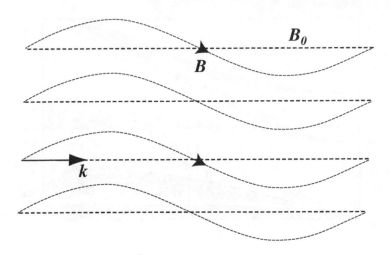

Figure 4.11: Sketch of the field distortion for a shear-Alfvén eigenmode propagating along the magnetic field.

where

$$V_A^2 = \frac{B^2}{\mu_0(m_i + m_e)n} \qquad (4.104)$$

is the squared *Alfvén speed*. Note that

$$\frac{V_A^2}{c^2} = \frac{\Omega_i \Omega_e}{\omega_{pe}^2} = \frac{\mu B^2}{\mathcal{N}},$$

so that the ratio of of the Alfvén speed to the speed of light parametrizes the parabolas corresponding to constant plasma parameters in the CMA diagram. In the most common case where $V_A \ll c$ and $\mu \ll 1$, we find

$$\omega = \pm k_\parallel V_A. \qquad (4.105)$$

This wave, called the shear-Alfvén wave, is sketched in Fig.4.11.

The second mode has an electric field lying along $\mathbf{e}_y$, perpendicular to both $\mathbf{k}$ and $\mathbf{B}_0$. This is the compressional Alfvén wave. Its eigenfrequency is given by

$$\omega^2 = \frac{k^2 V_A^2}{1 + \mu + \frac{V_A^2}{c^2}}, \qquad (4.106)$$

For $V_A \ll c$ and $\mu \ll 1$, we find

$$\omega = \pm k V_A. \qquad (4.107)$$

For $k_\perp = 0$, the compressional Alfvén wave becomes indistinguishable from the shear-Alfvén wave. In the opposite limit, $k_\parallel = 0$, the compressional Alfvén wave does not bend the field lines but merely compresses them, in a way similar to a sound wave in which the magnetic field provides the restoring force.

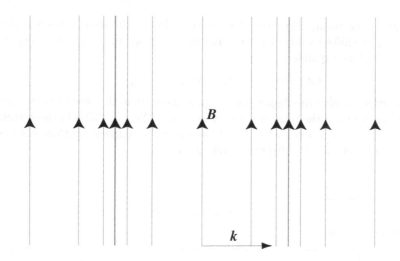

Figure 4.12: Sketch of the field distortion for a compressional Alfvén eigenmode propagating perpendicular to the magnetic field.

Note that the shear-Alfvén wave resonance occurs for $\theta \simeq 90^o$ at very low frequency. We have noted before how the cold plasma model fails in the vicinity of resonances. In the next chapter, we will show how the low-frequency assumption forms the basis for the MHD closure of the moment equations. The MHD closure improves upon the cold plasma model by describing the effects of finite pressure, such as sound waves.

## 4.9  Stability in the cold plasma model

In this section we consider two important instabilities described by the cold plasma model. Instabilities follow from either of two circumstances: the streaming of different components of the plasma with respect to each other, or spatial variation of the fields. We consider both cases in turn.

### The two-stream instability

The two-stream instability arises in a plasma consisting of two interpenetrating components in relative motion. It is the archetypal kinetic instability, driven by energy contained in the velocity distribution rather than in the spatial gradients. The dielectric for a two-stream distribution is obtained by summing the contributions of each beam to the plasma current (compare to (4.28):

$$K(k,\omega) = 1 - \frac{\omega_p^2}{\omega^2} - \frac{\omega_b^2}{(\omega - kv_b)^2},$$

where $\omega_p^2 = n_0 e^2 / m\epsilon_0$ is the plasma frequency for the component at rest in the lab frame, $\omega_b^2 = n_b e^2 / m\epsilon_0$ is the plasma frequency of the beam, and $v_b$ is the velocity of the beam. Note that the contribution of the second beam is simply the Doppler-shifted value of the coefficient $\mathcal{N}$ defined in (4.17). The dispersion relation is thus a quartic,

$$(\omega - kv_b)^2\omega^2 - (\omega - kv_b)^2\omega_p^2 - \omega_b^2\omega^2 = 0.$$

In order to solve the dispersion relation, we assume that the beam density is much smaller than the plasma density, $\omega_b \ll \omega_p$. We may think of the system as composed of two nearly independent plasma waves carried by their respective plasma component with dispersion relations

$$\omega = \omega_p$$

for the plasma and

$$\omega = kv_b \pm \omega_b \simeq kv_b$$

for the beam wave.

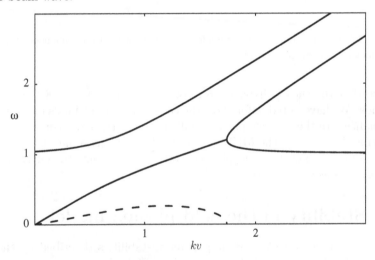

Figure 4.13: Wave frequency as a function of $kv$ for $\omega_b/\omega_p = 0.3$. Both axes are normalized to the plasma frequency of the stationary component, $\omega_p$. The solid line represent the real part of frequency for the three positive frequency roots, and the dashed line represents the growth rate for the two-stream instability.

The unstable domain can be obtained from the first order corrections to the dispersion relations. For the beam wave we find

$$\omega = kv_b \pm \frac{\omega_b}{(k^2v^2 - \omega_p^2)^{1/2}}kv_b.$$

The coupling between the beam and the rest plasma waves will thus cause instability in the wavenumber range

$$k < \omega_p/v_b.$$

The dispersion curves and growth rate for the two-stream instability are shown in Fig. 4.13.

## The sausage instability

In nature as well as in the laboratory, plasmas are generally confined by magnetic fields bent into bounded volumes. The magnetic tension associated with the bent magnetic fields constitutes a source of free energy for instabilities. We consider the two most important examples.

The simplest confinement geometry is the Z-pinch. In a Z-pinch, the current is made to run through a plasma between two electrodes. The magnetic field generated by the current provides an force that can accelerate the plasma inwards.

If a Z-pinch plasma is squeezed somewhere between the two electrodes, however, the current density and therefore the confining magnetic field will increase at that point. This has the effect of squeezing the plasma further. The resulting instability, called the sausage instability, can be eliminated by applying a longitudinal magnetic field $B_z$.

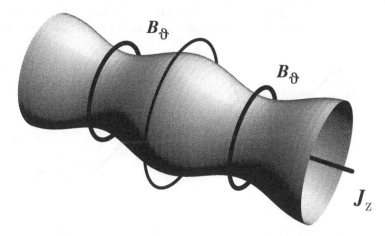

Figure 4.14: The sausage instability in a Z-pinch plasma.

Because of conservation of flux, the perturbation of the longitudinal field for a constriction $\delta a$ of the radius of the column, $a$, is

$$\delta B_z = 2B_z \delta a/a.$$

The azimuthal field perturbation is

$$\delta B_\theta = B_\theta \delta a/a.$$

The column is stable if the restoring pressure caused by the longitudinal field compression exceeds the increase in the azimuthal field pressure, or if

$$\delta(B_z^2) > \delta(B_\theta^2).$$

The stability criterion is thus

$$B_z^2 > \frac{B_\theta^2}{2}.$$

We see that a longitudinal field has a strong stabilizing influence. This stabilizing influence is ineffective, unfortunately, to prevent shear-Alfvén instabilities. This is a consequence of the fact that shear-Alfvén waves do not modify the magnetic field amplitude. They are thus much more difficult to stabilize, and much more common in nature. We next consider the most common shear-Alfvén instability, the kink mode.

## The kink instability

Consider the case of a long thin, current-carrying plasma wire immersed in a longitudinal magnetic field.[6] Twisting this plasma into a corkscrew introduces an angle between the external magnetic field and the plasma current. This results in a $\mathbf{J} \times \mathbf{B}$ force (Fig. 4.15). If the pitch of the twisted plasma matches that of the unperturbed magnetic field, the perturbation will amplify as the plasma tries to align the magnetic field produced by the current it is carrying with the externally applied magnetic field. The thin current carrying plasma is thus unconditionally unstable. This is called the kink or "firehose" instability.

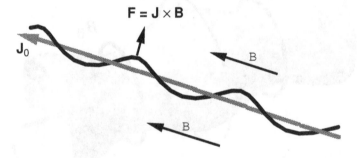

Figure 4.15:  Kink instability of a current-carrying plasma filament.

We next investigate the effect of the plasma cylinder's thickness.[6] The azimuthal field at a radius $r$ from the axis of the cylinder is given by

$$B_\theta = \frac{1}{2r} \int J_z \, dr.$$

This field reaches a maximum at the surface of the plasma. Consider two nearby cross-sections: As can be seen from Fig. 4.16, if the magnetic field twist is tighter than that of the perturbation, the perturbation will have the effect of increasing the twist of the field, or of increasing the strength of the azimuthal magnetic field outside the plasma column. The perturbation will then amplify. In finite-length cylinders, there is a minimum pitch allowed by the boundary conditions for a rigid kink, so that this mode will be stable provided that the azimuthal field is not too large. The corresponding stability condition,

$$q = \frac{rB_z}{RB_\theta} > 1$$

is called the Kruskal-Shafranov condition. A kinked cylindrical plasma is shown in Fig. 4.17

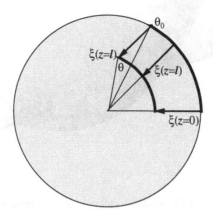

Figure 4.16: Kink instability of a cylindrical plasma. The displacement vector $\boldsymbol{\xi}$ for the instability is followed along the cylinder from $z = 0$ to $z = l$. The thick arcs are the projections of the initial and perturbed field lines. If the pitch of the field line is larger than that of the instability, so that the field line spans a larger angle than $\boldsymbol{\xi}$, the perturbation will increase the field-line pitch further. This increases the azimuthal field amplitude outside the column and thereby reinforces the perturbation

## Additional reading

The standard reference on wave propagation in the cold plasma model is the book by Stix.[68] Two less widely available but insightful references are the books by Allis, Buchsbaum and Bers[3], and Clemmow and Dougherty[18]. Whitham's book, "Linear and Nonlinear Waves," is an excellent general reference for the theory of wave propagation.[78] The WKB or eikonal theory was first extended to anisotropic media and applied to the cold plasma model by S. Weinberg.[75] A lively presentation of the theory of magnetohydrodynamic stability is given in a book by Bateman.[6]

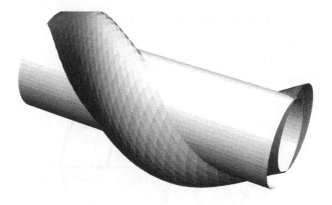

Figure 4.17: Kink instability of a cylindrical plasma.

# Exercises

1. **Ionospheric CMA diagram**: Write the dielectric permittivity tensor and sketch the CMA diagram for the ionosphere assuming that the plasma consists of a mixture of Hydrogen and Helium.

2. **Multi-component plasma**: How many waves (roots of the dispersion relation) will exist in general in an $N$-component plasma?

3. Working in the circular basis (where $\mathbf{K}$ is diagonal), solve for the components $E_R, E_L, E_\|$ in terms of $\mathbf{k} \cdot \mathbf{E}$. Substitute your result in $\mathbf{k} \cdot \mathbf{E}$ to show that the equation for the wave-normal surface may be written

$$\frac{\cos^2\theta}{v_{\text{ph}}^2 - v_0^2} + \frac{1}{2}\frac{\sin^2\theta}{v_{\text{ph}}^2 - v_R^2} + \frac{1}{2}\frac{\sin^2\theta}{v_{\text{ph}}^2 - v_L^2} = 0, \qquad (4.108)$$

   where

$$v_0^2 = \frac{c^2}{P}, \quad v_R^2 = \frac{c^2}{R}, \quad v_L^2 = \frac{c^2}{L}.$$

   are the phase velocities of the ordinary, right, and left waves. Note that the phase velocity of the fourth principal wave, the $X$ wave, is obtained by equating the sum of the coefficients of $\sin^2\theta$ to zero:

$$2v_X^2 = v_R^2 + v_L^2.$$

   The phase velocity of the $X$ wave is thus always between that of the $R$ and $L$ waves.[3]

4. Under what conditions is the phase velocity of the $X$ wave equal $c$, the speed of light in vacuum? Can any other principal wave have $v_{\text{ph}} = c$?

5. Derive the Åström and Allis form of the dispersion relation,

$$\tan^2 \theta = \frac{-P(n^2 - R)(n^2 - L)}{(Sn^2 - RL)(n^2 - P)}.$$   (4.109)

and describe the nature of the poles and zeroes.

6. Solve the ray equations for an ordinary wave in an unmagnetized plasma with a linearly varying density. Do the same for an electron cyclotron wave propagating parallel to the magnetic field in a plasma with a linearly varying magnetic field. What is the behavior of the rays near cutoff?

7. Describe the index, wave-normal and ray surfaces for a shear-Alfvén wave.

# Chapter 5

# MHD and the Drift Model

## 5.1 Magnetized fluid dynamics

Imagine a sequence of plasmas, all having the same density and temperature, but each subject to a stronger external magnetic field than its predecessor. At the beginning of the sequence is an unmagnetized, presumably isotropic, plasma; at the other extreme is a plasma in which a typical particle gyroradius is extremely small compared to the system size: $\delta \equiv \rho/L \ll 1$. Of course the strongly magnetized plasma is anisotropic: the helical orbits of its constituent particles are tightly wound around field lines. Note in particular that as the gyroradius decreases the perpendicular drift motions associated with both fluid flow (such as the diamagnetic drift) and with particle orbits (such as the curvature drift) also tend toward zero.

The present chapter attempts to describe the strongly magnetized plasma by fluid equations. Two questions underlie the discussion:

1. Is perpendicular motion forbidden, making plasma dynamics essentially one-dimensional, in the zero-gyroradius limit?

2. Does the strong magnetic field allow a general and rigorous fluid description—that is, does it provide an asymptotic closure of fluid equations in the sense of Chapter 3?

As is already clear from Chapter 3, the answer to both questions is no. With regard to the first question, we recall that perpendicular motion at vanishing $\delta$ is permitted because the $E \times B$ drift (unlike other particle or fluid drifts) need not vanish in this limit. Since this drift is measured by $E_\perp/B$, and since the relevant velocity scale is the thermal speed, keeping the electric drift finite in the $\delta \to 0$ limit is equivalent to assuming

$$\frac{E_\perp}{B} \sim v_t.$$ 

(5.1)

The plasma description that begins with this so-called *MHD ordering*, and then neglects all other finite-gyroradius effects, is called magnetohydrodynamics, or

MHD. We remark that the name "MHD" is also sometimes used in reference to general fluid descriptions; our usage includes the theories, based on (5.1), that use kinetic as well as fluid equations.

When the electric drift is assumed to vanish like other particle and fluid drifts as $\delta \to 0$—the so-called *drift ordering*—plasma motion indeed approaches a one-dimensional limit. In this case, however, one usually retains finite-$\delta$ effects to describe the dynamics across field lines. The resulting *drift model* is more complicated than MHD, including in particular terms of various orders in $\delta$. It is also more realistic in many cases of interest.

The answer to the second question is negative—small gyroradius does not in itself provide a simple fluid closure—because parallel motion is unaffected by the magnetic field. Even in the extreme case of a one-dimensional gas one needs some additional small parameter to close the chain of fluid equations. We have noted that such closure has been accomplished only for a collision-dominated plasma, for which the short mean-free path provides a small parameter.

The result is a Hobson's choice. If one finds (as we do in this book) the restriction to short mean-free path too confining, one can either abandon rigor, using some simplifying *ansatz* for the parallel dynamics, or reach beyond the fluid domain, supplementing moment equations with a kinetic description of the parallel plasma response. We adopt the first of these approaches for simplicity. However we outline the rigorous kinetic treatment in Section 5.6.

## 5.2   MHD

### Physical basis

The basic physical idea of MHD is that a magnetized plasma can move quickly across the magnetic field, with flow velocities near the thermal speed. Indeed, the characteristic MHD velocity is the Alfvén speed [recall (4.104)], which typically exceeds the ion thermal speed. The form of the magnetized, perpendicular flow, (3.70), shows that such rapid flow across $\mathbf{B}$ requires strong electric fields, as in (5.1). Other contributions to $\mathbf{V}_\perp$, such as the pressure gradient, could not yield motion this fast without contradicting the assumption of a magnetized plasma: $\delta \equiv \rho/L \ll 1$.

An electric field described by (5.1) is large in the following sense. If $\Phi$ is the electrostatic potential, we expect a plasma near equilibrium to have $e\Phi/T \sim 1$, implying

$$E/B \sim \Phi/(BL) \sim T/(eBL) \sim \delta v_t \qquad (5.2)$$

Equation (5.2) characterizes the electric field in the drift ordering. Comparing it to (5.1), we see why the MHD electric field is sometimes considered to be *minus-first* order in the gyroradius.

Note that (5.2) also describes the largest *parallel* electric field permitted in MHD: we recall from Chapter 2 that MHD requires

$$E_\parallel \sim \delta E_\perp. \qquad (5.3)$$

In other words MHD allows only the perpendicular field to be large.

MHD is designed to study rapid, even violent, motions of a magnetized plasma. Solar and magnetospherical phenomena, especially as they involve shocks and related disturbances, are well suited to the MHD description. Certain laboratory experiments, such as the pinch discharges explored in early plasma confinement experiments, are also realistically described by MHD. On the other hand MHD violence rarely occurs on present magnetic confinement devices, which are more accurately described by the drift ordering. (For example, even the so-called disruptions seen in some confinement experiments involve speeds too slow for the MHD ordering.) Indeed, the ability of confinement researchers to avoid rapid perpendicular flow is a scientific victory that owes much to MHD analysis.

Of course in using the MHD description one does not always deal with rapid flows or accelerations. Studies of MHD stability, for example, treat perturbations with weak flow, as well as MHD equilibria, where $\mathbf{E}$ may vanish. Investigating stability on this basis is a rewarding pursuit, allowing prediction and control of the worst plasma violence. But it is intrinsically limited, since (5.1) ignores certain avenues for slower evolution. Thus, if stability is predicted in MHD, one must next relax the MHD assumptions, seeking the possibility of instability on a longer time scale.

## MHD fluid

The simplest formal characterization of MHD is that, beginning with the exact fluid description (3.50) – (3.52), one allows the gyroradius to vanish, $\delta \to 0$, while using the MHD ordering (5.1) to preserve the electric drift. This procedure yields the MHD velocity of (3.81):

$$\mathbf{V} = \mathbf{V}_\| + \mathbf{V}_E, \tag{5.4}$$

thus capturing a cornerstone of the theory. But the statements of vanishing gyroradius and large flow are unfortunately insufficient to provide a system of equations that is both closed and physically interesting.

The first problem with (5.4) is that its naive interpretation would rule out any perpendicular plasma current: since $\mathbf{V}_E$ is common to all plasma species,

$$\sum_s e_s n_s \mathbf{V}_\perp = 0,$$

for a quasineutral plasma. In fact perpendicular current enters the equation of motion, (3.62), in a crucial way and must be included in any physical account. MHD addresses this weakness straightforwardly. Noting that the definition of current density contains a factor of electric charge $e$, we interpret that factor as an inverse power of $\delta = mv_t/eB$, and conclude that the *zeroth* order $\mathbf{J}$ is determined by the *first* order flow:

$$\lim_{\delta \to 0} \mathbf{J} = \sum_s e_s (n_s \mathbf{V}_s)_1, \tag{5.5}$$

where the subscript on the right indicates first order in $\delta$. In other words, it is fortunate that the electric drift does not contribute current: if it did, under the MHD ordering the plasma current would be unrealistically huge. On the other hand $(n_s \mathbf{V}_s)_1$ brings essential physical processes, such as the diamagnetic drift, into the perpendicular current.

Notice that a similar argument requires the (zeroth-order) parallel flow in (5.4) to be common to all species,

$$\mathbf{V}_{\|s} = \mathbf{V}_\|. \tag{5.6}$$

Although (5.5) is a key element of the theory, MHD uses it only implicitly; the first order velocity is not explicitly calculated. Instead a suitably approximate version of the equation of motion, (3.62) is incorporated into the MHD closed set, and used to compute $\mathbf{J}$.

In order to follow pressure evolution, the energy conservation law must enter the closed system. This observation leads to the second, rather more serious, issue in MHD closure: if the pressure tensor is a simple scalar, then the energy conservation law (with or without heat conduction) is sufficient to track its evolution; but any more complicated form for the pressure tensor entails too many unknowns for such a simple fluid closure.

It is at this point that MHD reaches a theoretical schism. Among the various proposed paths to closure, four are most representative and important:

1. *Kinetic MHD* is a fluid-kinetic hybrid that allows the pressure tensor to be gyrotropic. As noted in the Section 3.5, this path is rigorously consistent with the $\delta \to 0$ limit. The pressure components $p_\|$ and $p_\perp$ are computed from the lowest order kinetic equation, allowing the system to be arbitrarily far from thermal equilibrium. The consistency and lack of *ad hoc* assumptions in kinetic MHD is appealing: kinetic MHD is the only rigorous description of a collisionless plasma that is significantly simpler than straight-ahead kinetic theory. Furthermore, when the distribution is strongly non-Maxwellian—for example, due to a population of supra-thermal particles—kinetic MHD provides the only believable description. For this reason we outline its derivation in Section 5.6. On the other hand kinetic MHD is not a fluid closure. Its rather complicated structure is awkward to implement nonlinearly, while its linear predictions are often similar to those of simpler theories.

2. *Two-fluid theory* describes a magnetized plasma in which the collision frequency exceeds the parallel transit frequency. It uses the short mean-free-path ordering of conventional Chapman-Enskog theory to effect closure. (Strictly speaking, it is not a branch of MHD because it does not require the $\mathbf{E} \times \mathbf{B}$ drift to dominate the perpendicular motion. Thus its most common versions are closely related to the drift model considered in Section 5.4 below.) The two-fluid point of view is thoroughly explored in a famous article by Braginskii; [13] its name reflects the fact that it is usually applied to a plasma with only one ion species. Two-fluid theory has

the advantages of rigor, within its collisional domain, and of its a purely fluid methodology. Its disadvantage is that interesting plasma physics often occurs in situations where the collisional mean-free-path is long and Chapman-Enskog theory not pertinent.

3. *Double–adiabatic theory* allows for anisotropic stress, positing relatively simple fluid equations for $p_{\parallel}$ and $p_{\perp}$. It was proposed in the pioneering work of Chew, Goldberger and Low,[17] who introduced the gyrotropic tensor and proposed its evolution on the basis of effectively vanishing heat flow. This version of MHD closure is now rarely used, but it has had fruitful application.

4. *Fluid MHD* is the version of MHD emphasized in the remainder of this book. It is conceptually simple, combining the rapid MHD flow with an assumption of local thermal equilibrium (LTE). Thus the lowest-order distribution is taken to be a Maxwellian, moving at the MHD velocity:

$$f_{\text{MHD}} = f_M(\mathbf{v} - \mathbf{V}_{\text{MHD}}) \qquad (5.7)$$

The most important implication of this LTE *ansatz* is that the plasma pressure is scalar, described by a single parameter; all $O(\delta)$ terms in the fluid moments, such as the heat flow, are neglected. The enormous literature associated with fluid MHD includes numerous comparisons, both linear and nonlinear, to experimental observations.[26]

In the remainder of this chapter we refer to the fluid version simply as "MHD," or, when its dissipationless character is emphasized, "ideal MHD." This usage is common. (A modified version which allows dissipation is considered briefly at the end of the Section.) Ideal MHD is evidently cruder than double-adiabatic theory; it is in fact the simplest plausible context in which to explore the consequences of rapid perpendicular flow.

MHD is sometimes said to describe a magnetized, collision-dominated plasma, and indeed (5.7) is most plausible when $\nu$ is large. However, the neglect of dissipation becomes less credible as $\nu$ increases, so that a parameter regime in which (ideal) MHD is asymptotically valid is hard to identify. Furthermore, as we have noted, many plasmas of interest are magnetized but nearly collisionless; when MHD is used to study such plasmas, as it very often is, the justification appeals to simplicity rather than to rigor.

Note that MHD describes one class of phenomena with justifiable accuracy: the rapid evolution of a *low beta* plasma. (Recall that beta measures the ratio of plasma thermal energy or pressure to the magnetic field energy.) However complicated its form, the plasma stress tensor must vanish at vanishing pressure. Hence, if the latter is sufficiently small, the details of the stress tensor can affect fluid evolution only as minor corrections, and the simplifications of fluid MHD become harmless. Since low-beta evolution is an important topic, it is comforting to know that MHD describes it well. On the other hand MHD has also become a major tool for understanding high-beta regimes, where its application is less clearly reliable.

The derivation of MHD fluid equations involves only minor extension of the Chapter 3 analysis. One combines the center-of-mass particle conservation and momentum conservation laws, (3.60) and (3.62), with the LTE energy conservation law, (3.118). Equation (5.7) is used to omit heat flow,

$$\mathbf{q} = 0$$

as well as anisotropic stress,

$$\boldsymbol{\pi} = 0.$$

It is customary to omit the source terms $i_n$, and to use a simplified notation, omitting various ($cm$- and $m$-) subscripts. The results follow quickly from Chapter 3 and are summarized in the following section.

## MHD closure

The MHD fluid is described by the equations

$$\frac{d\rho}{dt} + \rho\nabla\cdot\mathbf{V} = 0, \tag{5.8}$$

$$\rho\frac{d\mathbf{V}}{dt} + \nabla p - \mathbf{J}\times\mathbf{B} = 0, \tag{5.9}$$

$$\frac{dp}{dt} + \frac{5}{3}p\nabla\cdot\mathbf{V} = 0. \tag{5.10}$$

where the time derivatives abbreviate

$$\frac{d}{dt} = \frac{\partial}{\partial t} + \mathbf{V}\cdot\nabla. \tag{5.11}$$

The closed system of fluid MHD combines (5.8) – (5.10) with

$$\mathbf{E} + \mathbf{V}\times\mathbf{B} = 0, \tag{5.12}$$

and the Maxwell relations

$$\nabla\times\mathbf{B} = \mu_0\mathbf{J}, \tag{5.13}$$

$$\nabla\times\mathbf{E} + \frac{\partial\mathbf{B}}{\partial t} = 0. \tag{5.14}$$

Equation (5.12), sometimes called the "MHD Ohm's law," neatly summarizes the flow specified in (5.4). We omit the Maxwell term in Ampere's law because MHD processes are slow on the scale of the light speed; in that case $\nabla\cdot\mathbf{B} = 0$ is guaranteed by (5.14) if it is presumed to hold initially. Similarly the assumption of quasineutrality makes Coulomb's law for $\nabla\cdot\mathbf{E}$ irrelevant.

Equations (5.8) – (5.14) evidently provide 14 (scalar) relations for the 14 unknowns $\rho$, $\mathbf{V}$, $p$, $\mathbf{J}$, $\mathbf{E}$ and $\mathbf{B}$. A more compact version uses (5.12) and (5.13) to eliminate, respectively, $\mathbf{E}$ and $\mathbf{J}$. In this regard, note that

$$\nabla\times\mathbf{E} = -\nabla\times(\mathbf{V}\times\mathbf{B}) = -(\mathbf{B}\cdot\nabla\mathbf{V} - \mathbf{V}\cdot\nabla\mathbf{B} - \mathbf{B}\nabla\cdot\mathbf{V})$$

whence

$$\frac{\partial \mathbf{B}}{\partial t} = \mathbf{B} \cdot \nabla \mathbf{V} - \mathbf{V} \cdot \nabla \mathbf{B} - \mathbf{B} \nabla \cdot \mathbf{V}.$$

Similarly,

$$\mu_0 \mathbf{J} \times \mathbf{B} = \mathbf{B} \cdot \nabla \mathbf{B} - \frac{1}{2} \nabla B^2.$$

Hence MHD is expressed by

$$\frac{d\rho}{dt} + \rho \nabla \cdot \mathbf{V} = 0, \tag{5.15}$$

$$\frac{dp}{dt} + \frac{5}{3} p \nabla \cdot \mathbf{V} = 0, \tag{5.16}$$

$$\rho \frac{d\mathbf{V}}{dt} + \nabla \left( p + \frac{B^2}{2\mu_0} \right) - \frac{1}{\mu_0} \mathbf{B} \cdot \nabla \mathbf{B} = 0, \tag{5.17}$$

$$\frac{d\mathbf{B}}{dt} - \mathbf{B} \cdot \nabla \mathbf{V} + \mathbf{B} \nabla \cdot \mathbf{V} = 0. \tag{5.18}$$

## MHD physics

MHD fluid evolution is adiabatic in the thermodynamic sense: there is no heat flow. This fact is clear from the derivation, and manifest in the two even moments, (5.15) and (5.16). After eliminating $\nabla \cdot \mathbf{V}$ between these relations we obtain

$$\frac{d \log p}{dt} - \frac{5}{3} \frac{d \log \rho}{dt} = 0,$$

or

$$\frac{d}{dt} \left( p \rho^{-5/3} \right) = 0, \tag{5.19}$$

the familiar adiabatic law for an ideal gas. Indeed the three MHD *fluid* equations can be said to describe the adiabatic evolution of an ideal gas that conducts electricity.

Some versions of MHD replace the adiabatic law,

$$p \rho^{-5/3} = \text{constant}$$

by an alternative "equation of state" $p = p(\rho)$. We do not pursue such alternatives, in part because their consequences often resemble those of (5.19), and in part because the justification for an arbitrary equation of state is problematic in typical MHD contexts—plasmas far from thermodynamic equilibrium. The adiabatic version has the advantage of a clearly identified origin in the LTE *ansatz* (5.7).

The most important feature of MHD is the relation (5.12) between $\mathbf{V}$ and $\mathbf{E}$. Recall from the Lorentz transformation law that the field $\mathbf{E}'$ measured in a frame moving with velocity $\mathbf{V}$ is related to the rest frame field $\mathbf{E}$ approximately by

$$\mathbf{E}' \cong \mathbf{E} + \mathbf{V} \times \mathbf{B}.$$

Hence the MHD Ohm's law states that $\mathbf{E}' = 0$, consistent with the picture of a moving, infinitely conducting fluid. [This interpretation explains the term "Ohm's law." Note however that (5.12), which simply reflects the MHD ordering in a magnetized plasma, is not based on any calculation of conductivity.] It turns out that requiring the fluid-frame electric field to vanish effectively binds the plasma fluid to the magnetic field lines.

To verify this remarkable fact, we consider the magnetic flux, $\Psi$, through a surface $S$ that moves with the local flow velocity of the plasma:

$$\Psi = \int_{S(t)} d\mathbf{S} \cdot \mathbf{B}(t)$$

The change in flux after some time $\Delta t$ is

$$\Delta\Psi = \int_{S(t+\Delta t)} d\mathbf{S} \cdot \mathbf{B}(t+\Delta t) - \int_{S(t)} d\mathbf{S} \cdot \mathbf{B}(t)$$

$$= \left( \int_{S(t+\Delta t)} d\mathbf{S} \cdot \mathbf{B}(t) - \int_{S(t)} d\mathbf{S} \cdot \mathbf{B}(t) \right) + \Delta t \int_{S(t)} d\mathbf{S} \cdot \frac{\partial \mathbf{B}}{\partial t}$$

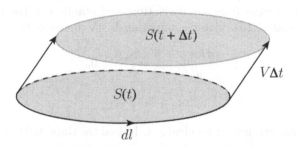

Figure 5.1: The disk-like surface $S(t)$ and its image under MHD advection, $S(t+\Delta t)$. The vector displacement between the two disks is $\mathbf{E}\Delta t$.

Here the two terms in parentheses combine to give an integral that can be evaluated by Gauss' law. We join the surfaces at $S(t)$ and $S(t+\Delta t)$ by a narrow ribbon forming a short hat-box shaped volume, as shown in Figure 5.1. Since the normal distance between the two surfaces is $\mathbf{V}\Delta t$, the outward-pointing surface element of the ribbon is

$$d\mathbf{S}_{\text{ribbon}} = d\mathbf{l} \times \mathbf{V}\Delta t$$

where $d\mathbf{l}$ denotes a small step around the border of the surface $S(t)$, in the right-hand-rule direction with respect to $d\mathbf{S}$. Note that $d\mathbf{l} \times \mathbf{V}$ points outward from the ribbon. Since the total magnetic flux out of the hat-box volume must vanish, we have

$$\int_{S(t+\Delta t)} d\mathbf{S} \cdot \mathbf{B}(t) - \int_{S(t)} d\mathbf{S} \cdot \mathbf{B}(t) = -\Delta t \oint d\mathbf{l} \times \mathbf{V} \cdot \mathbf{B}$$
$$= -\Delta t \oint d\mathbf{l} \cdot \mathbf{V} \times \mathbf{B}$$

and the flux change has become

$$\Delta \Psi = \Delta t \left( -\oint d\mathbf{l} \cdot \mathbf{V} \times \mathbf{B} + \int_{S(t)} d\mathbf{S} \cdot \frac{\partial \mathbf{B}}{\partial t} \right).$$

The two terms on the right-hand side of this equation precisely cancel, for (5.12) implies

$$\int_{S(t)} d\mathbf{S} \cdot \frac{\partial \mathbf{B}}{\partial t} = \int_{S(t)} d\mathbf{S} \cdot \nabla \times (\mathbf{E} \times \mathbf{B})$$

and Stokes' theorem gives

$$\int_{S(t)} d\mathbf{S} \cdot \nabla \times (\mathbf{E} \times \mathbf{B}) = \oint d\mathbf{l} \cdot \mathbf{E} \times \mathbf{B}.$$

We conclude that

$$\frac{d\Psi}{dt} = 0. \tag{5.20}$$

The magnetic flux through any surface that moves with the fluid cannot change.

Part of the importance of flux conservation, (5.20), is seen in the example of a *flux tube*—a cylindrical volume whose sides are defined by field lines. Suppose that, at some initial time, a flux tube is frozen to the plasma, each piece of it subsequently moving with the local fluid velocity. Thus the tube will become distorted by the flow; in general it will soon cease to be a flux tube, in that magnetic field lines will cross the tube's reformed sides. In an MHD fluid, however, (5.20) forbids such crossing: the MHD Ohm's law enforces precisely that magnetic field evolution which preserves the integrity of each flux tube.

Attaching the plasma to flux tubes has the effect of giving mass to field lines, which are forced to participate in the $\mathbf{E} \times \mathbf{B}$ motion of the plasma. This circumstance is seen clearly in shear-Alfvén waves, the propagating disturbances studied in Chapter 4. Recalling the discussion of Section 3.5, we treat the field line as a transversely vibrating string. If the string tension is $T$, then elementary argument shows its characteristic frequency and propagation velocity to be

$$\omega_{string} = \sqrt{\frac{2T}{L^2 \rho}}$$

and

$$V_{string} = L\omega_{string} = \sqrt{\frac{2T}{\rho}}. \tag{5.21}$$

where $L$ is the length of the string and $\rho$ its density per unit length. We showed in Chapter 3 that the field line tension is given by $T = B^2/2\mu_0$. Here we observe that locking the field to the fluid gives field lines an effective mass density $\rho_m$. Thus the string formula (5.21) implies that shear-Alfvén waves propagate at the Alfvén speed introduced in (4.104):

$$V_A = \sqrt{\frac{B^2}{\mu_0\rho_m}}. \tag{5.22}$$

The conventional inference from (5.20) is that individual field lines are frozen into the moving plasma; after all, field lines can be viewed as very thin flux tubes. It must be noted, however, that unique velocities cannot be ascribed to field lines. Thus, for example, an irrotational field can always be added to $\mathbf{V} \times \mathbf{B}$ without affecting (5.20). Also, of course, the parallel plasma motion does not enter (5.20).

One phenomenon ruled out by flux conservation is flux annihilation, as might occur if two flux tubes with oppositely directed fields were driven towards each other. Evidently (5.20) will keep such tubes apart. As a result, flux conservation limits magnetic-field topology change.

We next consider a fluid model in which the constraint of flux conservation is broken.

## 5.3   Resistive MHD and magnetic reconnection

Resistive MHD differs from the ideal version of (5.8) – (5.14) in only one respect: the MHD Ohm's law, (5.12), is replaced by

$$\mathbf{E} + \mathbf{V} \times \mathbf{B} = \eta\mathbf{J} \tag{5.23}$$

where $\eta$ is the plasma resistivity. Systematic analysis of short mean-free-path kinetic theory indeed produces a resistive term like the right hand side of (5.23); specifically one finds an anisotropic resistivity, with perpendicular and parallel resistivities differing by roughly a factor of 2. (The anisotropy is derived in Chapter 8.) Unfortunately the rigorous theory also produces a number of additional terms in Ohm's law and other MHD relations.

Thus resistive MHD should be viewed as a model which, without detailed mathematical justification, sheds light on real phenomena. Most importantly, resistive MHD includes a dissipative term with pronounced qualitative significance: even when $\eta$ is small, resistivity breaks the frozen flux constraint of ideal MHD, allowing the plasma fluid to slip through magnetic field lines. Thus, in particular, field annihilation and topology change become possible. The dissipative terms that resistive MHD leaves out usually have less dramatic effects.

Here we examine two elementary prediction of resistive MHD: magnetic field diffusion, and magnetic reconnection. After substituting the resistive Ohm's law into Faraday's law, we find that (5.18) is replaced by

$$\frac{d\mathbf{B}}{dt} - \mathbf{B} \cdot \nabla\mathbf{V} + \mathbf{B}\nabla \cdot \mathbf{V} = \frac{\eta}{\mu_0}\nabla^2\mathbf{B} \tag{5.24}$$

Here the last term evidently describes diffusion. A magnetic field with $\nabla \times \mathbf{J} \neq 0$ and scale length $L$ will diffuse on the time scale

$$\tau_R = \frac{\mu_0 L^2}{\eta}, \qquad (5.25)$$

called the resistive skin time. Of course the same diffusive process applies in any conducting material; (5.25) is used, for example, to estimate the penetration of an oscillating magnetic field into metals.

Since $\eta$ becomes very small in a hot plasma, resistive diffusion over macroscopic scales is a slow process. However the plasma can evolve into states with locally sharp variation of the field, corresponding to much smaller $L$ and more rapid diffusion. The resulting rapid rearrangement of the topology of the magnetic field is known as *magnetic reconnection*.

Magnetic reconnection is thought to be the cause of several observed phenomena such as coronal mass ejections, solar flares, magnetic storms, and disruptions in laboratory plasma. The theory is divided into two parts. The first part is concerned with the mechanism responsible for the formation of localized, sharply peaked current structures. The second part describes the dynamics within these current structures and the rate at which magnetic field lines "tear" and reconnect.

## Current ribbons

The narrow current structures that cause magnetic reconnection result from the resonance in the shear Alfvén dispersion relation (see Sections 4.4 and 4.8),

$$\omega = \mathbf{k} \cdot \mathbf{V}_A :$$

for $\mathbf{k} \cdot \mathbf{V}_A \to 0$, the index of refraction $n = kc/\omega \to \infty$. an estimate of the width $w$ of the resonance is

$$w = \frac{\omega}{|\nabla (\mathbf{k} \cdot \mathbf{V}_A)|}.$$

That is, the resonance width vanishes with the wave frequency. In order to demonstrate the formation of localized current structures, we are thus led to consider the response of a plasma to slow perturbations with $\mathbf{k} \cdot \mathbf{V}_A = 0$.

The geometry of the localized current structures that result from such perturbations was first demonstrated by Syrovatskii[71]. Syrovatskii considered the evolution of a planar magnetic field embedded in a pressureless plasma. Such a magnetic field can be described in terms of the flux function $\psi(x,y) = -\hat{\mathbf{z}} \cdot \mathbf{A}$ according to

$$\mathbf{B} = \hat{\mathbf{z}} \times \nabla \psi$$

where $\hat{\mathbf{z}}$ is the unit vector perpendicular to the plane containing $\mathbf{B}$. All quantities are assumed to be independent of $z$. Note that the surfaces of constant $\psi$ contain the lines of force of the magnetic field. The equilibrium condition $\mathbf{J} \times \mathbf{B} = 0$ yields $\mathbf{J} = 0$, or

$$\nabla^2 \psi = 0. \qquad (5.26)$$

The flux is thus a harmonic function, and the results of complex analysis can be applied. In particular, Cauchy's theorem implies that $\psi$ is uniquely determined throughout the plasma by its values along the boundary. This is precisely the situation that exists in the solar corona, where the "footpoints" of the magnetic field lines are anchored in the much denser photosphere. The magnetic flux through the bounding surface is thus determined at all time by the slow thermal convection of the photospheric fluid.

When the plasma contains a *neutral line* where $\mathbf{B} = 0$, however, the solutions of the Cauchy problem generally violate the frozen-flux constraint

$$\frac{d\psi}{dt} = 0.$$

This violation of the frozen-flux constraint is tied to the presence of an inductive electric field $\mathbf{E} = -\partial\psi/\partial t\,\hat{\mathbf{z}}$. Such an electric field will clearly drive a current along the neutral line. It is thus necessary to abandon the assumption $\mathbf{J} = 0$. Allowing for a line current at the magnetic null, however, again violates the frozen-in property, as seen in Fig. 5.2(a).

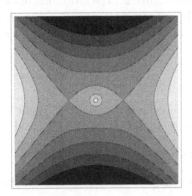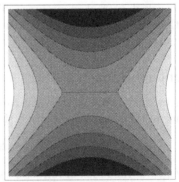

Figure 5.2: Magnetic flux close to a null line when (a) a line current is placed on the null line to satisfy the boundary conditions and (b) a branch cut is introduced so as to distribute the required current along a ribbon. The parameters chosen here are $a = I/\pi\mu_0$.

The solution proposed by Syrovatskii is to allow $\psi$ to have a branch cut along the line joining the two undesirable neutral points in Fig. 5.2. This leads to a family of solutions

$$\psi = \mathrm{Re}\left[\frac{a}{2}z\sqrt{z^2+1} + \frac{I}{2\pi\mu_0}\log(z + \sqrt{z^2+1})\right] \tag{5.27}$$

such that the magnetic field is parallel to the branch cut and reverses direction on either side of it. Here $I$ is the total current induced in the ribbon and $a$ is a parameter that must satisfy $I \leq \pi\mu_0 a$. The jump in the magnetic field defines the amplitude of the current density,

$$\int_{-w}^{w} dy\, J_z = B_x(x,w) - B_x(x,-w).$$

These solutions, shown in Fig. (5.2) are in good agreement with the results of numerical simulations, as shown in Fig. (5.3). They have also been shown to apply to more complicated geometries.

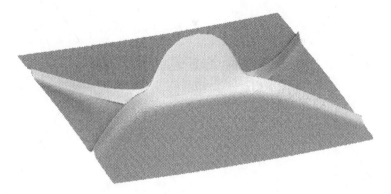

Figure 5.3: Current ribbon associated with magnetic reconnection in the Sweet-Parker regime. The current ribbon shown here developed during the coalescence of an array of flux tubes. The simulation uses an adaptive mesh to maintain appropriate resolution. Courtesy of C. Marliani.

## Sweet-Parker reconnection

The rate of magnetic reconnection in the presence of current ribbons was first described by Sweet and Parker[56]. The principal consideration in the Sweet-Parker model is momentum balance along the current ribbon (Fig. 5.4). Since flux tubes are subjected to a magnetic pressure $B_x^2(0)/2\mu_0$ in the center and zero at the end of the ribbon, the flux tubes are accelerated along the ribbon and exit with an outflow velocity equal to the Alfvén speed in the center of the ribbon,

$$v_{\text{out}}^2 = B_x^2(0)/\mu_0 n m_i.$$

The reconnection rate is computed by combining the above result with an equation expressing conservation of mass,

$$v_{\text{out}} w = v_{\text{in}} L,$$

where $w$ and $L$ are the width and length of the current ribbon, and with Ohm's law

$$v_{\text{in}} B_x(0) = \eta J_z \simeq \frac{\eta B_x(0)}{\mu_0 w}.$$

There follows

$$\frac{v_{\text{in}}}{V_A(0)} = \left(\frac{L}{V_A(0)\tau_R}\right)^{1/2}.$$

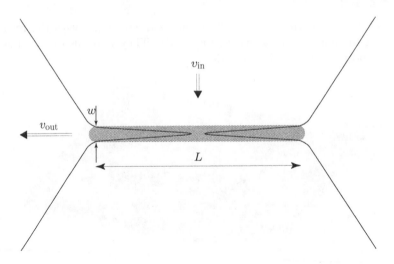

Figure 5.4: Geometry of the current ribbon during Sweet-Parker reconnection.

This is known as the Sweet-Parker rate of reconnection. It is in good agreement with numerical simulations.

The Sweet-Parker rate of reconnection, although fast by diffusive standards, is orders of magnitude too slow to account for most observations of reconnection. The two most promising explanations for the discrepancy are that the rate of reconnection is enhanced by the effects of drifts (discussed in the next section) or by the effects of turbulence.

## 5.4   The drift model

### Physical basis

In this Section we study magnetized plasma evolution in the case of moderate electric fields: when electric forces are comparable to the pressure gradient. This so-called drift ordering, discussed in Chapter 2 and described by (5.2),

$$E/B \sim \delta v_t \tag{5.28}$$

describes plasma behavior in modern magnetic confinement experiments, as well as a host of astrophysical and space-physical phenomena. Its most obvious implication is that diamagnetic drifts become comparable to the electric drift.

It is consistent to further assume that time evolution is relatively slow; otherwise the induced electric field from Faraday's law could contradict (5.28). Thus, while the time scale of MHD is characterized by the transit frequency $\omega_t$ or Alfvén frequency, here the characteristic frequency is

$$\omega \sim \delta \omega_t \sim \delta^2 \Omega. \tag{5.29}$$

The *drift model* considered in this section uses both (5.28) and (5.29).

We stress that the drift model applies only to a magnetized plasma: $\delta$ remains a small parameter. What distinguishes the drift model from MHD is that, without the dominating $O(1)$ electric drift of the former, $O(\delta)$ terms must now be taken into account. Thus the new processes that enter in the drift model are often called "finite Larmor-radius" or *FLR* effects: they are simply the terms—in the plasma velocity, pressure tensor, heat flow and elsewhere—that MHD labels as $O(\delta)$ and ignores.

The physics underlying the drift model, as shown in Section 3.5, is simply gyration. Each guiding center in a magnetized plasma represents a gyrating particle: a tiny spinning hoop or magnetic dipole whose rotation axis coincides with the local **B**. Spatial variation in the distribution or energy of such hoops induces electric current, heat flow and momentum flow.

There are two broad categories of physical processes that the drift model takes into account. The first category is associated with the so-called *drift frequency,*

$$\omega_{*s} = \frac{1}{e_s nB}\mathbf{k} \cdot \mathbf{b} \times \nabla p_s \tag{5.30}$$

where **k** denotes the wave vector of some plasma perturbation. This frequency, given by the product of wave number and diamagnetic drift, is evidently consistent with (5.29). *Drift waves*, a characteristic feature of magnetized plasmas with pressure or density gradients, have frequencies near $\omega_*$. Many other varieties of (sufficiently slow) plasma modes have their frequencies displaced by, or otherwise related to, the drift frequency. We stress that, despite the appearance of the wave vector in (5.30), the drift frequency can pertain in fully nonlinear situations, including large-amplitude coherent motions and turbulence.

The second category of drift-model physics involves an effective blurring of the electromagnetic field. The point is that a magnetized plasma is affected, not by the instantaneous field, but by its average over many Larmor periods. Because of the finite extent of Larmor orbits, this temporal average samples a spatial domain of dimension $\rho$, effectively smoothing any field variation over smaller length scales. A representative effect is the appearance, in linear theory, of the smoothed electrostatic potential $\Phi$, according to

$$\Phi \to (1 - \rho^2\nabla^2)\Phi. \tag{5.31}$$

Any model of FLR physics is complicated, with more independent variables and equations than MHD. In particular, the disparity between gyroradii of different plasma species, with

$$\delta_i \gg \delta_e \tag{5.32}$$

is taken into account: usually electron gyroradius corrections are ignored. As a result, the drift model must account separately for different plasma species; it begins with the individual species moment equations, rather than the summed equations used by MHD. (The drift model is sometimes classified as "two-fluid theory" for this reason.) However, the major complication of the drift model comes, not from its additional fluid variables, but rather from the many FLR terms that enter virtually all its equations.

It is not hard to see how fluid closure is to be obtained in the drift model. In Section 3.5, we noted that perpendicular moments could be expressed in a way that made minimal demands on kinetic theory: one could compute the desired moment through order $\delta^n$ using a kinetic theory accurate only through order $\delta^{n-1}$. This device was used to construct MHD; it is just as applicable in the drift model.

Thus, beginning with (3.50) — (3.52), one uses (3.79) for the perpendicular velocity and (3.94) for the perpendicular heat flux. An extension of the same procedure supplies the off diagonal, gyro-viscous, terms in the ion stress tensor; [electron gyroviscosity is negligible because of (5.32)], and the collisional moments are not hard to compute directly. There is the usual problem of parallel dynamics—including gyro-anisotropy as well as the parallel flows of particle and heat—which can be addressed, as in kinetic MHD, by recourse to low-order kinetic theory. Such complications aside, the FLR-closure based on Section 3.5 is straightforward and well-founded.

It is not simple, however, and the set of equations produced is so complicated as to be rarely used. For this reason we are content to sketch the derivation of FLR-closure while presenting only a simplified version of its results.

## Gyroviscosity

Flow velocities in the drift model are of order $\delta$ compared to the thermal speed; frequencies are of order $\delta$ compared to the transit frequency. Hence the acceleration term in the equation of motion is measured by

$$mn\frac{d\mathbf{V}}{dt} \sim \delta^2 mn\omega_t v_t. \tag{5.33}$$

A consistent description of plasma motion must keep, in the equation of motion, all terms of this order. The MHD ordering would put no factors of $\delta$ on the right hand side of (5.33), leading to the simple MHD rule: throw out all $O(\delta)$ terms. Here the rule—keep $O(\delta^2)$—requires more elaborate calculation.

The ugliest consequence of (5.33) is that it compels us to keep gyroviscous terms in the pressure tensor. Gyroviscosity, which is also called "magnetoviscosity," was introduced in Chapter 3, and identified as the solution, $\boldsymbol{\pi}_{gv}$, to

$$\mathbf{b} \times \boldsymbol{\pi}_{gv} + (\mathbf{b} \times \boldsymbol{\pi}_{gv})^\dagger = \frac{m}{\Omega}\nabla \cdot \mathbf{M}_3 \tag{5.34}$$

where

$$\mathbf{M}_3 = \int d^3v f\mathbf{v}\mathbf{v}\mathbf{v} \tag{5.35}$$

is the third-order moment of $f$. Now, by a familiar argument, the $1/\Omega$ factor in (5.34) corresponds to a factor of $\delta$, allowing us to compute $\boldsymbol{\pi}_{gv}$ through second order, if $\mathbf{M}_3$ is known through first order. Indeed, we see $\mathbf{M}_3$ cannot be larger than $O(\delta)$, since the zeroth order distribution is Maxwellian and isotropic. The difficulty is that here, for the first and only time in our exploration of fluid closure, we need the $O(\delta)$ correction to this distribution.

The required correction is computed systematically in Chapter 4, but the following, somewhat oversimplified, argument will serve here. In order to contribute perpendicular components to (5.35), the relevant distribution must depend upon the gyrophase, $\gamma$. But a distribution depending on gyrophase will vary on the scale of the gyrofrequency, $d\gamma/dt = \Omega$, unless its $\gamma$-dependence occurs solely through the slowly varying guiding-center position,

$$\mathbf{x}_{gc} = \mathbf{x} - \boldsymbol{\rho}$$

where

$$\boldsymbol{\rho} = \frac{1}{\Omega}\mathbf{b} \times \mathbf{v}_\perp$$

is the vector gyroradius introduced in Chapter 2. In other words the desired distribution should have the form

$$F(\mathbf{x} - \boldsymbol{\rho}) \cong F(\mathbf{x}) - \boldsymbol{\rho} \cdot \nabla F \tag{5.36}$$

Here the first term allows us to identify the function $F$ by comparison with (3.76):

$$F = f_M.$$

Hence the desired first-order correction is given simply by

$$\tilde{f}_1 = -\boldsymbol{\rho} \cdot \nabla f_M. \tag{5.37}$$

where the tilde reminds us that this distribution has vanishing gyro-phase average:

$$\oint d\gamma \tilde{f} = 0.$$

The more careful derivation of Chapter 8 will uncover an additional term in $\tilde{f}_1$, reflecting energy change due to electromagnetic work. The extra term has a very simple effect: it combines with the electrostatic field $\nabla\Phi$ to produce the full $\mathbf{E} \times \mathbf{B}$ drift, as in (5.45), below. Thus it is implicitly included in the following gyroviscosity calculation.

Two relevant components of $\mathbf{M}_3$ depend on the gyro-phase averaged part of $f$,

$$\bar{f} = \oint d\gamma f$$

and contribute whenever $\bar{f}$ includes terms odd in $v_\parallel$. Thus we need a distinct correction to the Maxwellian, denoted by $\bar{f}_1$. For the present purpose of computing gyro-viscosity it is appropriate to assume that the terms odd in $v_\parallel$ are simply those corresponding to a slow parallel drift, $V_\parallel \sim \delta v_t$. Thus we expand the moving Maxwellian,

$$f_M(\mathbf{v} - \mathbf{V}_\parallel) = f_M(\mathbf{v}) \left(1 + \frac{2V_\parallel v_\parallel}{v_t^2}\right) + O(\delta^2)$$

and find that

$$\bar{f}_1 = \frac{2V_\| v_\|}{v_t^2} f_M \tag{5.38}$$

Equation (5.38) has consequences beyond the context of gyroviscosity that deserve emphasis. Because it represents a moving Maxwellian, and because the heat flux and viscosity tensor are defined in the moving frame, (3.78) remains operative. Thus, as can be verified by straightforward integration, (5.38) yields vanishing parallel heat flow,

$$\mathbf{q}_\| = 0. \tag{5.39}$$

and isotropic pressure: $p_\| = p_\perp$. In other words the pressure tensor is given by

$$\mathbf{p} = \mathbf{I}p + \boldsymbol{\pi}_{gv}. \tag{5.40}$$

Both of these results lack physical justification; the parallel heat flow and gyrotropic pressure should be computed from kinetic analysis rather than the *ansatz* (5.38). Note in particular that collisional transport theory applies only for short mean-free path and does *not* supply a general formula for $q_\|$.

With (5.37) and (5.38), it is a straightforward matter to evaluate the nonvanishing components of $\mathbf{M}_3$, and then unravel (5.34) for the gyroviscosity. The result can be expressed as

$$\boldsymbol{\pi}_{gv} = \boldsymbol{\pi}_\perp + \mathbf{b}\boldsymbol{\pi}_\| + \boldsymbol{\pi}_\|\mathbf{b} \tag{5.41}$$

where $\boldsymbol{\pi}_\perp$ is the tensor,

$$\boldsymbol{\pi}_\perp = \frac{p}{4\Omega}\left[(\mathbf{b}\times\nabla)\mathbf{V}_\perp + \nabla(\mathbf{b}\times\mathbf{V}_\perp) + \text{transpose}\right]$$
$$+ \frac{1}{10\Omega}\left[(\mathbf{b}\times\nabla)\mathbf{q}_\perp + \nabla(\mathbf{b}\times\mathbf{q}_\perp) + \text{transpose}\right], \tag{5.42}$$

and $\boldsymbol{\pi}_\|$ is the vector,

$$\boldsymbol{\pi}_\| = \frac{p}{\Omega}\left[\mathbf{b}\times\nabla V_\| + \mathbf{b}\cdot\nabla(\mathbf{b}\times\mathbf{V}_\perp)\right] + \frac{1}{5\Omega}\mathbf{b}\cdot\nabla(\mathbf{b}\times\mathbf{q}_\perp). \tag{5.43}$$

These formulae suffice for construction of the drift model. Note that

$$\pi_{gv} \sim \delta^2 nT$$

as anticipated.

Equation (5.42) is complicated enough to justify more exposition. In a right-handed $(\mathbf{b},\mathbf{e}_2,\mathbf{e}_3)$ coordinate system, we consider the component

$$\pi_{gv23} = \frac{p}{4\Omega}\left[\epsilon_{2\beta\gamma}b_\beta\frac{\partial V_3}{\partial x_\gamma} + \frac{\partial(\epsilon_{3\beta\gamma}b_\beta V_{\perp\gamma})}{\partial x_2} + \epsilon_{3\beta\gamma}b_\beta\frac{\partial V_2}{\partial x_\gamma} + \frac{\partial(\epsilon_{2\beta\gamma}b_\beta V_{\perp\gamma})}{\partial x_3}\right] + \cdots$$

where the ellipsis indicates terms of the same form involving $\mathbf{q}$, and where $\epsilon_{\alpha\beta\gamma}$ is the unit antisymmetric tensor. Since $b_\alpha = \delta_{1\alpha}$, we find

$$\pi_{gv23} = \frac{p}{2\Omega}\left(-\frac{\partial V_3}{\partial x_3} + \frac{\partial V_2}{\partial x_2}\right) + \cdots \tag{5.44}$$

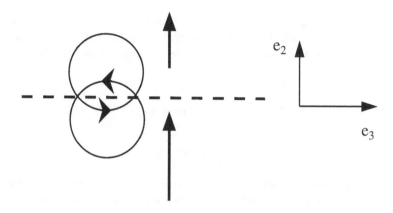

Figure 5.5: Gyroviscosity. Two gyrating particles are shown, each carrying $e_3$-momentum in the $e_2$ direction, across the dashed surface indicated. Because of the displacement of the two guiding centers, each carries oppositely directed $e_3$-momentum; if they had the same vertical speed, there would be no net momentum transport. In the presence of velocity shear, however, one of the particles has a larger vertical speed and outweighs the other, giving a momentum flux density proportional to the shear.

Note that magnetic field variation is ignored here for simplicity.

The physics underlying gyroviscosity is easily understood. Figure 5.5 shows two identical particles, drifting vertically, at slightly different vertical positions. (For simplicity the pressure is taken to be constant, so that the vertical motion stems from an electric drift.) Gyration implies that each carries horizontal momentum, but the net flux of horizontal momentum can be seen to vanish unless the two particles have different vertical speeds. Thus gyroviscous momentum transport is proportional to the velocity gradient: in this case, $\partial V_2 / \partial x_2$, as in (5.44).

## Ion drift dynamics

Gyroviscosity has considerable intrinsic interest, but its primary role is played in the ion equation of motion. Consider the first-order velocity,

$$\mathbf{V} = \mathbf{V}_{\|i} + \mathbf{V}_E + \mathbf{V}_{di} \tag{5.45}$$

where, we recall, $\mathbf{V}_{di}$ is the ion diamagnetic drift of (3.80):

$$\mathbf{V}_d = \frac{1}{eBn}\mathbf{b} \times \nabla p_i.$$

Note that all subscripts on the left-hand side of (5.45) are suppressed: from now on, $\mathbf{V}$ will refer to the lowest order ion velocity.

The ion flow enters the equation of motion explicitly, through the acceleration term and $\pi_{gv}$, and also implicitly, through the convective derivative

$$\frac{d}{dt} = \frac{\partial}{\partial t} + \mathbf{V} \cdot \nabla.$$

This would seem to generate a plethora of terms, and indeed it does—but not quite as many as one might fear. For a remarkable cancellation occurs between certain acceleration terms and the gyroviscosity, with the result that

$$m_i n \frac{d\mathbf{V}_{di}}{dt} + (\nabla \cdot \boldsymbol{\pi}_{gv})_\perp = -\nabla_\perp \chi_g, \tag{5.46}$$

$$(\nabla \cdot \boldsymbol{\pi}_{gv})_\| = -m_i n (\mathbf{V}_{di} \cdot \nabla) V_{\|i} - 2\nabla_\| \chi_g, \tag{5.47}$$

where

$$\chi_g = \frac{p_i}{2\Omega_i} \mathbf{b} \cdot \nabla \times \mathbf{V}_\perp,$$

Actually these results, sometimes referred to as the *gyroviscous cancellation*, are not exact, and depend upon certain simplifications (such as small compressibility, $\nabla \cdot \mathbf{V} \ll V/L$).

The quantity $\chi_g$ involves the parallel *vorticity*, $\mathbf{b} \cdot \nabla \times \mathbf{V}_\perp$. It is not hard to interpret, since it ultimately appears as a correction to the ion pressure:

$$\nabla_\perp p_i \to \nabla_\perp p_i \left( 1 - \frac{\mathbf{b} \cdot \nabla \times \mathbf{V}}{2\Omega_i} \right)$$

with a similar correction to the parallel pressure gradient. Notice, from (5.45), that the vorticity will involve $\nabla^2 p_i$, so that the correction includes terms of the form, $1 - \rho^2 \nabla^2$, anticipated in (5.31). In other words, the appearance of vorticity in the modified pressure manifests the blurring or local averaging induced by finite ion gyroradius.

Because of (5.47), drift acceleration does not enter the equation of motion, (3.62). The equation of motion is also simplified by approximating $\mathbf{V}_{cm} \cong \mathbf{V}$ and omitting all electron FLR corrections. The result is, for the perpendicular acceleration,

$$m_i n \frac{d\mathbf{V}_E}{dt} + \nabla_\perp (p_T - \chi_g) = \mathbf{J} \times \mathbf{B},$$

and for the parallel acceleration,

$$m_i n \frac{\partial \mathbf{V}_\|}{\partial t} + (\mathbf{V}_E + \mathbf{V}_\|) \cdot \nabla \mathbf{V}_\| + \nabla_\| (p_T - 2\chi_g) = 0.$$

## Electron drift dynamics

The perpendicular motion of magnetized electrons, because of their much smaller mass, is simpler than that of ions: such effects as electron gyroviscosity are proportional to $\rho_e$ and consistently neglected. (On the other hand the electron diamagnetic drift is mass-independent: $V_{de} \sim V_{di}$.) In constructing the drift model description of electrons, it is their parallel motion that requires attention.

Consider first the electron equation of motion, (3.51). If terms small in the electron mass are neglected, it assumes the form

$$\mathbf{E} + \mathbf{V}_e \times \mathbf{B} = -\frac{\nabla p_e}{en} + \frac{\mathbf{F}_e}{en}, \tag{5.48}$$

a relation that generalizes the MHD Ohm's law, (5.12). We have included the friction force on the right-hand side to allow for dissipation, as in resistive MHD. However the dissipative terms are often very small; the most important new feature in (5.48) is the appearance of the electron pressure gradient.

It is sometimes convenient to write (5.48) in terms of the ion velocity. Since only two species are present we have

$$\mathbf{V}_e = \mathbf{V} - \frac{\mathbf{J}}{en}$$

and

$$\mathbf{E} + \mathbf{V} \times \mathbf{B} + \frac{1}{en}\left(\nabla p_e - \mathbf{J} \times \mathbf{B}\right) = \frac{\mathbf{F}_e}{en}$$

The last term on the left-hand side here is called the *Hall term*.

In the perpendicular directions, the pressure-gradient term in (5.48) simply adds diamagnetic motion to the $\mathbf{E} \times \mathbf{B}$ drift of MHD—as in the ion case. Its effect on the parallel dynamics is more important: (5.48) allows electrons to respond to electric fields in a manner forbidden by MHD. The essential point is that the electrons can equilibrate, in the Maxwell-Boltzmann sense, to a parallel electrostatic field; they are said to respond adiabatically to it. Here the word "adiabatic" describes rapid equilibration to an electrostatic potential, $\Phi$, that changes adiabatically slowly compared to the electron transit frequency, $v_{te}/L$.

To derive the adiabatic response we need an additional equation: the energy-flow version of (5.48). In Chapter 3 we took note of the exact moment

$$\frac{\partial \mathbf{Q}}{\partial t} + \nabla \cdot \mathbf{R} - \frac{e}{2m}\left[\mathbf{E}\,Tr(\mathbf{P}) + 2\mathbf{E} \cdot \mathbf{P}\right] + \Omega \mathbf{b} \times \mathbf{Q} = \mathbf{C}_3$$

In the electron case the first, time-derivative term can be neglected; it is smaller than subsequent terms by the same ratio as the electron inertia term is smaller than the terms retained in (5.48). Also as in (5.48), we can evaluate the other terms in lowest $\delta_e$-order, assuming scalar pressure and using the LTE result, again from Chapter 3, $\mathbf{R}_M = \mathbf{I}(5/2)pT/m$. The resulting expression of electron thermal balance,

$$\frac{5}{2m_e}\left[\nabla(p_e T_e) + ep_e\mathbf{E}\right] - \frac{eB}{m_e}\mathbf{b} \times \mathbf{Q}_e = \mathbf{C}_3$$

is conveniently expressed in terms of $\mathbf{q}_e = \mathbf{Q}_e - (5/2)p_e\mathbf{E}_e$ and combined with (5.48) to obtain

$$\mathbf{b} \times \mathbf{q}_e - \frac{5}{2}\frac{p_e}{eB}\nabla T_e = -\frac{m_e\mathbf{G}_e}{eB} + \frac{5}{2}\frac{T_e}{eB}\mathbf{F}_e \qquad (5.49)$$

where $\mathbf{G}_e$ is an alternative notation for $\mathbf{C}_{3e}$,

$$\mathbf{G}_e \equiv \int d^3v \frac{1}{2}m_e v^2 \mathbf{v} C_e.$$

Note the close analogy between (5.48) and (5.49), which result from applying identical approximations to, respectively, the first and third moments of the kinetic equation. Corresponding equations do *not* pertain to the ions because of their larger mass.

The perpendicular components of (5.49) reproduce the electron heat flux discussed in Chapter 3,

$$\mathbf{q}_{de} = -\frac{5}{2}\frac{p_e}{eB}\mathbf{b} \times \nabla T_e,$$

plus additional terms proportional to the collision frequency (the heat-flow analog to resistivity). Its parallel component,

$$-\frac{5}{2}p_e\nabla_\| T_e = -m_e G_{\|e} + \frac{5}{2}T_e F_{\|e}$$

shows that parallel electron temperature gradients can be sustained only in the presence of collisions—an easily understood consequence of rapid electron parallel streaming.

The collisionless case,

$$\nu_e \ll \omega_{te} \qquad (5.50)$$

characterized by vanishing electron temperature gradient along the field, is especially important; the adiabatic electron response pertains only in this limit. For the case of an electrostatic field $\mathbf{E} = -\nabla\Phi$, we can use (5.50) to neglect the parallel temperature gradient in the momentum law (5.48) and quickly find

$$\nabla_\| \left( ne^{-e\Phi/T_e} \right) = 0. \qquad (5.51)$$

Thus the electrons relax, along the magnetic field, to a Maxwell-Boltzmann distribution in the effectively static potential.

In linear theory, where the adiabaticity requirement takes the form

$$\omega \ll k_\| v_{te},$$

(5.51) forces the linear perturbation of the electron density to be proportional to that of $\Phi$:

$$\frac{n_1}{n_0} = \frac{e\Phi_1}{T_{e0}} \qquad (5.52)$$

It is this relation that is generally associated, in the plasma stability literature, with the phrase "adiabatic electrons." Note that it only applies electrostatically; electromagnetic terms would enter through the more general form of $E_\|$ as well as through the electromagnetic electron temperature perturbation, given in linear theory by

$$\mathbf{B}_0 \cdot \nabla T_{e1} = -\mathbf{B}_1 \cdot \nabla T_{e0}. \qquad (5.53)$$

## Parallel electron heat flow

Parallel electron thermal balance, (5.49), raises issues of fluid closure that deserve comment. Consider the electron energy conservation law; making the usual small-$\rho_e$ approximations in the general energy conservation law of Chapter 3, (3.52), we obtain

$$\frac{3}{2}\frac{\partial p_e}{\partial t} + \frac{3}{2}\mathbf{V}_e \cdot \nabla p_e + \frac{5}{2}p_e\nabla \cdot \mathbf{V}_e + \nabla \cdot \mathbf{q}_e = 0. \tag{5.54}$$

Evidently a formula for $\mathbf{q}_e$ is required for closure; what is the drift model for the electron heat flow?

The perpendicular heat flow is reasonably approximated by $\mathbf{q}_{de}$, as noted previously:

$$\mathbf{q}_{\perp e} = \mathbf{q}_{de}. \tag{5.55}$$

One is tempted to assume further that $q_{\|e} = 0$, as in MHD. The corresponding physical assumption, that parallel energy flow occurs mainly through convection rather than thermal conduction, is not obviously unreasonable. But neglecting parallel heat flow will generally contradict the thermal balance equation: this simplest truncation leads to a fluid description that is not self-consistent.

The difficulty is most obvious in the collisionless limit, when (5.49) requires $\nabla_\| T_e = 0$. A pressure evolution law without $q_{\|e}$ will not in general maintain constant temperature along field lines: in the drift model, fluid flow is not tied to the field. Hence a system using (5.49) cannot also require $q_{\|e} = 0$ without self-contradiction.

Having to retain the parallel electron heat flux is especially awkward because a closed-form expression for $q_{\|e}$ is available only at short mean-free path. (The short mean-free path theory is presented in Chapter 8.) At long mean-free path—a regime of prominent interest—$q_{\|e}$ cannot be expressed in terms of local gradients at all. Of course the kinetic equation "knows" the true parallel heat flux which, among other things, makes all the moments consistent; our difficulty is the classic one of representing long mean-free path physics in fluid form.

In linear theory, an expression for $q_{\|e}$ is not needed for closure: the temperature perturbation is fixed by (5.49) alone, with the unperturbed temperature prescribed on each field line. But nonlinear theory—including numerical simulation of the drift model—requires an accounting of parallel heat flux that preserves (5.49) at each time. Therefore we briefly digress to show how a closed description of electron energy conservation can be obtained, in principle, without recourse to kinetic theory. Our discussion assumes, for simplicity, that the field lines are closed.

Noting that

$$\nabla \cdot \mathbf{q}_{\|e} = \mathbf{B} \cdot \nabla(q_{\|e}/B),$$

or, in terms of the distance $s$ along the field-line,

$$\nabla \cdot \mathbf{q}_{\|e} = B\frac{\partial}{\partial s}\frac{q_{\|e}}{B}.$$

we can express the energy conservation law as

$$B\frac{\partial}{\partial s}\frac{q_{\|e}}{B} = -\dot{p}_e,$$

where $\dot{p}_e$ abbreviates all the other terms in (5.54). We next introduce the field line average,

$$\langle X \rangle = \frac{\oint ds\, X/B}{\oint ds/B} \tag{5.56}$$

where X is any function of position and the integrals extend over the full (closed) line. Since the field line average of $B\partial F/\partial s$ vanishes for any single-valued $F$, we must have $\langle \dot{p}_e \rangle = 0$, or

$$\langle \frac{3}{2}\frac{\partial p_e}{\partial t} + \frac{3}{2}\mathbf{V}_e \cdot \nabla p_e + \frac{5}{2}p_e \nabla \cdot \mathbf{V}_e + \nabla \cdot \mathbf{q}_{\perp e} \rangle = 0. \tag{5.57}$$

This field-line averaged energy conservation law, which does not depend upon parallel heat flow, leads to a consistent model for electron fluid evolution and can be incorporated into the drift model.

To obtain a closed fluid description from (5.57), we must relate $\langle p_e \rangle$ to the unaveraged electron pressure, $p_e$—the quantity that appears elsewhere in the drift model equations. This relation is provided in principle by the two parallel balance laws, (5.48) and (5.49). We do not pursue the required, quite lengthy analysis (nor is it available in the literature); it is complicated by the fact that derivatives, including time-derivatives, do not generally commute with the averaging operator, and also by the fact that $T_e$, rather than $p_e$, is constant along the field.

It is fortunate that linear applications of the drift model, at least, do not use electron energy conservation.

## Drift model equations

The drift model advances nine quantities in time: the density $n = n_i = n_e$, the two pressures $p_i$ and $p_e$, the three flow components in $\mathbf{E}_E$ and $\mathbf{E}_\parallel$, and the three components of $\mathbf{B}$. Its nine evolution equations are given by:

$$\frac{dn}{dt} + n\nabla \cdot \mathbf{V} = 0, \tag{5.58}$$

$$\frac{3}{2}\frac{dp_i}{dt} + \frac{5}{2}p_i \nabla \cdot \mathbf{V} + \nabla \cdot \mathbf{q}_{di} = 0, \tag{5.59}$$

$$\left\langle \frac{3}{2}\frac{\partial p_e}{\partial t} + \frac{3}{2}\mathbf{V}_e \cdot \nabla p_e + \frac{5}{2}p_e \nabla \cdot \mathbf{V}_e + \nabla \cdot \mathbf{q}_{de} \right\rangle = 0, \tag{5.60}$$

$$m_i n \frac{d\mathbf{V}_E}{dt} + \nabla_\perp (p_i + p_e - \chi_g) = \mathbf{J} \times \mathbf{B}, \tag{5.61}$$

$$m_i n \frac{\partial \mathbf{V}_\parallel}{\partial t} + (\mathbf{V}_E + \mathbf{V}_\parallel) \cdot \nabla \mathbf{V}_\parallel + \nabla_\parallel (p_i + p_e - 2\chi_g) = 0, \tag{5.62}$$

$$\frac{\partial \mathbf{B}}{\partial t} + \nabla \times \mathbf{E} = 0. \tag{5.63}$$

We have in addition two constraint equations:

$$E_\| + \frac{1}{en}\nabla_\| p_e = \frac{F_{\|e}}{en}, \tag{5.64}$$

and

$$\frac{5}{2}p_e\nabla_\| T_e = m_e G_{\|e} - \frac{5}{2}\frac{p_e}{n}F_{\|e}, \tag{5.65}$$

Finally, the current is determined by Ampere's law,

$$\nabla \times \mathbf{B} = \mu_0 \mathbf{J},$$

and we use the familiar abbreviations

$$\frac{d}{dt} = \frac{\partial}{\partial t} + \mathbf{V}\cdot\nabla,$$

$$\mathbf{V} = \mathbf{V}_\| + \mathbf{V}_E + \mathbf{V}_d,$$

$$\mathbf{V}_d = \frac{1}{eBn}\mathbf{b}\times\nabla p_i,$$

$$\mathbf{V}_e = \mathbf{E} - \frac{\mathbf{J}}{en},$$

$$\mathbf{q}_{ds} = \frac{5}{2}\frac{p_s}{e_s B}\mathbf{b}\times\nabla(p_s/n),$$

$$\chi_g = \frac{p_i}{2\Omega_i}\mathbf{b}\cdot\nabla\times\mathbf{V}_\perp.$$

The field line average appearing in (5.60) is defined by (5.56).

The two constraint equations are needed, first, to determine the parallel electric field, which does not enter the perpendicular evolution equation (5.61); and, second, to fix the relation between $\langle p_e \rangle$ and the variable $p_e$. The point is that (5.60) advances only the former.

Before looking at specific applications of this model we comment generally on the role ion diamagnetic drifts play in it. What must be pointed out is that $\mathbf{V}_d$ almost disappears from the evolution equations, due to cancellations: it ultimately enters only through magnetic field variation. Consider first the particle conservation law. A simple calculation shows that

$$\nabla\cdot\mathbf{V}_d = -\frac{\mathbf{b}\times\nabla T_i}{eBn}\cdot\nabla n + \frac{\nabla p_i}{en}\cdot\nabla\times\left(\frac{\mathbf{b}}{B}\right).$$

Hence

$$n\nabla\cdot\mathbf{V}_d + \mathbf{V}_d\cdot\nabla n = \frac{\nabla p_i}{e}\cdot\nabla\times\left(\frac{\mathbf{b}}{B}\right), \tag{5.66}$$

and density evolution is affected by diamagnetic drifts only through the factor $\nabla\times(\mathbf{b}/B)$.

The cancellation reflects more than the solenoidal character of magnetization flow, as in (3.86); the same cancellation occurs in the ion energy conservation law, where one finds

$$\frac{5}{2}p_i\nabla\cdot\mathbf{V}_d + \nabla\cdot\mathbf{q}_i = -\frac{5}{2e}T_i\nabla p_i\cdot\nabla\times\left(\frac{\mathbf{b}}{B}\right). \tag{5.67}$$

Thus the occurrence of the drift frequency

$$\omega_{*i} = \mathbf{k}\cdot\mathbf{V}_d$$

in drift-model dispersion relations is not a simple consequence of diamagnetic convection. Indeed, as we find in the next section, $\omega_*$ enters even without (ion) diamagnetic drift.

## 5.5   Applications of the drift model

### Drift waves

The archetypal application of the drift model, the *drift wave*, does not involve $\mathbf{E}_d$ and propagates when $T_i = 0$. The simplest drift-wave dispersion relation involves only $\omega_{*e}$, which enters through $E\times B$ convection across the equilibrium density gradient.

The classical drift wave, sometimes called the "electron drift wave," is an electrostatic perturbation,

$$\mathbf{E} = -\nabla\Phi$$

whose frequency satisfies

$$k_\| v_{ti} \ll \omega \ll k_\| v_{te}. \tag{5.68}$$

We show that such a disturbance propagates, using the simplest version of the drift model, with vanishing ion temperature,

$$T_i = 0.$$

The more general case is considered presently. We use Cartesian coordinates $(x, y, z)$, with constant magnetic field in the $z$-direction:

$$\mathbf{B} = \hat{\mathbf{z}}B = \text{constant}.$$

The equilibrium is assumed to vary only with $x$; in that case perturbations will depend on $y$ and $z$ through the familiar eikonal, so that we have, in particular,

$$n = n_0(x) + n_1(x)e^{-i\omega t + ik_\perp y + ik_\| z},$$

where $n_1$ is the linear perturbation. We suppose that $\mathbf{E}_\|$ and $\mathbf{E}_E$ vanish in equilibrium and for simplicity omit collisional effects.

Propagation of the electron drift wave depends upon the nature of the electron response to an electrostatic potential: its relaxation to a Maxwell-Boltzmann equilibrium. This so-called "adiabatic response," resulting from fast electron motion through the nearly stationery potential ($\omega \ll k_\parallel v_{te}$), is described by the parallel Ohm's law, (5.64). In the absence of collisions the electron temperature is constant along field lines; then (5.64) can be seen to imply

$$\frac{n_1}{n_0} = \frac{e\Phi}{T_e}. \tag{5.69}$$

Here and below we suppress the 1-subscript on $\Phi$ (since $\Phi_0 = 0$) and the 0-subscript on $T_e$ (since $T_{e1} = 0$). We emphasize that there is no counterpart to this relaxation—no allowance for adiabatic electron response—in MHD.

The ions, on the other hand, do not see a quasi-static potential and respond very differently. To understand their (quasineutral) density perturbation, note first that

$$\nabla \cdot \mathbf{V}_E = 0,$$

for a uniform magnetic field. Hence the (nonlinear) particle conservation law becomes simply

$$\frac{\partial n}{\partial t} + \left(\mathbf{V}_E + \mathbf{V}_\parallel\right) \cdot \nabla n + n\nabla_\parallel V_\parallel = 0,$$

or, after linearization,

$$-i\omega n_1 + ik_\parallel n_0 V_{\parallel 1} + i\omega_{*ne} n_0 \frac{e\Phi}{T_e} = 0, \tag{5.70}$$

Here we noted that

$$\mathbf{V}_E \cdot \nabla n_0 = i\omega_{*ne} n_0 \frac{e\Phi}{T_e}.$$

where

$$\omega_{*ne} = -\frac{k_\perp T_e}{eB} \frac{d\log n_0}{dx}$$

is the drift frequency defined with the density gradient, rather than the pressure gradient. Thus, as remarked previously, the drift frequency enters through $E \times B$ motion.

Our description is completed by the parallel equation of motion, whose linear version is

$$-i\omega m_i n_0 V_{\parallel 1} + ik_\parallel p_{e1} = 0. \tag{5.71}$$

Equations (5.70) and (5.71) combine to yield

$$\left(\omega - \frac{k_\parallel^2 c_s^2}{\omega}\right) \frac{n_1}{n_0} = \omega_{*ne} \frac{e\Phi}{T_e}, \tag{5.72}$$

where

$$c_s = \sqrt{T_e/m_i} \tag{5.73}$$

is called the "ion sound speed." Equation (5.72) is sometimes said to describe a *hydrodynamic* density response. The parenthesized factor on the left-hand side evidently describes sound waves propagating along the magnetic field.

Finally we combine (5.72) with electron adiabaticity, (5.69), to obtain the drift-wave dispersion relation

$$\omega - \omega_{*ne} = \frac{k_\parallel^2 c_s^2}{\omega},$$  (5.74)

which implies, in the regime of (5.68), that

$$\omega \simeq \omega_{*ne} + \frac{k_\parallel^2 c_s^2}{\omega_{*ne}}.$$

Drift waves can be destabilized by a variety of effects, usually involving dissipation. The most common drift instabilities are characterized by a modified electron response, including a contribution out of phase with $\Phi$. Thus (5.69) is replaced by

$$\frac{n_1}{n_0} = \frac{e\Phi}{T_e}(1 - i\Delta)$$

where $\Delta$ reflects some dissipative process, such as resistivity, in Ohm's law. It is clear from (5.72) that positive $\Delta$ corresponds to instability.

## Ion temperature gradient instability

The stability of electron drift waves is a fragile and occasionally controversial issue, depending on such kinetic considerations as Landau damping or magnetic trapping. On the other hand, when ion temperature variation is allowed the drift model predicts a simple and robust instability, driven by the ion temperature gradient. The evidence that ion-temperature-gradient (ITG) modes affect magnetic confinement experiments is strong; they are likely to be important in other contexts as well.

The ITG dispersion relation is derived similarly to the electron drift wave, now allowing for finite ion temperature. In particular we retain Cartesian geometry with uniform, constant magnetic field. However some simplifying notations are helpful. We first suppress subscripts to write

$$\omega_* = -\frac{k_\perp T_e}{eB}\frac{d\log n_0}{dx}$$

instead of $\omega_{*ne}$. Note that this is the only drift frequency that enters the particle conservation law. Next, the ion drift frequency,

$$\omega_{*i} = \frac{k_\perp T_i}{eB}\left(\frac{d\log n_0}{dx} + \frac{d\log T_i}{dx}\right)$$

is expressed in terms of $\omega_*$ and the ratio of gradients,

$$\eta_i \equiv \frac{d\log T_i/dx}{d\log n_0/dx} = \frac{d\log T_i}{d\log n_0}.$$

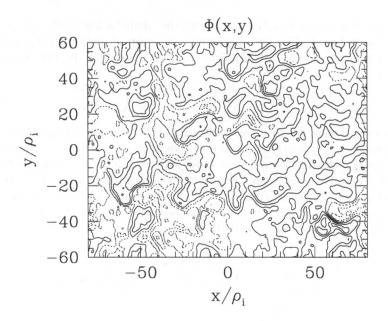

Figure 5.6: Contours of constant electrostatic potential in a plasma subject to turbulence driven by the ITG instability. These contours may be viewed as stream lines for the $\mathbf{E} \times \mathbf{B}$ velocity. The simulation is based on the gyrofluid model,[32] a fluid closure devised to include some of the features of the Vlasov model (Chapter 6) in an approximate way. Courtesy of M. Beer.

Thus

$$\omega_{*i} = -\tau \omega_* (1 + \eta_i).$$

Here we have introduced our final abbreviation,

$$\tau \equiv \frac{T_i}{T_e}.$$

The key effect of finite $T_i$ is to change the relation between ion pressure and density:

$$p_{i1} = n_0 T_{i1} + n_1 T_i.$$

Thus we have four variables, $n_1, T_{i1}, V_{\|1}$ and $\Phi$, instead of the three describing electron drift waves. The necessary additional information is provided by the ion energy conservation law, (5.59), which is very simple in the case of a uniform magnetic field. Recalling (5.67), we see that

$$\frac{5}{2} p_i \nabla \cdot \mathbf{V}_d = -\nabla \cdot \mathbf{q}_i;$$

since, in addition, the electric drift remains solenoidal and $\mathbf{E}_d \cdot \nabla p_i = 0$, the linearized energy equation is simply

$$-\frac{3}{2} i\omega p_{i1} - \frac{3}{2} i\omega_{*i} n_0 e\Phi + \frac{5}{2} i n_0 T_i k_\| V_{\|1} = 0$$

It remains only to combine this result with the adiabatic law, (5.69); the particle conservation law, (5.70); and the parallel equation of motion (5.71), in which the ion pressure must now be included. We thus obtain the ITG dispersion relation

$$\omega(\omega - \omega_*) = k_\parallel^2 c_s^2 \left[ \frac{5}{3}\tau + 1 + \tau\frac{\omega_*}{\omega}\left(\eta - \frac{2}{3}\right)\right] \qquad (5.75)$$

As a cubic equation with real coefficients, (5.75) has one solution $\omega(k_\parallel)$ that is real and two solutions that are complex conjugates of each other. The real solution generalizes the electron drift wave, satisfying $\omega(0) = \omega_*$, and confirming what was found at constant temperature: the electron drift wave is stable without dissipation. The two complex solutions generalize the ion acoustic wave, being proportional to $k_\parallel$ at large $k_\parallel$ and vanishing at $k_\parallel = 0$. Thus the ITG instability corresponds to a modified sound wave.

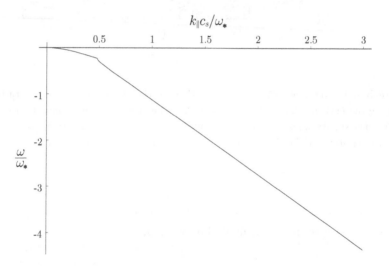

Figure 5.7: The real part of the frequency corresponding to ITG instability, normalized by $\omega_*$ and plotted as a function of normalized parallel wave number, $k_\parallel c_s/\omega_*$.

But the sound wave solutions are not always complex: for instability the temperature gradient must be sufficiently large:

$$\eta > \eta_c$$

where, for the present simplified model, $\eta_c = 2/3$. This value is obviously critical to the form of (5.75), since the equation is quadratic at that point. Its stability significance is most easily seen in the limit

$$\omega \sim k_\parallel c_s \ll \omega_*$$

in which the balance is between the second, $\tau$−term on the left-hand side of

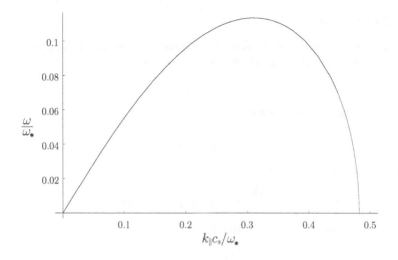

Figure 5.8: The ITG growth rate, for $\eta = 1$, normalized and plotted as in Figure 5.7.

(5.75) and the inverse-$\omega$ term on the right. Thus we find in lowest order that

$$\omega^2 \cong -\frac{k_\parallel^2 c_s^2}{\tau}\left(\eta - \frac{2}{3}\right).$$

More detailed descriptions of the instability (including, for example, magnetic field curvature) yield different values for $\eta_c$, but the main conclusion pertains generally: an ion temperature gradient that is sufficiently sharp compared to the density gradient is unstable.

The nature of the exact (unstable) solution to (5.75) is shown in Figures 5.7 and 5.8. Note the limited $k_\parallel$ range for instability; as $\eta \to \eta_c$, this unstable range (as well as the growth rate magnitude) shrinks.

## 5.6  Kinetic MHD

Here we outline the derivation of a rigorously asymptotic closure that combines moment equations with a lowest-order kinetic equation. Kinetic MHD is so-named because it uses the key physical assumptions of MHD, allowing the perpendicular electric field to be large as in (5.1), and omitting all finite-gyroradius corrections. Kinetic MHD differs from the fluid version discussed previously in this chapter in that it eschews the LTE assumption, instead determining the distribution from a zero-gyroradius version of the kinetic equation.

The basic ordering assumptions of the theory are therefore

$$\delta \ll 1. \tag{5.76}$$
$$V \cong v_{ti}. \tag{5.77}$$

where $\delta = \rho/L$ is the usual gyroradius parameter.

## Fluid equations

We start with the center-of-mass moment equations derived in Chapter 3. Omitting source terms and using quasineutrality to set

$$\rho_c = 0, \tag{5.78}$$

we recall the mass conservation law,

$$\frac{d\rho_m}{dt} + \rho_m \nabla \cdot \mathbf{V}_{cm} = 0 \tag{5.79}$$

and the equation of motion

$$\rho_m \frac{d\mathbf{V}_{cm}}{dt} + \nabla \cdot \mathbf{p}_{cm} = \mathbf{J} \times \mathbf{B}. \tag{5.80}$$

These are supplemented by the MHD *ansatz*

$$\mathbf{V}_{cm} = \mathbf{V}_\| + \mathbf{V}_E$$

as in (5.4). From here on we simplify notation by omitting *cm* subscripts; in particular the MHD velocity is denoted simply by $\mathbf{V}$.

We close our system by providing a kinetic prescription for the pressure tensor, $\mathbf{p}$. If we could complete this prescription *ab initio*, giving once and for all an expression for the pressure tensor in terms of lower-order moments and the electromagnetic field, then our procedure would be a true fluid closure. Instead however we derive a suitably approximate kinetic equation that must be solved in each application of kinetic MHD. Thus our result is a hybrid fluid-kinetic description.

## Kinetic equation

Our kinetic equation is the collisionless version of (3.22),

$$\frac{\partial f_s}{\partial t} + \mathbf{v} \cdot \nabla f_s + (\mathbf{E} + \mathbf{v} \times \mathbf{B}) \cdot \frac{\partial f_s}{\partial \mathbf{v}} = 0, \tag{5.81}$$

simplified in two ways. First, it is expressed in terms of velocity variables appropriate to a magnetized, MHD-ordered plasma: magnetic moment, kinetic energy and gyrophase angle. Second, it is averaged over the qyrophase, omitting all terms of first or higher order in the gyroradius parameter, $\delta$. The resulting kinetic equation is indeed much simpler than (5.81): because of adiabatic invariance, it involves only a single velocity variable.

Consider first the coordinate transformation

$$(\mathbf{x}, \mathbf{v}) \rightarrow (\mathbf{x}, \mathcal{K}, \mu, \gamma) \tag{5.82}$$

where $\mathcal{K}$ is the kinetic energy, $\mu$ the magnetic moment and $\gamma$ the gyrophase angle. Since our kinetic equation retains only lowest order terms, the adiabatic invariance of $\mu$ will allow us to neglect

$$\frac{d\mu}{dt} \frac{\partial f}{\partial \mu},$$

thereby converting $\mu$ from a velocity variable into a parameter. (Species subscripts are suppressed wherever they are not essential.)

This contraction of phase space represents of course a huge simplification. Note however that it pertains only in the $V_E$-drifting frame: when $\mathbf{V}_E$ is large, as we have assumed, the magnetic moment measured in terms of the laboratory velocity is not approximately invariant. If $\mathbf{v}$ represents the conventional laboratory-frame velocity, then the conserved quantity is

$$\mu = \frac{m}{2B}\left(\mathbf{v}_\perp - \mathbf{V}_E\right)^2$$

or, in the notation we will use,

$$\mu = \frac{m\mathbf{w}_\perp^2}{2B} \tag{5.83}$$

with

$$\mathbf{w} \equiv \mathbf{v} - \mathbf{V} \tag{5.84}$$

Hence we require first a coordinate transformation into the moving frame:

$$(\mathbf{x},\ \mathbf{v}) \to (\mathbf{x},\ \mathbf{w})$$

Because of the space and time dependence of $\mathbf{E}$, the moving frame is not inertial. The resulting inertial forces appearing in the transformed equation represent the main technical complication of kinetic MHD.

We introduce a familiar triad of unit vectors,

$$(\mathbf{e}_1,\ \mathbf{e}_2,\ \mathbf{b}),$$

as in Chapter 2, to express the velocity in the moving frame as

$$\mathbf{w} = \mathbf{b}w_\| + w_\perp(\mathbf{e}_1 \sin\gamma + \mathbf{e}_2 \cos\gamma)$$

The appropriate energy variable is

$$\mathcal{K} = mw^2/2.$$

Hence we have

$$w_\| = \sqrt{(2/m)(w - \mu B)} \tag{5.85}$$

$$w_\perp = \sqrt{(2/m)\mu B} \tag{5.86}$$

$$\gamma = \arctan\frac{w_1}{w_2} \tag{5.87}$$

Expressing the kinetic equation (5.81) in terms of the variable $(\mathbf{x}, \mathcal{K}, \mu, \gamma)$ we obtain the form

$$\frac{\partial f}{\partial t} + (\mathbf{V} + \mathbf{w}) \cdot \nabla f + \dot{\mathcal{K}}\frac{\partial f}{\partial \mathcal{K}} + \dot{\mu}\frac{\partial f}{\partial \mu} + \dot{\gamma}\frac{\partial f}{\partial \gamma} = 0, \tag{5.88}$$

where the coefficients are to be computed from (5.85) – (5.87). This computation parallels the guiding center derivation of chapter 2; it yields

$$\dot{\mathcal{K}} = -m\mathbf{w} \cdot \left[ \frac{\partial \mathbf{V}}{\partial t} + (\mathbf{V} + \mathbf{w}) \cdot \nabla \mathbf{V} - \frac{e}{m} \mathbf{b} E_{\parallel} \right] \tag{5.89}$$

$$\dot{\mu} = \frac{\partial \mu}{\partial t} + (\mathbf{V} + \mathbf{w}) \cdot \nabla \mu - m \frac{\mathbf{w}_{\perp}}{B} \left[ \frac{\partial \mathbf{V}}{\partial t} + (\mathbf{V} + \mathbf{w}) \cdot \nabla \mathbf{V} \right] \tag{5.90}$$

$$\dot{\gamma} = \Omega \left[ 1 + O(\delta) \right] \tag{5.91}$$

We omit the detailed expression for $\dot{\gamma}$ because only its leading term is needed.

As we have emphasized, kinetic MHD is an asymptotic theory, valid in the limit of vanishing gyroradius. Hence we consider (5.88) in the $\delta \to 0$ limit. Expanding the distribution in powers of $\delta$,

$$f = f_0 + f_1 + O(\delta^2)$$

and recalling (5.3), we find in lowest order that only gyration, as expressed by (5.91), survives:

$$\Omega \frac{\partial f_0}{\partial \gamma} = 0$$

In other words the zeroth-order distribution is isotropic in gyrophase:

$$\frac{\partial f}{\partial \gamma} = O(\delta) \tag{5.92}$$

This circumstance is clear physically: when viewed on times scales much longer than the gyroperiod, each particle is smeared over its gyroorbit, and phase information is lost.

In next order we have

$$-\Omega \frac{\partial f_1}{\partial \gamma} = \mathcal{L} f_0 \tag{5.93}$$

in terms of the abbreviation

$$\mathcal{L} f_0 \equiv \frac{\partial f_0}{\partial t} + (\mathbf{V} + \mathbf{w}_{\parallel}) \cdot \nabla f_0 + \langle \dot{\mathcal{K}} \rangle \frac{\partial f_0}{\partial \mathcal{K}} + \langle \dot{\mu} \rangle \frac{\partial f_0}{\partial \mu}$$

The $\gamma-$derivative term has been omitted from $\mathcal{L}$ because of (5.92). Since the distribution must be periodic in $\gamma$ to all orders, (5.93) has a solution only if

$$\langle \mathcal{L} f_0 \rangle \equiv \oint \frac{d\gamma}{2\pi} \mathcal{L} f_0 = 0.$$

Thus we obtain the kinetic equation

$$\frac{\partial f_0}{\partial t} + (\mathbf{V} + \mathbf{w}_{\parallel}) \cdot \nabla f_0 + \langle \dot{\mathcal{K}} \rangle \frac{\partial f_0}{\partial \mathcal{K}} = 0, \tag{5.94}$$

which is a cornerstone of kinetic MHD.

Notice that

1. Only the lowest order distribution appears; in this sense, kinetic MHD is a zero-gyroradius theory. Yet we will see that key finite gyroradius effects enter through the moment equations.

2. The quantity $w_\parallel$ is not a variable but an abbreviation, as in (5.85).

3. There is no $\mu$–derivative term; a calculation parallel to the adiabatic invariance demonstration in chapter 2 shows that $\langle \dot\mu \rangle = O(\delta)$. On the other hand, the coefficients in the kinetic equation depend on $\mu$ as will, generally, the distribution $f_0$. Thus the relevant phase space is five-dimensional.

Finally we complete our kinetic description by computing, through straightforward if lengthy calculation, the gyrophase-averaged energy change:

$$\langle \dot{\mathcal{K}} \rangle = \frac{e}{m} w_\parallel E_\parallel - w_\parallel \mathbf{b} \cdot \frac{d\mathbf{V}}{dt} - \frac{\mu B}{m} \nabla \cdot \mathbf{V} - \left( w_\parallel^2 - \frac{\mu B}{m} \right) \mathbf{b} \cdot (\mathbf{b} \cdot \nabla)\mathbf{V}. \quad (5.95)$$

## Closed system

If the exact distribution were known it could be used to compute all currents and charge densities and thus to close Maxwell's equations. Kinetic MHD, however, attempts only to compute the lowest-order distribution $f_0$, which is too crude for even a lowest-order calculation of the current: we have already noted, in (5.5), that the current must be computed from $f_1$. Thus the solution of kinetic MHD's kinetic equation does not by itself provide closure. Closure is obtained from the judicious combination of this solution with fluid equations.

In fact the kinetic part of kinetic MHD is used only to compute the lowest-order plasma pressure tensor,

$$\mathbf{P}_{cm} = \sum_s \int d^3w f_{0s} m_s \mathbf{w}\mathbf{w} \quad (5.96)$$

Since the pressure tensor contributes to the perpendicular current through (5.80), it is clear that kinetic MHD uses a familiar manipulation of moments to study $O(\delta)$ physics using the $\delta \to 0$ distribution. On the other hand the closed set of kinetic MHD, like that of MHD itself, is most simply written without *explicit* reference to $\mathbf{J}$.

Note also that the pressure tensor of (5.96) is gyrotropic, since $f_0$ is independent of gyrophase; recall the discussion of (3.98). Thus our kinetic equation has the sole purpose of evaluating, for each plasma species, the two quantities

$$p_\parallel = \int d^3w f_0 m w_\parallel^2$$

and

$$p_\perp = \int d^3w f_0 \mu B,$$

in terms of the electromagnetic field.

Kinetic MHD advances the magnetic field using the same version of Faraday's law as ordinary MHD:

$$\frac{d\mathbf{B}}{dt} - \mathbf{B} \cdot \nabla\mathbf{V} + \mathbf{B}\nabla \cdot \mathbf{V} = 0. \tag{5.97}$$

Here, again as in fluid MHD, the velocity is determined by the equation of motion, (5.80), together with mass conservation, (5.79). However the energy conservation law of fluid MHD, (5.16), is based on the scalar-pressure assumption of LTE and not generally credible. Thus the main distinguishing feature of kinetic MHD is to replace the fluid energy law by kinetic calculation of the perpendicular and parallel pressures.

There is one other fundamental difference between the fluid and kinetic systems, pertaining to the parallel electric field. Both theories agree that this field must be small, but only fluid MHD is able to neglect it entirely. Thus the electric field enters the fluid system (5.15) – (5.18) only through the $E \times B$ drift, which is obviously independent of $E_\parallel$. But the parallel field enters the MHD kinetic equation (5.94), through (5.95), because it occurs multiplied by the large ratio $e/m$. This additional variable is covered in kinetic MHD by explicit use of the charge-neutrality relation, (5.78).

In summary, kinetic MHD advances the ten coordinate-space variables

$$\mathbf{V}, \mathbf{B}, E_\parallel, \rho_m, p_\parallel, p_\perp$$

using eight coordinate-space equations—(5.80), (5.97), (5.79), and (5.78)—together with the phase-space kinetic equation (5.94) for the two pressure components.

## Role of kinetic MHD

Kinetic MHD is rigorously asymptotic in the sense discussed in chapter 3: it identifies a small parameter (the gyroradius parameter $\delta$), allows for MHD-sized perpendicular flow, and proceeds to derive a closed system without further assumptions. All approximations are in this sense controlled; refinements to the theory can be obtained in principle by proceeding to higher order in $\delta$. There are no swindles hidden in the closet, no simplifications swept under the rug.

But kinetic MHD is not a fluid closure: it is set in a five-dimensional phase space rather than in coordinate space. One should not underestimate the resulting loss, compared to fluid MHD, of tractability and simplicity—especially with regard to numerical solution. It makes sense to ask, when attacking a specific problem in MHD dynamics, whether replacing fluid MHD by its more rigorous kinetic version is likely to reward the increased effort: whether, in particular, the problem is likely to be affected by pressure anisotropy.

The complexity of kinetic MHD is one reason it is not often used. Another reason pertains to plasma stability: it can be shown by variational means that kinetic MHD is always more optimistic regarding stability than it fluid counterpart. Thus demonstrating MHD stability using the fluid version is sufficient.

# Additional reading

Two helpful texts on MHD are by Freidberg [26] and Bateman [6]; the former considers a number of fusion confinement systems in detail. Alfvén's treatment of MHD [2] and Kadomtsev's stability monograph [40] are also useful. A recent presentation of nonlinear magnetohydrodynamics, including the theory of magnetic reconnection and turbulence, is given by Biskamp.[10] Discussions of kinetic MHD more thorough than ours, especially with regard to plasma stability, can be found in the original literature: Kruskal and Oberman [47], and Rosenbluth and Rostoker [62].

The original formulations of the drift model were presented by Rosenbluth and Simon [63] and Roberts and Taylor [59]. A more detailed derivation of the equations, including the gyroviscous cancellation, may be found in [34]. Drift-waves (including the ITG instability) are discussed in detail by Krall [42]; their nonlinear consequences are reviewed by Horton [36].

# Problems

1. In the neighborhood of a sunspot, the solar magnetic field might reach 0.5 Tesla. The typical plasma density in this region is $10^{22}$ ions/m$^3$, the temperature is roughly 6000°K and plasma flows (including perpendicular flows) near sunspots are in the kilometer/sec range. Suppose an understanding of sunspot structure on a length-scale of hundreds of kilometers is desired. Is MHD an appropriate model for this purpose?

2. Show that the Alfvén speed is a natural speed of MHD, in the following way. Express each MHD variable, $X$, in terms of its characteristic value $\bar{X}$ and its dimensionless measure, $\hat{X} = X/\bar{X}$; thus $p = \hat{p}\bar{p}$ and so on. Similarly introduce the characteristic length $\bar{L}$ and time $\bar{t}$, with $\bar{V} = \bar{L}/\bar{t}$ and

$$\nabla = \frac{1}{\bar{L}}\hat{\nabla}, \quad d/dt = \frac{1}{\bar{t}}d/d\hat{t}.$$

   Show that the equation of motion is simplified by the choice

$$\bar{V} = v_A,$$

   and that only the dimensionless parameter

$$\beta \equiv \frac{2\mu_0\bar{p}}{\bar{B}^2}$$

   remains.

3. Consider a cylindrically symmetric plasma, using cylindrical coordinates $(r, \theta, z)$, with symmetry in the $\theta$- and $z$-directions. A purely axial current, $\mathbf{J} = \hat{z}J(r)$, maintains the azimuthal magnetic field $\mathbf{B} = \hat{\theta}B(r)$. Use MHD to derive the equilibrium relation between $B(r)$, $J(r)$, and the pressure $p(r)$. Also compute the magnetic curvature, $\boldsymbol{\kappa} = \mathbf{b} \cdot \nabla \mathbf{b}$, for this configuration, and show that

$$\hat{r}\frac{d}{dr}\left(p + \frac{B^2}{2\mu_0}\right) = \frac{B^2}{\mu_0}\boldsymbol{\kappa}.$$

   In the plasma confinement literature, this configuration is called a *z-pinch*, since a plasma current along the *z*-axis creates the confining magnetic field.

4. Express the Alfvén speed in terms of the ion thermal speed $v_{ti}$ and the ion beta, $\beta_i = 2\mu_0 p_i/B^2$.

5. Demonstrate that, in the absence of collisional dissipation, the drift-model equations imply

$$-\mathbf{V} \times \mathbf{B} = \mathbf{E} - \frac{\nabla p_i}{en}.$$

   Then show that an equation of state $p_i = p_i(n)$ yields a collisionless drift model that satisfies the same frozen-flux condition as MHD.

6. Verify in detail the relation (5.66) for the divergence of the diamagnetic drift.

7. Show that the two constraint equations of the drift model, (5.64) and (5.65), force collisionless electrons to assume (nonlinearly) a Maxwell-Boltzmann distribution along the magnetic field. This circumstance is at the core of drift-model physics.

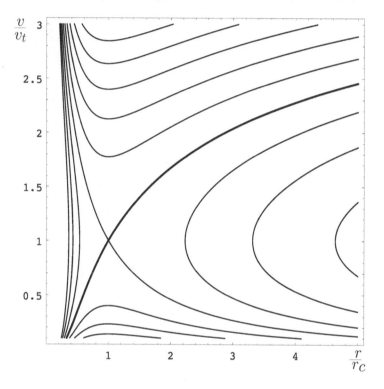

Figure 5.9: Solution curves for the velocity of the solar wind in the isothermal model of Parker.

8. **Solar wind:** Use the equations of continuity and momentum conservation to relate the velocity and density of the solar wind to the distance from the sun. Assume that the expansion is isothermal, and that the plasma acceleration is dominated by the pressure gradient and gravitational force. Justify the features and discuss the admissibility of the various solutions plotted in Fig. 1.

   The solution curve drawn in bold was proposed by E. Parker in 1958 and was later confirmed by satellite observations.[57] It predicts a transition to supersonic flow and a finite asymptotic solar wind velocity. Calculate the distance from the sun where the flow becomes supersonic, and compare it to the position of the planets.

9. Use the conservation of flux property to calculate the form of the solar magnetic field carried by the solar wind. Assume that the field is purely radial on the surface of the sun at $r = r_0$ and of magnitude $B_r(r_0, \theta, \phi)$. Show that the field lines in the ecliptic plane describe Archimedes' spirals. Assuming a solar wind velocity of 400 km/s, calculate the solar latitude of the photospheric footpoint of the field line that connects to the earth. Give your answer by reference to the sun-earth line.

# Chapter 6

# Vlasov Description of a Plasma

The aim of Vlasov theory is to describe phenomena for which the differences in the behavior of particles moving at different velocities play an important role. Recall that fluid theory describes the plasma in terms of the low-order moments of the distribution function. This is appropriate when all the particles at a given location have a nearly uniform response to perturbations—when the plasma response, while local in coordinate space, is global in velocity space. The fluid description is useful because some disturbances, by the nature of their propagation, produce similar responses in nearby particles, even when their velocities differ.

When the phase velocity of a wave is comparable to the plasma thermal velocity, however, the response of particles moving at different velocities is highly nonuniform: the particle response becomes local in velocity space as well as coordinate space. The nonuniformity results from the fact that a significant fraction of particles are nearly at rest with respect to the wave and thus affected by it with particular force. These particles and the wave are said to experience *resonant interaction*. The fluid characterization of the plasma by a few low-order moments of the distribution function is then inadequate. In order to describe the plasma properly, it is necessary to follow the evolution of the distribution function itself.

The Vlasov equation,

$$\frac{\partial f_s}{\partial t} + \mathbf{v} \cdot \nabla f_s + \mathbf{a}_s \cdot \frac{\partial f_s}{\partial \mathbf{v}} = 0, \tag{6.1}$$

describes the evolution of the single-particle distribution function in the absence of collisions. It is obtained from the kinetic equation (3.22) simply by setting $C_s(f) = 0$. Recall that this is equivalent to neglecting correlations, in the sense of (3.6). Here

$$\mathbf{a}_s = \frac{e_s}{m_s}(\mathbf{E} + \mathbf{v} \times \mathbf{B}) \tag{6.2}$$

is the Lorentz acceleration and the subscript $s$ denotes the particle species. Note that the velocity $\mathbf{v}$ appearing in Vlasov's equation is an *independent variable* labeling points of the six-dimensional phase space $(\mathbf{x}, \mathbf{v})$. It should not be confused with the velocity of a particle (a function of time) or with the local fluid velocity (a function of time and position).

Note also that the resonant interaction of particles with waves propagating at the same velocity involves a departure from the Maxwellian distribution function. This departure is opposed by collisions, so that it is natural to begin the study of wave-particle interactions by neglecting collisions.

The Vlasov equation is formally identical to Liouville's equation for a single particle system, but has a quite different meaning. The Liouville equation is an exact description of the dynamics, which serves as a starting-point for devising closures. The Vlasov equation is an approximate description, resulting from one such closure: the neglect of correlations. While the Liouville equation describes the evolution of a microscopic probability distribution function, the Vlasov equation describes the evolution of the density of particles in phase space. (Recall that the phase-space density is the ensemble average of the microscopic probability distribution.) Similarly the electromagnetic field in the Vlasov equation is an ensemble-averaged field, rather than the exact microscopic field of the Liouville equation.

The formal identity between Vlasov's and Liouville's equations makes it frequently useful to refer to properties of single-particle motion to gain insight into the behavior of the solutions of the Vlasov equation. One should bear in mind, when doing so, that particle discreteness has been erased by the neglect of correlations. Thus, the purpose of the Vlasov equation is to describe the evolution of a *self-interacting phase-space fluid*.

## 6.1   Properties of the Vlasov equation

The formal identity between the Vlasov and Liouville equations allows some of the known techniques for solving the Liouville equation to be applied to the Vlasov equation. We briefly summarize the most pertinent results.

The Vlasov equation is a first-order partial differential equation: its structure may be brought forward by writing it as

$$\partial_t f + \dot{\mathbf{z}} \cdot \nabla_6 f = 0, \tag{6.3}$$

where $\partial_t = \partial/\partial t$, $\mathbf{z} = (\mathbf{x}, \mathbf{v})$, $\dot{\mathbf{z}} = (\mathbf{v}, \mathbf{a})$, and

$$\dot{\mathbf{z}} \cdot \nabla_6 = \sum_{n=1}^{6} \dot{z}_n \frac{\partial}{\partial z_n}. \tag{6.4}$$

Equation (6.3) states that the surfaces $f =$ constant are everywhere tangent to the vector field $(1, \mathbf{v}, \mathbf{a})$. It follows that the lines of force of this vector field, called characteristic curves or simply characteristics, lie on surfaces of constant

$f$. The characteristics are defined by

$$\dot{\mathbf{x}} = \mathbf{v} \tag{6.5}$$
$$\dot{\mathbf{v}} = \mathbf{F}(\mathbf{x}, \mathbf{v}, t)/m \tag{6.6}$$
$$\dot{t} = 1 \tag{6.7}$$

where the dot represents derivation with respect to an arbitrary parameter along the characteristics. These equations are easily recognized as those of the trajectories of the single-particle motion.

Complete solutions to (6.5)-(6.6) can only be found in special cases, such as when the fields enjoy certain symmetries. In these special cases the initial position of the particle, $(\mathbf{x}_0, \mathbf{v}_0)$, can be expressed in terms of the position, $(\mathbf{x}, \mathbf{v})$. The complete solution of the Vlasov equation may then be written

$$f(\mathbf{x}, \mathbf{v}, t) = f(\mathbf{x}_0(\mathbf{x}, \mathbf{v}, t), \mathbf{v}_0(\mathbf{x}, \mathbf{v}, t), 0), \tag{6.8}$$

where $f(\mathbf{x}, \mathbf{v}, 0)$ is the initial distribution function. Note that we have assumed in (6.3) and (6.8) that the forces are conservative—that they can be derived from a Hamiltonian—so that $d\mathbf{x}\, d\mathbf{v} = d\mathbf{x}_0\, d\mathbf{v}_0$. Otherwise the Jacobian of the transformation from $(\mathbf{x}, \mathbf{v})$ to $(\mathbf{x}_0, \mathbf{v}_0)$ would appear in (6.8).

In general, the trajectories cannot be integrated explicitly, but particular solutions may nevertheless be obtained when the symmetry or invariance of a system makes it possible to identify constants of the motion. Specifically, if $C_j(\mathbf{x}, \mathbf{v}, t)$, $j = 1, ..., n$ are constants of the motion,

$$\frac{dC_j}{dt} = 0,$$

then

$$f(\mathbf{x}, \mathbf{v}, t) = g(C_1(\mathbf{x}, \mathbf{v}, t), ..., C_n(\mathbf{x}, \mathbf{v}, t)) \tag{6.9}$$

is a solution of the Vlasov equation for any function $g$. Perhaps the most important application of this type of solution is to stationary systems, for which the energy is conserved. In the presence of symmetries, the conjugate momenta also provide useful constants of the motion. The solution (6.9) is the basis for the nonlinear solution of the self-consistent Vlasov-Poisson system. It is also important in many other contexts.

## The shuffling property

We have said that the fundamental property of MHD is the constraint of frozen flux. The Vlasov theory has a similar constraint: the conservation of generalized entropy or, more pictorially, the shuffling property. To demonstrate this property we invoke the conservation of phase-space volume,

$$\nabla_6 \cdot \dot{\mathbf{z}} = 0, \tag{6.10}$$

to express the kinetic equation, including collisions, as

$$\partial_t f + \nabla_6 \cdot (\dot{\mathbf{z}} f) = C. \tag{6.11}$$

Here $C$ is the collision operator and we use the notation of (6.3). Since the collision operator conserves particles, we conclude that

$$\frac{d}{dt} \int d^6 z \, f = \int d^6 z \, \frac{\partial f}{\partial t} = 0. \tag{6.12}$$

The Vlasov equation, however, allows a much stronger conclusion. Suppose that $G(f)$ is a function of $f$ (such as $G(f) = f^2$) and consider

$$\frac{d}{dt} \int d^6 z \, G(f) = \int d^6 z \, \frac{\partial G(f)}{\partial t}. \tag{6.13}$$

For a Vlasov plasma, the integrand on the right-hand side of (6.13) is given by

$$\frac{\partial G(f)}{\partial t} = G'(f)\frac{\partial f}{\partial t} = -G'(f)\nabla_6 \cdot (\dot{z}f)$$

where the prime indicates a derivative with respect to argument. In view of (6.10), we can write the integrand as a divergence,

$$\begin{aligned} G'(f)\nabla_6 \cdot (\dot{z}f) &= \dot{z} \cdot \nabla_6(G(f)) \\ &= \nabla_6 \cdot (\dot{z}G(f)), \end{aligned}$$

whence

$$\frac{d}{dt} \int d^6 z \, G(f) = 0. \tag{6.14}$$

Thus we have an uncountable infinity of conservation laws.

This extremely strong result applies only to the Vlasov equation. For the particular choice $G = -f \log f$, (6.14) expresses entropy conservation; it may thus be characterized as a generalized entropy conservation law.

To understand the significance of generalized entropy conservation, consider a histogram of $f$: a set of columns, giving the value of $f$ for each increment of its variables. The columns, of course, evolve in time. Now we may choose for $G$ a function equal to 1 for all the columns on which $f = 7$, for example, and equal to zero for all other columns; this function counts the number of columns that are seven chips high. The conservation law (6.14) states that the number of 7-chip-high columns remains constant at all times! This implies that the columns cannot disappear or change height under the motion; they can only be rearranged. Vlasov evolution consists of a shuffling in phase-space of histogram columns of fixed height.

The shuffling property has important consequences for the stability of Vlasov plasmas. First, it implies that a vast class of perturbed states are simply not accessible from a given equilibrium state, since accessing these states would require violating (6.14). When investigating the stability of an equilibrium, one may thus restrict attention to *dynamically accessible* states that satisfy (6.14). Of course, collisions are always present in practice and will lead eventually to violation of the shuffling property. But because such collisional dissipation occurs only after a time much longer than the growth time of a typical Vlasov instability, (6.14) has important practical relevance.

The concept of dynamical accessibility introduced above can be used to derive a very general stability condition. Consider the kinetic energy of a plasma,

$$W_K = \int d^3x \int d^3v \, \frac{mv^2}{2} f(\mathbf{x}, \mathbf{v}, t) \tag{6.15}$$

Clearly an instability can only arise if a state of lower energy is accessible to the plasma. The state of lowest energy compatible with the shuffling property can be constructed by taking the tallest column and moving it to the state of lowest energy, namely $\mathbf{v} = 0$, then moving the next tallest column right besides it, and so on, until all the columns of the distribution function have been rearranged into a sequence such that their height decreases monotonically as one moves away from the origin in velocity. This argument leads to the result that the kinetic energy of distributions that are monotone decreasing about the origin cannot be further reduced, so that these distributions are absolutely stable. This result, known as Gardner's theorem,[29] is notable for being free of any linearization.

## 6.2   The wave-particle resonance

The most important application of the Vlasov equation is to describe the effects of the resonant interaction between a wave and the particles with velocity close to its phase velocity. There are two methods for evaluating these effects, both of which are concerned with times long compared to the transit time $1/kv_{\text{th}}$ (the time it takes a particle to travel one wavelength). The first method is based on the observation that for $v_t \ll \omega/k$, there will be few resonant particles and the solution should be close to that predicted by fluid theory. The effects of the resonance can then be evaluated perturbatively. The second method uses the Laplace transformation and is more general in scope. These two methods yield different insights and we will describe each in turn.

For simplicity we restrict consideration in the first part of this chapter to the case of unmagnetized plasma ($\mathbf{B}_0 = 0$). We will describe waves in a magnetized plasma in Section 6.6. In the unmagnetized case, fluid theory predicts two types of solutions: transverse electromagnetic waves with $\omega^2 = \omega_{pe}^2 + k^2c^2$, and longitudinal electrostatic oscillations with $\omega^2 = \omega_{pe}^2$. We note that the electromagnetic waves have $\omega/k \equiv v_{ph} > c > v_{te}$. That is, their phase velocity is greater than the speed of light, so that they do not experience wave-particle interaction. We thus focus attention on the longitudinal, electrostatic oscillation.

We look for solutions of the Vlasov equation that vary only in the $x$ direction, and denote the $x$ component of the velocity by $u$. The remaining components of the velocity are passive parameters, and we omit them for clarity. We consider small-amplitude, plane-wave perturbations of the electron distribution of the form,

$$f_e(x, u, t) = f_0(u) + f_k(u, t) \exp(ikx), \tag{6.16}$$

where $f_0(u)$ represents the distribution function for the homogeneous equilibrium and $f_k(u, t)$ is the amplitude of the perturbation. Recall that the unperturbed plasma density $n_0$ is given by $n_0 = \int du\, f_0(u)$. The corresponding linearized Vlasov equation is

$$\frac{\partial f_k(u, t)}{\partial t} + iku f_k(u, t) = \frac{eE_k(t)}{m_e} f_0'(u), \qquad (6.17)$$

where $f_0'(u) = df_0/du$. We are interested in high frequency oscillations with $\omega \gg \omega_{pi}$, so that the ions do not participate in the oscillation and their role is limited to that of a neutralizing background. The linearized system is completed by Poisson's equation,

$$ik\epsilon_0 E_k(t) = -e \int_{-\infty}^{\infty} du\, f_k(u, t). \qquad (6.18)$$

The usual comments concerning plane waves in homogeneous media apply here: First, Fourier's theorem assures us that any three-dimensional perturbation may be represented by a linear superposition of plane waves of the form $f_k(\mathbf{v}, t)\exp(i\mathbf{k} \cdot \mathbf{x})$. Second, solutions of the form (6.16) may be generalized to inhomogeneous plasma by the use of the WKB theory provided that $kL \ll 1$, where L represents a typical length-scale for the variation of the equilibrium quantities.

We next describe the perturbative solution of (6.17)-(6.18).

## Perturbative solution

In order to treat the wave-particle resonance perturbatively, we separate the equilibrium distribution function into two parts corresponding to particles travelling faster and slower than a cutoff velocity $u_c$, chosen such that $v_t \ll u_c \ll v_{ph}$. That is, we take

$$f_0(u) = f_{0b}(u) + f_{0t}(u), \qquad (6.19)$$

where $f_{0b}(u) = f_0(u)\Theta(u_c - u)$ describes the particles constituting the bulk of the plasma and $f_{0t}(u) = f_0(u)\Theta(u - u_c)$ describes the particles populating the tail of the distribution function. Here $\Theta$ is the Heaviside step-function. We may then expand the solution using the ordering

$$f_{0t}(u) \ll f_{0b}(u) \sim n_0/v_t.$$

We begin by integrating the lowest-order Vlasov equation, obtained by neglecting $f_{0t}$ from (6.17):

$$f_k(u, t) = f_k(u, 0)e^{-ikut} + \frac{ef_{0b}'(u)}{m_e} \int_0^t d\tau\, E_k(t - \tau)e^{-iku\tau}. \qquad (6.20)$$

Here the first term represents the free streaming of the initial perturbation and the second term represents the response of the bulk particles under the effect of the electric field caused by the perturbation. More specifically, this second term can be seen to result from the acceleration of the background particles as a result of the impulse they acquire from the electric field along their unperturbed trajectories, $x = x_0 + u\tau$.

We are interested in solutions describing a plasma oscillation with

$$E_k(t) = \hat{E}_k(t)e^{-i\omega t},$$

where the amplitude of the oscillation, $\hat{E}_k$, may vary slowly in time as a result of wave-particle interaction:

$$\frac{1}{\hat{E}_k}\frac{d\hat{E}_k}{dt} \sim \omega \frac{v_t f_{0t}}{n_0} \ll \omega. \tag{6.21}$$

We evaluate the integral in (6.20) by parts and use (6.21) to neglect the term proportional to the time derivative of $\hat{E}_k$. There follows

$$f(x,u,t) = f_0(u) + \left( f_k(u,0) - \frac{ek\hat{\phi}_k(0)}{m_e}\frac{f'_{0b}(u)}{\omega - ku} \right) e^{ik(x-ut)}$$

$$+ \frac{ek\hat{\phi}_k(t)}{m_e}\frac{f'_{0b}(u)}{\omega - ku} e^{i(kx-\omega t)}, \tag{6.22}$$

where we have substituted $\hat{E}_k = ik\hat{\phi}_k$ in order to show that the perturbation of the distribution is in phase with the electrostatic potential, as one would expect from energy conservation. We have written the result in terms of the complete distribution $f(x,u,t)$, rather than the Fourier components $f_k(u,t)$, in order to display the form of its spatio-temporal dependence, which we discuss next.

Equation (6.22) shows that the response to an electric field oscillating at constant amplitude consists of three parts. The first part, proportional to $f_k(u,0)$, is the initial perturbation of the distribution function that creates the electric field. This part of the distribution evolves through the free-streaming of the corresponding particles, giving rise to the factor $e^{ik(x-ut)}$. The next two terms describe the response of the background plasma to the imposed electrostatic potential. To interpret these two terms, we note that in a frame moving with the wave, the electrostatic potential is constant, so that the energy of the particles $m(\delta u)^2/2 - e\phi(x)$ is conserved (here $\delta u = u - \omega/k$ is the velocity of a particle in the frame moving with the wave). The background distribution may be separated into a part that is constant on surfaces of constant energy, and a small, spatially oscillating part that results from the difference between surfaces of constant energy and velocity. The former part is constant in time, since energy is conserved, whereas the latter part evolves ballistically. After transforming back to the laboratory frame, these two parts yield respectively the fourth and third terms in (6.22).

## The ballistic response and phase mixing

The two terms in (6.22) proportional to $e^{ik(x-ut)}$,

$$b_k(u) \equiv f_k(u,0) - \frac{ek\phi_k(0)}{m_e} \frac{f'_{0b}(u)}{\omega - ku}. \tag{6.23}$$

are collectively referred to as the ballistic response of the plasma. It is interesting to consider the part of the electric field caused by the ballistic response,

$$ik\epsilon_0 E_{k,b} = -e \int_{-\infty}^{\infty} du\, b_k(u) e^{-ikut}, \tag{6.24}$$

We see that this field is the Fourier transform with respect to velocity of the ballistic distribution $b_k(u)$. It will thus decay at a rate $k\Delta u$, where $\Delta u$ is the shortest scale of velocity-variation of $b_k$. The rigorous version of this assertion is given by the Riemann-Lebesgue theorem.

The decay of the ballistic electric field is a manifestation of phase-mixing, one of the most important phenomena in plasma physics. A physical interpretation of the decay is given in Fig. 6.1. This Figure shows the evolution of two contours of constant $b$ defined by $b(x,u,t) = b_k \sin(kx - ut) = \pm b_k/2$. Ballistic evolution stretches these contours continuously, while the periodicity of the perturbation effectively folds them back into the fundamental period. The resulting transformation is known as the baker's map. After a few iterations, the two contours of constant $b$ cover the entire phase-space area included between its initial maximum and minimum extensions in velocity. We emphasize that this transformation is fully consistent with the shuffling property.

The time $\Delta t$ needed to stretch a contour by a factor of two is inversely proportional to its width $\Delta u$: $\Delta t = 2\pi/k\Delta u$. For a perturbation of width $\Delta u_0$, the electric field will thus initially experience a rate of decay of the order of $k\Delta u_0$, as we noted earlier.

## The Langmuir wave

Having established that the ballistic response leads to a decaying electric field, we next restrict consideration to initial conditions such that the ballistic response is absent from the lowest-order terms. Examination of (6.22) shows that we can do this by choosing the initial conditions

$$f_k(u,0) = \frac{ek\phi_k(0)}{m_e} \frac{f'_{0b}(u)}{\omega - ku}.$$

The resulting solution is an *eigenmode* $f_k(u,t) = \hat{f}_k(u,t)e^{-i\omega t}$ where

$$\hat{f}_k(u,t) = \frac{ek\hat{\phi}_k(t)}{m_e} \frac{f'_{0b}(u)}{\omega - ku}. \tag{6.25}$$

In order for the solution (6.25) to be consistent we must verify that Poisson's equation is satisfied. This requires that

$$K_b(\omega,k) \equiv 1 + \frac{\omega_{pe}^2}{kn_0} \int du \frac{f'_{0b}}{\omega - ku} = 0, \tag{6.26}$$

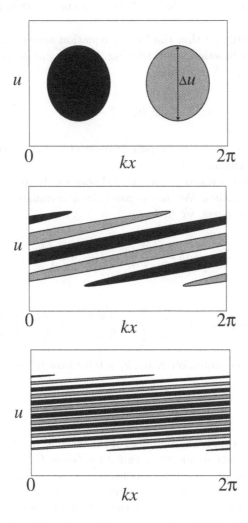

Figure 6.1: Illustration of the phase mixing process brought about by the ballistic propagation. The two ellipses in the top frame represent surfaces of constant $b(u, x, t) = \pm b_k/2$ at time $t = 0$ containing an excess (black) and a deficit (gray) of electrons. The perturbation is assumed to have an extent $\Delta u$ in velocity space. The second frame shows the same surfaces after a time $2\pi/k\Delta u$, and the third after a time $t = 12\pi/k\Delta u$. The electric field at a given point $x$ is the integral of the space charge over all velocities, and thus decays at a rate $\gamma \sim k\Delta u$.

where $K_b(\omega, k)$, the dielectric constant of the bulk plasma, measures the Debye shielding of the initial charge perturbation by the bulk electrons. Equation (6.26) is a dispersion relation that can be used to determine the frequency $\omega$ of the wave. We next compare the resulting frequency to that predicted by fluid theory.

We begin by verifying that the Vlasov equation agrees with fluid theory for a cold distribution function. Substituting $f_0(u) = n_0\delta(u)$ into the expression for the bulk dielectric $K_b$, we find

$$K_b(\omega, k) = 1 - \frac{\omega_{pe}^2}{\omega^2}.$$

We see that $K_b(\omega, k) = 0$ has the pair of roots $\omega = \pm\omega_{pe}$ corresponding to the plasma oscillation eigenmodes.

We next consider a warm plasma and choose the frame of reference so that the fluid velocity vanishes. We may expand the denominator in powers of $ku/\omega$. After integration by parts, we find

$$K_b(\omega, k) = 1 - \frac{\omega_{pe}^2}{\omega^2} \int du \, \frac{f_{0b}(u)}{n_0} \left[ 1 + 2\frac{ku}{\omega} + 3\left(\frac{ku}{\omega}\right)^2 + O\left(\frac{ku}{\omega}\right)^3 \right].$$

or

$$K_b(\omega, k) = 1 - \frac{\omega_{pe}^2}{\omega^2} \left[ 1 + \frac{3}{2}\frac{k^2 v_t^2}{\omega^2} + O\left(\frac{kv_t}{\omega}\right)^3 \right]. \tag{6.27}$$

Solving the dispersion equation $K_b(\omega, k) = 0$ iteratively for $kv_t \ll \omega$, we find

$$\omega^2 = \omega_{BG}^2 \equiv \omega_{pe}^2 + k^2\frac{3T}{m_e}. \tag{6.28}$$

Equation (6.28), describing the effect of the thermal spread of the equilibrium distribution on the wave frequency, is called the Bohm-Gross dispersion relation. The frequency determined by this dispersion relation is given to lowest order by the plasma frequency. Thus we have found that thermal motion transforms the plasma oscillation into a wave, called the Langmuir wave, propagating at the group velocity

$$v_g = \frac{3}{2}\frac{kv_t^2}{\omega_{pe}} \ll v_t.$$

The asymptotic method used above to determine the frequency of Langmuir waves can also be used to determine the growth rate of the two-stream instability when the relative velocity of the beams exceeds their thermal spread. In the general case, however, the question of the existence of roots to the dispersion relation with $\text{Im}(\omega) > 0$ is most easily answered using contour integration techniques described later in this chapter.

## Unstable eigenmodes

We are now ready to determine the effects of the wave-particle interaction by calculating the corrections due to the resonant particles in the tail of the distribution function. To determine these corrections, we substitute $f_k(u,t) = \hat{f}_k(u,t)e^{-i\omega t} + \delta f_k(u,t)$ into the Vlasov equation, where $\hat{f}_k(u,t)$ is the lowest-order solution (6.25). There follows

$$\frac{\partial \delta f_k(u,t)}{\partial t} + iku\,\delta f_k(u,t) = \hat{S}_k(u,t)e^{-i\omega t}, \tag{6.29}$$

where

$$\hat{S}_k(u,t) = -\frac{ike}{m_e}\left(\delta\hat{\phi}_k(t)f'_{0b}(u) + \hat{\phi}_k(t)f'_{0t}(u) - i\frac{f'_{0b}(u)}{\omega - ku}\frac{d\hat{\phi}_k(t)}{dt}\right)$$

is a slowly varying function of time. Note that $S_k(u,t)$ is a well-behaved function of velocity, since $f'_{0b}(u)$ vanishes at the resonance.

Treating $\hat{S}_k(u,t)$ as constant in time, we solve (6.29):

$$\delta f_k(u,t) = \delta f_k(u,0)e^{-ikut} - i\hat{S}_k(u,t)\frac{e^{-ikut} - e^{-i\omega t}}{\omega - ku}. \tag{6.30}$$

This solution is regular at $u = \omega/k$, since the numerator vanishes at this point.

We would like to choose the initial conditions so as to eliminate the ballistic response, as we did for the lowest-order solution. Inspection of (6.30), however, reveals that this calls for a singular initial perturbation,

$$\delta f_k(u,0) = \frac{i\hat{S}_k(u,t)}{\omega - ku}. \tag{6.31}$$

We may sidestep this problem by allowing the frequency to have a small imaginary part, $\omega = \omega_r + i\gamma$, describing the slow growth or decay of the mode:

$$\phi_k(t) = \hat{\phi}_k(0)e^{(\gamma - i\omega_r)t} = \hat{\phi}_k(0)e^{-i\omega t}.$$

This has the effect of regularizing the wave-particle singularity, thereby allowing us to choose the initial condition as in (6.31), so as to entirely eliminate the ballistic response. The solution is then

$$\delta f_k(u,t) = \frac{ke}{m_e}\left(\delta\hat{\phi}_k(t)f'_{0b}(u) + \hat{\phi}_k(0)f'_{0t}(u) - i\frac{f'_{0b}(u)}{\omega - ku}\gamma\hat{\phi}_k(0)\right)\frac{e^{-i\omega t}}{\omega - ku}. \tag{6.32}$$

Substituting the complete solution in Poisson's equation and using the zeroth-order dispersion relation (6.26) to eliminate the $\delta\hat{\phi}_k$ terms, we find

$$K_b(\omega_r + i\gamma, k) = i\gamma K'_b(\omega_{pe}, k) = -\frac{\omega_{pe}^2}{kn_0}\int du\,\frac{f'_{0t}(u)}{\omega_r + i\gamma - ku}, \tag{6.33}$$

For small $\gamma$ the integral may be evaluated by using

$$\lim_{\gamma \to 0} \int dz \, \frac{g(z)}{\omega_r + i\gamma - z} = \lim_{\gamma \to 0} \int dz \, g(z) \frac{(\omega_r - z) - i\gamma}{(\omega_r - z)^2 + \gamma^2}$$

$$= P \int dz \, \frac{g(z)}{\omega_r - z} - i\pi \, \text{sign}(\gamma) g(\omega_r) \quad (6.34)$$

The above, frequently useful result is conveniently summarized by Plemelj's formula,

$$\lim_{\gamma \to 0} \frac{1}{\omega \pm i\gamma - z} = P \frac{1}{\omega - z} \mp i\pi \delta(\omega - z). \quad (6.35)$$

Substituting (6.34) into (6.33) and neglecting the principal part integral of the tail distribution, there follows

$$|\gamma| = \frac{\pi}{2} \frac{\omega_{pe}^3}{k^2 n_0} f_0' \left( \frac{\omega_{pe}}{k} \right). \quad (6.36)$$

We see that solutions exist only for $f_0'(v_{ph}) > 0$. Distributions satisfying this condition are said to display population inversion. They may occur, for example, in the presence of a beam of suprathermal particles. Since either sign of $\gamma$ yields an acceptable solutions, there is both a growing and a damped eigenmode. The growing mode is called the bump on tail instability.

In the more common case where $f_0'(v_{ph}) < 0$, (6.36) shows that there are no eigenmode solutions to first order in our expansion. In order to describe the evolution of the Langmuir waves, it is then necessary to retain the ballistic response. We consider this next.

## Landau damping

We have shown that for $f_0'(v_{ph}) < 0$, wave propagation with resonant particles must excite a ballistic response. In order to determine the effect of the ballistic excitation, we return to the general solution (6.30). Since we are now looking for solutions that are not eigenmodes, we revert to the more general description of the amplitude evolution through $\hat{\phi}_k(t)$ with $\text{Im}(\omega) = 0$. We may simplify the subsequent analysis by choosing the initial condition so as to eliminate the ballistic response of the *bulk* plasma,

$$\delta f_k(u, 0) = \frac{ke}{m_e} \left( \frac{\delta \hat{\phi}_k(t)}{\omega - ku} - \frac{i}{(\omega - ku)^2} \frac{d\hat{\phi}_k(t)}{dt} \right) f_{0b}'(u). \quad (6.37)$$

The corresponding solution is

$$\delta f_k(u, t) = -\frac{ke}{m_e} \hat{\phi}_k(t) f_{0t}'(u) \, R[(\omega - ku)t] \, t e^{-i\omega t}$$

$$- \frac{ke}{m_e} \left( \delta \hat{\phi}_k(t) - \frac{i}{\omega - ku} \frac{d\hat{\phi}_k(t)}{dt} \right) \frac{f_{0b}'(u)}{\omega - ku} e^{-i\omega t}. \quad (6.38)$$

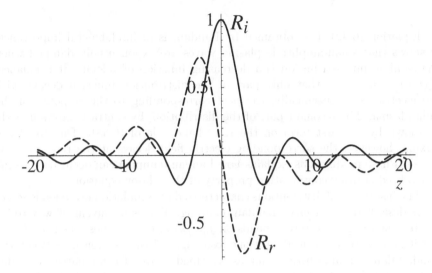

Figure 6.2: Shape of the perturbed distribution function in the resonance region for a weakly damped plasma wave. The solid line is the imaginary part and the dashed line the real part of the function $R(z)$ defined in (6.39)

Here the function

$$R(z) = \frac{e^{iz} - 1}{z}, \tag{6.39}$$

shown in Fig. 6.2, describes the shape of the response in the resonant region. The evolution of the response in the resonant region can be visualized by imagining the scale of the function $R$ in velocity space being compressed at a constant rate, while the amplitude of the response grows linearly in time.

We may now obtain an evolution equation for the amplitude of the mode by substituting the solution (6.38) into Poisson's equation. Making use of the dispersion relation $K_b(\omega, k) = 0$, we find

$$\frac{d\hat{\phi}_k(t)}{dt} \int du \, \frac{f'_{0b}(u)}{(\omega - ku)^2} = -i\hat{\phi}_k(t) \int du \, f'_{0t}(u) \, t \, R[(\omega - ku)t]. \tag{6.40}$$

The first integral is proportional to $\partial K_b(\omega, k)/\partial \omega \simeq 2/\omega_{pe}$. For $t \gg 1/kv_t$, we may evaluate the second integral by approximating $R$ in terms of a Dirac delta function according to

$$\lim_{t \to \infty} tR(zt) = i\pi\delta(z).$$

It follows that the field amplitude evolves according to

$$\frac{d\hat{\phi}_k(t)}{dt} = \gamma_L \hat{\phi}_k(t), \tag{6.41}$$

where

$$\gamma_L = \frac{\pi}{2} \frac{\omega_{pe}^3}{k^2 n_0} f'_0 \left(\frac{\omega_{pe}}{k}\right). \tag{6.42}$$

Equation (6.42), first obtained by Landau, is of fundamental importance. It shows that small-amplitude plasma waves are exponentially damped when the equilibrium distribution is a decreasing function of velocity at resonance, $f_0'(v_{ph}) < 0$. Note that only part of the distribution function decays with the electric field, specifically, the part corresponding to the response of the bulk plasma. The resonant part of the distribution, by contrast, grows linearly as shown by the first term on the right-hand side of (6.38). Due to phase-mixing, however, the corresponding electric field decays at the rate given by (6.42). Phase mixing thus explains how Landau damping can occur despite the generalized entropy conservation property of the Vlasov equation.

The instability of distributions characterized by population inversion is sometimes described as "inverse" Landau damping. This is a convenient way to indicate that the growth rate depends on $f_0'(v_{ph})$ in the way described by (6.42), but it is potentially misleading if one loses sight of the differences between the Landau-damped and unstable modes described above. To summarize these differences, for $f_0'(v_{ph}) > 0$ there is a pair of exact eigenmodes, such that the perturbation grows or decays at the same rate $\gamma$ for all velocities. For $f_0'(v_{ph}) < 0$, by contrast, the solutions are only eigenmodes to lowest order in $f_0'(v_{ph})$ and the decay affects only the perturbation of the bulk particles: the ballistic response in the tail, by contrast, grows during the evolution of the mode.

A disturbing aspect of Landau damping is that it seems to violate the known time-reversibility of the Vlasov equation. Since the Vlasov equation is invariant under time reversal, a perturbation with an exponentially growing electric field can in principle be obtained by taking a damped wave and reversing the velocity of every particle. One could then argue that for every initial condition such that the electric field decays for some time, there is exactly one other initial condition such that it will grow exponentially for the same amount of time. This suggests that growth and decay are equally likely. What then, is the meaning of the statement that the electric oscillations in a plasma are exponentially damped?

The answer is twofold: first, the electric perturbation constructed by the above velocity reversal will only grow for a finite amount of time, until the smooth original state is restored. After this it will indeed decay. Second, while any smooth initial distribution will decay, constructing an initial perturbation that will momentarily amplify requires replicating in exquisite detail the very fine structure of the velocity-reversed version of the resonant response of a damped wave. Although this difficult task can in fact be achieved experimentally, as described below, the corresponding states clearly cannot be thought of as "equally likely" as the smooth initial conditions leading to damped oscillations.

The resonant ballistic response can be observed experimentally as follows. A plasma is excited at time $t_1$ by a wave with wavevector $k_1$. At a subsequent time $t_2$, the excited wave is modulated by a second perturbation with wavevector $k_2$. The modulation results in a perturbation of the form

$$\delta f(x, u, t) = h(x, u) \exp[i k_1 u(t - t_1) + i k_2 u(t - t_2)].$$

At the later time

$$t_{\text{echo}} = \frac{k_1 t_1 + k_2 t_2}{k_1 + k_2},$$

phase-mixing is nullified and the ballistic response produces an electrostatic pulse known as a plasma echo. Comparisons between theoretical predictions and experimental observations of plasma echoes have provided a rigorous test of the validity of Vlasov theory.

Another counterintuitive aspect of Landau damping is the dependence of the damping rate on the *local* gradient of $f_0$ at the resonant velocity $v_{ph}$. Taken literally, (6.42) seems to assert that the damping depends exclusively on particles that have infinitesimal velocity relative to the wave. This is an artifact of the long-time asymptotic limit used to derive (6.42): the dependence of the damping on the distribution of neighboring particles as well as on time is best understood in terms of the energetics of the plasma oscillation, which we consider next.

## Energy

In order to interpret the dependence of Landau's result on the derivative of the distribution function at the resonant velocity, it is enlightening to consider the energetics of the Vlasov dynamics. There are two different forms for the energy in a plasma described by the Vlasov equation. The first, called the Vlasov energy, is exact. It is expressed in terms of the distribution function. The second, called the wave energy, is a product of the perturbation analysis. It is expressed in terms of the electric field amplitude, and describes only the energy contained in slowly evolving eigenmodes. (it is incapable, for example, of describing the ballistic response.) The wave energy is much simpler than the Vlasov energy, and is the most frequently used. Here we derive and compare both forms.

The exact energy is

$$\mathcal{W} = \frac{1}{2} \int d^3x \left( \sum_s \int d^3v \, m_s \mathbf{v}^2 f_s + \epsilon_0 |\mathbf{E}|^2 \right). \tag{6.43}$$

This energy is conserved by the full, nonlinear Vlasov equation. A quadratic version of the energy that is conserved by the linearized system, however, is not immediately evident. The difficulty is due to the fact that the first term in $\mathcal{W}$ depends linearly on the distribution function. As a result the expansion of the energy to second order in the perturbation amplitude, as required by linear theory, involves the second order correction to the distribution function.

To avoid calculating this correction, we use the energy conservation equation, obtained by multiplying the electrostatic version of Maxwell's equation by the electric field:

$$\frac{\epsilon_0}{2} \frac{\partial}{\partial t} |\mathbf{E}|^2 = -\text{Re}(\mathbf{J}^* \cdot \mathbf{E}). \tag{6.44}$$

This shows that the change in electrostatic energy density is given by Ohmic power density. We evaluate the electric field appearing in the Ohmic power term by using Vlasov's equation:

$$E_k = \frac{m_e}{e} \left( \frac{\partial f_k}{\partial t} + iku f_k \right) \frac{1}{f_0'(u)}. \tag{6.45}$$

We may limit consideration to perturbations that satisfy the accessibility constraint, or such that the initial perturbation satisfies the shuffling property

$$f(x, u, t) = f_0(u + \xi(x, u)) = f_0(u) + \xi(x, u) f_0'(u),$$

where $\xi(x, u)$ is the displacement of the phase-space fluid element at $(x, u)$. It follows that $f_k(u, t)$ vanishes where $f_0'(u) = 0$, so that the above expression for the electric field is well defined. When (6.44) is applied to a plane-wave perturbation of the form (6.16), the Ohmic power density takes the form

$$\text{Re}(J_k^* E_k) = -\frac{\partial}{\partial t} \int du \, \frac{m_e u |f|^2}{2 f_0'(u)}. \tag{6.46}$$

The energy conservation equation (6.44) thus takes the form

$$\frac{d\mathcal{W}_\mathbf{k}}{dt} = 0, \tag{6.47}$$

where

$$\mathcal{W}_\mathbf{k} = \frac{1}{2} \left( \epsilon_0 |\mathbf{E_k}|^2 - \int du \frac{m_e u}{f_0'(u)} |f(u, t)|^2 \right) \tag{6.48}$$

is called the spectral energy density. The spectral energy density $\mathcal{W}_\mathbf{k}$ is exactly conserved by the *linear* Vlasov-Poisson equations.

We next consider the implications of the conservation of energy for the evolution of a plasma oscillation. We first separate the energy of the particles, the second part of (6.48), into the contribution of the bulk and tail of the distribution. Using (6.38), we find that the tail contribution to the energy density is

$$\begin{aligned}
\mathcal{W}_{\mathbf{k},t} &= -\frac{e^2}{2m_e} |\hat{E}_k(t)|^2 \int du \, u f_{0t}'(u) \, t^2 |R[(\omega - ku)t]|^2 \\
&\to -\pi \frac{e^2}{m_e k^2} |\hat{E}_k(t)|^2 f_0'(\omega/k) \omega t. \tag{6.49}
\end{aligned}$$

where the second line refers to the large-$t$ limit. The function $|R[(\omega - ku)t]|^2$ appearing in the energy integral can be interpreted as describing the diffraction pattern between the coherent and the ballistic responses to the electric field.

Equation (6.49) shows that the energy associated with the resonant response depends on the sign of the gradient of the distribution function in the resonance region. The role of the distribution of particles in the resonant region is unsurprising: by virtue of the shuffling property, the distribution can only change through the interchange of particles in equal volume elements of phase-space. Thus, particles gaining kinetic energy must be replaced by others losing kinetic energy. When there are more particles in volume elements with low kinetic energy than in those with high kinetic energy, an interchange can only take place if energy is supplied by the wave.

Turning next to the bulk contribution, we insert (6.25) into the energy density integral to find

$$\mathcal{W}_{k,b} = -\frac{e^2}{2m_e}|\hat{E}_k(t)|^2 \int du \, \frac{u f_0'(u)}{(\omega - ku)^2} = \frac{\epsilon_0}{2}|\hat{E}_k(t)|^2 \left( \omega \frac{\partial K}{\partial \omega} - 1 \right). \quad (6.50)$$

Substituting the two parts of the energy density contained in the particles into (6.47)-(6.48), we recover Landau's result (6.42).

Equation (6.49) shows that the local dependence of the Landau damping rate on $f_0'(v_{ph})$ is an artifact of the long-time asymptotic limit. For finite time, the dominant energy exchange is between the wave (supported by the bulk particles) and the resonant particles with velocities $|u - v_{ph}| < \pi/kt$ lying in the central zone of the diffraction pattern described by $R^2$.

The sum of the energy densities of the field and of the bulk particles has a special significance:

$$\mathcal{W}_{kw} = \frac{\epsilon_0}{2}|\hat{E}_k(t)|^2 \omega \frac{\partial K}{\partial \omega}. \quad (6.51)$$

This quantity is called the wave energy. It is identical to the Vlasov energy for exact eigenmodes in homogeneous plasma. In particular, both forms of the energy vanish for unstable eigenmodes, since these may grow from a vanishingly small initial amplitude. The Vlasov energy applies to a wider class of disturbances but its application to inhomogeneous plasma is impractical at best. The wave energy, by contrast, is restricted to approximate eigenmodes but constitutes a convenient tool for the study of space- and time-varying problems. We note that the dielectric energy is considerably more general than suggested by the simple derivation presented here. A more general derivation has been described by Bernstein.[8]

An interesting feature of the wave energy is that it need not be positive. In fact, for an observer moving at a velocity $V_o$ the energy of a Plasma oscillation is

$$\mathcal{W}_k = \left( 1 - \frac{kV_o}{\omega_{pe}} \right) \epsilon_0 |E_k(t)|^2.$$

We see that the sign of the wave energy depends on the frame of reference. This simply reflects the fact that energy can be extracted from a moving plasma by exciting a wave, just as a windmill extracts energy from a breeze.

# 6.3 General solution of the Vlasov equation

In the previous section we applied the perturbation method to find solutions of the Vlasov equation describing waves similar to those predicted by the fluid model. Although these solutions are very important and constitute the most commonly encountered type, a more general method of solution is nevertheless desireable. In particular, a troubling defect of the perturbation method is its inability to address the existence of long-lived modes that lack a fluid analog. In this section we will describe a more general method, due to Landau, that provides a highly efficient and powerful description, at the cost of greater mathematical abstraction.

Landau's solution of the Vlasov equation combines the methods of Laplace transformation and analytic continuation. The Laplace transform of the distribution function, $\bar{f}_k$, is defined as the Fourier transform of the positive-time part of $f_k$:

$$\bar{f}_k(u,\omega) = \int_0^\infty dt\, e^{i\omega t} f_k(u,t). \tag{6.52}$$

The function $\bar{f}_k(u,\omega)$ is bounded whenever the imaginary part of $\omega$ is larger than the most rapid growth rate. Applying the Laplace transformation to the Vlasov and Poisson equations results in

$$(\omega - ku)\bar{f}_k(u,\omega) \;=\; if(u,0) + i\frac{e}{m_e} f_0'(u)\bar{E}_k(\omega), \tag{6.53}$$

$$ik\epsilon_0\bar{E}_k(\omega) \;=\; -e\int dv\, \bar{f}_k(v,\omega). \tag{6.54}$$

These equations are equivalent to an integral equation for $\bar{f}_k(u,\omega)$, as can easily be seen be eliminating $\bar{E}_k$. Solving instead (6.53) for $\bar{f}_k$ yields

$$\bar{f}_k(u,\omega) = \frac{if(u,0)}{\omega - ku} + i\frac{e}{m_e}\frac{f_0'(u)}{\omega - ku}\bar{E}_k(\omega). \tag{6.55}$$

The perturbed distribution function consists of two parts: the first part represents the evolution of the initial perturbation $f(u,0)$ in the absence of electric field: in linear theory, this evolution is determined by the free streaming of the electrons along their unperturbed trajectories. The second part represents the response of the background electrons $f_0(u)$ to the electric field perturbation. Both parts of the distribution function have a singularity at $u = \omega/k$, where the particle velocity matches the phase velocity of the wave. This singularity is a signature of the wave-particle resonance.

We next eliminate $\bar{f}_k$ from the equation for the electric field by substituting (6.55) in Poisson's equation (6.54). There follows

$$\bar{E}_k(\omega) = -\frac{e}{k\epsilon_0 K(\omega,k)} \int du\, \frac{f(u,0)}{\omega - ku}, \tag{6.56}$$

where

$$K(\omega,k) \equiv 1 + \frac{\omega_{pe}^2}{kn_0} \int du\, \frac{f_0'}{\omega - ku} = 0, \tag{6.57}$$

The function $K(\omega,k)$ is clearly an extension of the bulk dielectric constant that accounts for tail and resonant particles. It is an analytic function in the upper and lower half of the frequency plane, but suffers a discontinuity on the real axis

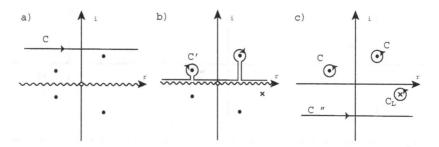

Figure 6.3: Sketch of the Laplace inversion path and the modified integration paths in the complex frequency plane. The diagrams correspond to a distribution that vanishes outside the interval $(-v_{max}, v_{max})$ and such that $f_0'(0) = 0$. They show a) the original Laplace inversion path $C$ lying above all the poles of $\bar{E}_k(\omega)$; b) the modified contours used to evaluate the Laplace inversion integral in the long-time asymptotic limit; c) same as b) but with $K(\omega, k)$ replaced by its analytic continuation $\kappa(\omega, k)$, allowing the contour to be moved to $C''$ and uncovering the Landau pole $\omega_L$. We have also removed the canceling segments, leaving the two contours $C_1$ and $C_2$ corresponding to the eigenmodes and the contour $C_L$ corresponding to the Landau pole.

caused by the wave-particle resonance. This discontinuity may be evaluated with Plemelj's formula (6.35). It is

$$K(\omega \pm i0) = 1 + \frac{\omega_{pe}^2}{kn_0} P \int du \, \frac{f_0'(u)}{\omega - ku} \mp \pi i \frac{\omega_{pe}^2}{k^2 n_0} f_0' \left( \frac{\omega}{k} \right). \tag{6.58}$$

We next determine the evolution of the electric field by inverting the Laplace transform so as to obtain $E_k(t)$ from (6.56).

## Time evolution

The evolution of the field amplitude $E_k(t)$ is given by the inverse Laplace transform of $\bar{E}_k(\omega)$,

$$E_k(t) = \frac{1}{2\pi} \int_C d\omega \, e^{-i\omega t} \bar{E}_k(\omega), \tag{6.59}$$

where the integrations path $C$ lies above all the singularities of $\bar{E}_k$. It is generally both impractical and unnecessary to evaluate the Laplace inversion integral (6.59) exactly. Since we are interested in coherent modes characterized by long-lived electric perturbations, we are content to evaluate the asymptotic form of the solution for long times. We may do this as follows.

Recall that the value of an integral in the complex plane is unaffected by a change of integration path, provided that the closed contour formed by joining the original and the new integration paths does not contain any singularities (poles or branch points) of the integrand. Inspection of (6.56) shows that for smooth initial conditions, the singularities of $\bar{E}_k(\omega)$ are of two types. The first type are poles corresponding to the roots $\omega_j$ of the dispersion relation, $K(\omega_j, k) = 0$. These are the unstable and damped eigenmmodes already discussed in Sec. 6.2. The second type of singularity corresponds to the discon-

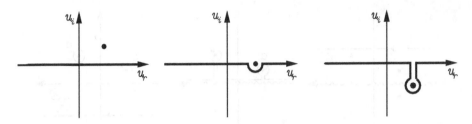

Figure 6.4: Integration paths in the complex velocity plane defining the analytic continuation of the dielectric: (a) original path along the real axis for $\text{Im}(\omega) > 0$. (b) Vlasov's integration path for $\text{Im}(\omega) = 0$ and (c) for $\text{Im}(\omega) < 0$.

tinuity of the dielectric $K(\omega, k)$ on the real axis caused by the wave-particle resonance.

The discontinuity of $K(\omega, k)$ prevents us from moving the Laplace inversion contour below the real axis in order to evaluate the asymptotic behavior of the electric field. The solution proposed by Landau is to eliminate the discontinuity by *analytic continuation* of the function $K(\omega, k)$. Landau begins by noting that the Laplace inversion integral is inchanged if $K(\omega, k)$ is replaced by its analytic continuation. Next, observing that the discontinuity is caused by the pole in the integrand of $K$ crossing the integration path, Landau found the desired analytic continuation by deforming the velocity integration contour so as to avoid the crossing by the pole. Note that the integration contour in the $u$ plane used to determine $K(\omega, k)$ should not be confused with the Laplace inversion contour in the $\omega$ plane. The contour shown in Fig. 6.4, called the Landau contour, is the most convenient choice for evaluating the integral explicitly. We will denote the analytic continuation of $K(\omega, k)$ to the lower half complex $\omega$ plane by $\kappa(\omega, k)$. From Fig. 6.4, we find

$$\begin{aligned} \kappa(\omega, k) &= K(\omega, k), & \text{Im}(\omega) > 0; \\ &= K(\omega, k) - 2\pi i \frac{\omega_{pe}^2}{k^2 n_0} f_0'\left(\frac{\omega}{k}\right), & \text{Im}(\omega) < 0. \end{aligned} \tag{6.60}$$

It is easy to see from (6.58) that $\kappa$ is continuous on the real axis.

Having adjusted the velocity integration contour, we may now resume pushing the contour downward, below the real $\omega$ axis. The long-time behavior of $E_k(t)$ is dominated by the root of $\kappa(\omega, k)$ with the largest $\text{Im}(\omega)$. If $K$ has an approximate real root $\omega = \omega_0$ such that $\text{Re}[K(\omega_0, k)] = 0$, we may determine the exact root of its analytic continuation $\kappa$ by expansion:

$$\kappa(\omega_0 + \delta\omega, k) = K(\omega_0 + i0, k) + \delta\omega \left.\frac{\partial K}{\partial \omega}\right|_{\omega_0} = 0. \tag{6.61}$$

There follows

$$\delta\omega = -\frac{K(\omega_0 + i0, k)}{\partial K/\partial \omega|_{\omega_0}}. \tag{6.62}$$

Substituting the form of $K$ for the Langmuir wave found in Sec. 6.2, one easily recovers the Landau damping rate (6.42). We emphasize that $\omega = \omega_0 + \delta\omega$ is a root of $\kappa(\omega, k) = 0$: although $\delta\omega$ is evaluated by extrapolating the function $K$, it is only a root of $K$ itself if $f_0'(v_{ph} > 0$.

The long-time evolution of the distribution function $f_k(u, t)$ can be determined in a way similar to that of $E_k(t)$. One thus finds that the residue of the pole at $\omega = ku$ caused by the wave-particle resonance accounts for the ballistic response (6.23), while the residue of the Landau pole gives rise to the bulk response (6.25) found earlier with the perturbation theory. Landau's method is clearly more general than the perturbation theory, being free of any assumption concerning the closeness of the solution to a mode described by fluid theory. Its most important advantage, however, is to bring the full arsenal of complex analysis to bear upon the problem of Vlasov dynamics. In particular, complex analysis provides convenient methods for investigating the existence of roots of the dispersion function in a given region of the $\omega$ plane. We describe these methods next.

## The Nyquist theorems

In the previous sections, we found that eigenmodes correspond to roots of $K(\omega) = 0$, or

$$\frac{\omega_{pe}^2}{n_0 k^2} \int du \, \frac{f_0'(u)}{u - \omega/k} = 1. \tag{6.63}$$

Here we derive a practical criterion for determining whether (6.63) has any roots in the upper half of the complex plane, corresponding to instabilities. A pair of theorems due to Nyquist form the basis of a powerful, general method for determining the existence of such roots. We begin by presenting these theorems.

The first theorem of Nyquist states that the number $N$ of zeroes of an analytic function $w(z)$ lying within a closed contour $C$ is given by

$$N = \frac{1}{2\pi i} \oint_C dz \, \frac{w'(z)}{w(z)}, \tag{6.64}$$

where $w'(z) = dw/dz$. Note that since $w$ is assumed to be analytic, its zeroes are the only singularities in the integrand. The first Nyquist theorem, (6.64), is thus a direct consequence of the residue theorem. The evaluation of the integral in (6.64), however, is impractical in the best of circumstances. Fortunately, Nyquist's second theorem makes this unnecessary.

The second theorem of Nyquist is concerned with the image of the contour $C$ in the $w$ plane under the mapping defined by the function $w(z)$. It states that the number of times that this image contour surrounds the origin of the $w$ plane is equal to the number of zeroes of $w(z)$. This follows by changing variables in (6.64),

$$\frac{1}{2\pi i} \oint_C dz \, \frac{w'(z)}{w(z)} = \frac{1}{2\pi i} \oint_{w(C)} \frac{dw}{w}, \tag{6.65}$$

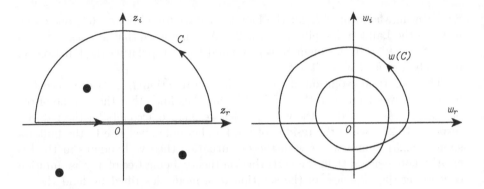

Figure 6.5: Contour $C$ used to determine the number of zeroes of a function $w(z)$ in the upper half of the complex z plane, and its image $w(C)$ in the $w$ plane. The number of times the image $w(C)$ surrounds the origin (2 in this example) is equal to the number of zeroes of $w$ enclosed by $C$.

where $w(C)$ denotes the image of $C$. The interpretation of the right hand side of (6.65) as the number of times the contour $w(C)$ surrounds the origin follows again from the residue theorem. The usefulness of the second Nyquist theorem follows from the fact that it is often possible to determine how many times a contour surrounds the origin from topological arguments, knowing only a few points of the contour. This is the case, in particular, for the electrostatic eigenmode problem.

## Penrose criterion

We now apply the Nyquist theorems to determine the existence of exponential instabilities for an equilibrium plasma.[58] We consider the contour $C_N$ formed by a straight line lying immediately above the real frequency axis and closed by a semi-circle at radius $\Omega \to \infty$. The number of zeroes of $K(\omega)$ in the upper half plane is then given by the number of times the image of $C_N$ surrounds the origin in the $K$ plane.

To determine this number, we first consider the image of the semi-circle $\omega = \Omega e^{i\phi}$,

$$K(\Omega e^{i\phi}) \sim 1 - \frac{\omega_{pe}^2}{\Omega^2} e^{-2i\phi}.$$

The semi-circle is mapped onto the point $K = 1 \pm i0$ as $\Omega \to \pm\infty$. We next consider the image of a line lying on top of the real axis and closing the semi-circle. The image of the line immediately above the real axis is given by (6.58): this shows that the image of the real $\omega$ axis will cross the real $K$ axis whenever $f_0'(\omega/k) = 0$, that is, at every extremum of the distribution function. Furthermore we know that the crossings will be in the downward direction at the maxima of $f_0$ and in the upward direction at the minima. The only remaining information needed to determine the wrapping number is the location of the crossings.

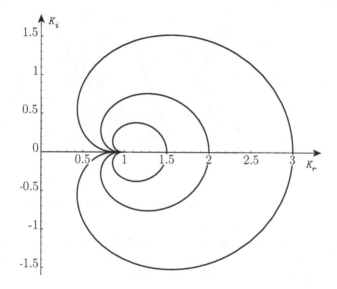

Figure 6.6: Nyquist diagrams for a Maxwellian distribution, for several values of wavenumber $k$, illustrating the scaling property of the Nyquist contours.

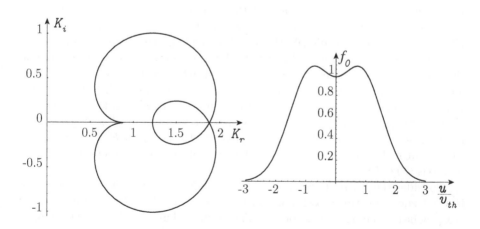

Figure 6.7: Nyquist diagram for a stable two-stream distribution.

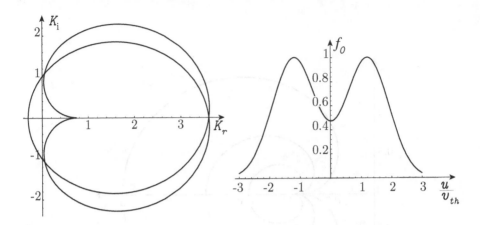

Figure 6.8: Nyquist diagrams for an unstable two-stream distribution.

The image of the real-$\omega$ axis crosses the real $K$ axis at $K(ku_j)$, where the $u_j$ represent the extrema of $f_0$. For $u = u_j$ a more perspicuous form of the principal value integral is obtained by integrating by parts, using

$$f_0'(u) = \frac{d}{du} (f_0(u) - f_0(u_j))$$

to obtain

$$K(ku_j) = 1 - \frac{\omega_{pe}^2}{n_0 k^2} \int du \, \frac{f_0(u) - f_0(u_j)}{(u - u_j)^2} \qquad (6.66)$$

It follows from (6.66) that if $f_0$ has only one maximum, $u_{\max}$, then $\text{Re}[K(ku_{\max})] > 1$ and the image of the Nyquist contour does not surround the origin. This is an explicit proof, within the context of linear theory, of Gardner's theorem that distributions with a single maximum are stable.

In the general case where the distribution function has one or more minima, the stability analysis relies in a crucial way on the fact that the Nyquist contours for different $k$ are copies of each other scaled about the point $K = 1$ (Fig. 6.6). If all the crossings of the real axis occur to the right of $K = 1$, then clearly the Nyquist winding number is zero and the system is stable for all $k$ (Fig. 6.7). If, however, a Nyquist contour crosses the real axis to the left of $K = 1$ for some $k$, then a range of $k$ can be found for which the Nyquist contours cross the negative real axis (Fig. 6.8).

Consider the leftmost crossing $\omega_l$ such that $K(\omega_l) < 1$, and choose the scale factor $k$ such that this crossing lies just to the left of the origin, while the next crossing lies to the right of the origin. The origin is thus encircled once (Fig. 6.8). If the leftmost crossing corresponded to a maximum of the distribution function, the Nyquist winding number would be $-1$ indicating a pole in $K(\omega)$. We know that this cannot be, and so conclude that if $K(\omega_l) < 1$, the leftmost crossing occurs at a minimum of $f_0$ and the system is unstable.

The results of the foregoing analysis form the

> **Penrose criterion**[58]: *An equilibrium distribution $f_0(u)$ has unstable eigenmodes if and only if it has a minimum $u_{min}$ such that*

$$\int du \, \frac{f_0(u) - f_0(u_{min})}{(u - u_{min})^2} > 0. \tag{6.67}$$

Interestingly, the Penrose criterion shows that a distribution function consisting of a pair of warm beams may be stable, even though the shuffling property allows the energy of such a distribution to be reduced by rearrangement of the "columns." The two beams must in fact be sufficiently well resolved in order for the two-stream instability to be excited, as illustrated in Figs. 6.7-6.8.

## 6.4 The plasma as a dielectric

In many applications, a great economy of argument is possible if the plasma can be treated as a conducting medium without concern for the details of the microscopic dynamics. This is the case, in particular, for all problems involving the coupling of a plasma to an external source by means of an antenna. In such problems it is useful to distinguish between the "free" charge $\rho_f$ in the antenna, that can be manipulated at will by the physicist, and the "bound" plasma charge $\rho_{bd}$, that must obey the Vlasov equation. The bound charge induces a *polarization* $\mathbf{P}$ given by

$$\nabla \cdot \mathbf{P} = i\mathbf{k} \cdot \mathbf{P} = -\rho_{bd}$$

In linear theory, the bound charge perturbation is proportional to the total electric field:

$$\mathbf{P} = \epsilon_0 \chi \mathbf{E}$$

The constant of proportionality $\chi$, called the dielectric susceptibility, is one of the most important quantities calculated by the Vlasov theory. The object of this section is to derive this constant for electrostatic oscillations in unmagnetized plasma and to describe its properties.

In the presence of an antenna exciting charge oscillations with wavelength $k$, the Laplace transform of Poisson's equation is

$$ik\epsilon_0 \bar{E}(\omega) = \bar{\rho}_{ext}(\omega) - e \int dv \, \bar{f}(v, \omega). \tag{6.68}$$

where $\bar{\rho}_{ext}(\omega)$ is the externally driven charge and the perturbed distribution function $\bar{f}$ is given, as before, by the Vlasov equation. The bound charge density can be evaluated from the solution (6.55) of Vlasov's equation. Assuming $f(u, 0) = 0$, we find

$$\bar{\rho}_{bd}(\omega) = i\epsilon_0 \frac{\omega_{pe}^2}{n_0} \int du \, \frac{f_0'(u)}{\omega - ku} \bar{E}(\omega) \tag{6.69}$$

The susceptibility $\chi$ is thus

$$\chi = \frac{\omega_{pe}^2}{n_0 k} \int du \, \frac{f_0'(u)}{\omega - ku} \tag{6.70}$$

The susceptibility measures the ability of the plasma to shield external (free) charges. Its most important property is that it is additive. That is, the susceptibility of a multi-component plasma is given by

$$\chi = \sum_s \chi_s.$$

This property may be used to obtain a simpler derivation of (6.70) by summing the susceptibility for a cold beam, $\chi_{\text{beam}} = -\omega_{pe}^2/(\omega - ku)^2$, over a continuous distribution of such beams $f_0(u)$:

$$\chi = -\frac{\omega_{pe}^2}{n_0} \int du \frac{f_0(u)}{(\omega - ku)^2}$$

Integration by parts immediately yields (6.70). Substituting the perturbed charge into Poisson's equation, we find the electric field

$$ik\epsilon(\omega)\bar{E}(\omega) = \rho_f(\omega) \tag{6.71}$$

where

$$\epsilon(\omega) \equiv \epsilon_0 \kappa(\omega) = \epsilon_0 \left[ 1 + \sum_s \chi_s(\omega) \right]. \tag{6.72}$$

The function $\epsilon(\omega)$ is the dielectric permittivity of the plasma. The above derivation confirms our previous identification of $\kappa$ with the dielectric constant or relative permittivity.

## The dielectric for a Maxwellian plasma

The Maxwellian plasma is of such importance as to merit special consideration. For a Maxwellian distribution the dielectric susceptibility takes the form

$$\chi(\omega, k) = \frac{2}{\sqrt{\pi}} \frac{\omega_p^2}{k^2 v_t^2} \int_{-\infty}^{\infty} ds \frac{se^{-s^2}}{s - \omega/kv_t} \tag{6.73}$$

It is conventional in the plasma physics literature to express the dielectric susceptibility in terms of the function

$$Z(\zeta) = \frac{1}{\sqrt{\pi}} \int_{-\infty}^{\infty} ds \frac{e^{-s^2}}{s - \zeta}, \qquad \text{Im}(\zeta) > 0, \tag{6.74}$$

called the plasma dispersion function. On the real axis and in the lower half plane, this function is defined by analytic continuation:

$$Z(\zeta) = \begin{array}{ll} i\sqrt{\pi}e^{-\zeta^2} + \frac{1}{\sqrt{\pi}}P\int_{-\infty}^{\infty} ds \frac{e^{-s^2}}{s-\zeta}, & \text{Im}(\zeta) = 0; \\ 2i\sqrt{\pi}e^{-\zeta^2} + \frac{1}{\sqrt{\pi}}\int_{-\infty}^{\infty} ds \frac{e^{-s^2}}{s-\zeta}, & \text{Im}(\zeta) < 0. \end{array}$$

The susceptibility is given in terms of $Z$ by

$$\chi(\omega, k) = 2\frac{\omega_p^2}{k^2 v_t^2}\left[1 + \frac{\omega}{kv_t}Z\left(\frac{\omega}{kv_t}\right)\right]. \tag{6.75}$$

The introduction of the symbol $Z$ is unfortunate, since $Z$ is simply proportional to the special function $w(\zeta)$,

$$Z(\zeta) = i\sqrt{\pi}w(\zeta)$$

the properties of which are described in standard reference works. Note also that

$$w(\zeta) = e^{-\zeta^2}\text{erfc}(-i\zeta),$$

where erfc is the complementary error function. It follows that

$$Z(0) = i\sqrt{\pi}.$$

An important property of $Z$ is that

$$\frac{dZ}{d\zeta} = -2[1 + \zeta Z(\zeta)].$$

This may be used to write the susceptibility as

$$\chi(\omega, k) = -\zeta_p^2\left(\frac{dZ}{d\zeta}\right)_{\zeta=\omega/kv_t} \tag{6.76}$$

where $\zeta_p = \omega_p^2/k^2 v_t^2$. The real and imaginary parts of $\chi/\zeta_p^2$ are shown in Fig. 6.9.

The asymptotic properties of $Z$ follow from the large argument expansion of $\kappa$ found in our derivation of the Bohm Gross dispersion relation,

$$Z(\zeta) \sim -\zeta^{-1} - \frac{1}{2}\zeta^{-3} - \frac{3}{4}\zeta^{-5} + O(\zeta^{-7}). \tag{6.77}$$

Rational approximations of $Z$ that reproduce the small-$\zeta$ as well as the large-$\zeta$ expansions can be very useful in practical applications.

We next apply the above results to the propagation of ion waves in an unmagnetized plasma.

## Ion-Acoustic waves

In Sec. 6.2, we considered only high-frequency waves for which the ion response was negligible. Here we consider the role of the ion response. In order to allow the ions to play a role while keeping ion Landau-damping small, we assume that $T_i \ll T_e$ so that there is a band of frequency where $kv_{ti} \ll \omega \ll kv_{te}$.

Using (6.73) to evaluate the electron susceptibility and (6.77) for the ion susceptibility, the dispersion relation takes the form

$$K(\omega, k) = 1 - \frac{\omega_{pi}^2}{\omega^2} + \frac{1}{k^2\lambda_{de}^2} + i\sqrt{\frac{\pi}{2}}\frac{\omega_{pi}^2\omega}{k^3 c_s^3}\left[\left(\frac{T_e}{T_i}\right)^{3/2}e^{-\omega^2/k^2 v_{ti}^2} + \sqrt{\frac{m_e}{m_i}}\right]$$

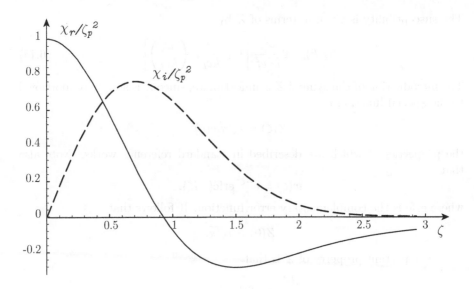

Figure 6.9: Real (solid line) and imaginary (dashed line) parts of the plasma susceptibility for real values of the frequency.

Assuming that the damping is small, a perturbative solution of this dispersion relation yields the mode frequency

$$\omega_r^2 = \frac{k^2 c_s^2}{1 + k^2 \lambda_{de}^2} \tag{6.78}$$

and the damping rate

$$\gamma = -\frac{\omega_r \sqrt{\pi/8}}{(1 + k^2 \lambda_{de}^2)^{3/2}} \left[ \left( \frac{T_e}{T_i} \right)^{3/2} \exp\left( -\frac{T_e/T_i}{2(1 + k^2 \lambda_{de}^2)} \right) + \sqrt{\frac{m_e}{m_i}} \right]. \tag{6.79}$$

The corresponding wave is called the ion-acoustic wave.

The ion-acoustic wave is the Vlasov analog of the sound wave found in fluid models, but its propagation speed $c_s$ differs from the fluid result

$$C_{s,\text{fluid}} = \sqrt{\frac{\gamma p}{n m_i}} \simeq \sqrt{\gamma}\, c_s$$

through the presence of the adiabatic index $\gamma = 5/3$ in the fluid sound speed. This difference is the result of the unjustified neglect of heat fluxes in the fluid model. More importantly, the Vlasov theory shows that the ion-acoustic wave is severely damped outside the frequency range $k v_{ti} \ll \omega \ll k v_{te}$, so that this wave only propagates in plasmas with $T_i \ll T_e$.

## 6.5 Case-Van Kampen modes

In quantum mechanics it is common practice to view the Hamiltonian operator as an infinite-dimensional matrix. A given problem is considered solved once its Hamiltonian has been "diagonalized," or represented in terms of a complete collection of eigenstates as

$$H = \sum_\nu E_\nu |\psi_\nu\rangle\langle\psi_\nu|. \tag{6.80}$$

Here $E_\nu$ and $|\psi_\nu\rangle$ are respectively the eigenvalues and eigenvectors of the Hamiltonian,

$$H|\psi_\nu\rangle = E_\nu|\psi_\nu\rangle.$$

The index $\nu$ describing the spectrum generally ranges over both discrete values, corresponding to bound states, and a continuum, describing scattering states. One advantage of this representation is that it provides an immediate solution of the initial value problem in the form

$$|\psi(t)\rangle = \sum_\nu |\psi_\nu\rangle\langle\psi_\nu|\psi(0)\rangle e^{-2\pi i E_\nu t/h}, \tag{6.81}$$

where $h$ is Planck's constant. The representation (6.80) is also useful in perturbation theory.

In order to develop a similar description for the Vlasov equation, we rewrite it in a form similar to Schrödinger's equation:

$$i\frac{\partial f_k}{\partial t} = V f_k, \tag{6.82}$$

where the Vlasov operator $V$ is defined by

$$V f_k = -ku f_k(u,t) + \frac{\omega_{pe}^2}{n_0 k} f_0'(u) \int dv\, f_k(v,t). \tag{6.83}$$

The Vlasov operator $V$, unlike the Hamiltonian, is asymmetric:

$$\langle h_k|V|g_k\rangle - \langle g_k|V|h_k\rangle = \frac{\omega_{pe}^2}{k} \iint du\, dv\, (f_0'(u) - f_0'(v))h_k(u)g_k(v) \neq 0.$$

Remarkably, a complete set of eigenfunctions can nevertheless be found. The discrete states are simply the eigenmodes found in (6.25),

$$|\hat{f}_{k,j}\rangle = \frac{\omega_{pe}^2}{n_0 k} \frac{f_0'(u)}{\omega_{k,j} - ku},$$

$$\langle\hat{f}_{k,j}| = \frac{1}{\omega_{k,j} - ku}, \tag{6.84}$$

where the index $j$ labels the roots of $K(\omega) = 0$. Note that the left eigenfunctions differ from the right eigenfunctions due to the asymmetry of $V$.

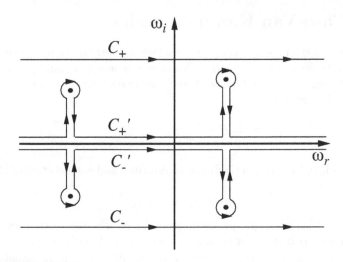

Figure 6.10: Integration paths for the inverse Laplace transform for positive ($C_+$) and negative ($C_-$) time, and modified paths $C'_\pm$ surrounding the singularities.

The modes corresponding to the continuous spectrum are called the Case-Van Kampen (CVK) modes. These modes may be obtained by rearranging the Laplace transform solution,

$$f_k(u,t) = \frac{i}{2\pi} \int d\omega \, \frac{e^{-i\omega t}}{\omega - ku} \left\{ f_k(u,0) - \frac{\omega_{pe}^2 f_0'}{kK(\omega)} \int dv \, \frac{f_k(v,0)}{\omega - kv} \right\}. \tag{6.85}$$

so as to bring it into the form (6.81). The solution for positive time corresponds to calculating the integral in (6.85) along the contour $C_+$ lying above all singularities. The solution for negative time, by contrast, is obtained by integrating the Laplace transform along the contour $C_-$ obtained by rotating the positive-time contour by $\pi$. The complete solution is thus obained simply by extending the integration contour in (6.85) to encircle all the singularities of the integrand, as shown in Fig. 6.10. We omit the algebra and simply list the results,

$$|f_{k,\omega}\rangle = \left( P \frac{f_0'(u)}{\omega - ku} + K_r(\omega)\delta(u - \omega/k) \right), \tag{6.86}$$

$$\langle f_{k,\omega}| = \left( P \frac{1}{\omega - ku} + \frac{K_r(\omega)}{f_0'(u)}\delta(u - \omega/k) \right), \tag{6.87}$$

where $K_r(\omega) = \mathrm{Re}[K(\omega)]$.

It is often stated that the CVK modes are unphysical due to their $\delta$-function singularity. To place this statement into the proper perspective, we may compare CVK modes to plane waves which have infinite extent and energy. Physically acceptable states may be constructed as linear superposition of plane waves as well as CVK modes, and both may be viewed as limits or idealizations of actual waves (in particular, we will see below that CVK modes are obtained as the small-amplitude limit of the nonlinear BGK modes).

The Case-van Kampen basis is principally used for addressing fundamental theoretical questions regarding Vlasov dynamics. For example, it can be applied to show that the perturbed distribution function is completely determined for all times by the knowledge of the electric field spectrum $E_k(\omega)$. It can further be used to show that the power density $|E_k(\omega)|^2$ and phase of the electric field constitute canonical action-angle variables for the linearized Vlasov system.

## 6.6 Waves in a magnetized plasma

Magnetization complicates the plasma response to electromagnetic perturbations in several ways. The most salient effects are the replacement of the straight-line trajectories of the unperturbed particle motion by spirals, and the consequent replacement of the isotropic dielectric permittivity by a tensor.

The dielectric tensor for a magnetized plasma, or more conveniently the conductivity tensor, can be obtained by a straightforward application of the method of characteristics presented in Sec. 6.1. For simplicity we consider an isotropic distribution, $f_0(\mathbf{v}) = f_0(v)$ where $v = |\mathbf{v}|$, and assume that this distribution is unperturbed at $t = -\infty$. Equation (6.8) then takes the form

$$f(\mathbf{x}, \mathbf{v}, t) = f_0(v_0(\mathbf{x}, \mathbf{v}, t)) \tag{6.88}$$

The initial velocity $v_0$ can be evaluated from the change in kinetic energy

$$\frac{m}{2} v^2 = \frac{m}{2} v_0^2 + e \int_{-\infty}^{t} dt' \, \mathbf{v}' \cdot \mathbf{E}(x', t').$$

where $(x', v')$ represents the position at time $t' < t$ of the particle that goes through the point $(x, v)$ at time $t$. Linearization yields

$$\Delta v = v_0(\mathbf{x}, \mathbf{v}, t) - v = -\frac{e}{mv} \int_{-\infty}^{t} dt' \, \mathbf{V}' \cdot \mathbf{E}(\mathbf{X}', t'), \tag{6.89}$$

where $(\mathbf{X}', \mathbf{V}')$ represents the particle coordinates along the unperturbed trajectory. This is merely the statement that to linear order, the change in energy is given by the work done by the perturbation along the unperturbed orbit.

Using the property that the parallel and perpendicular velocities are conserved by the unperturbed motion, the trajectories are

$$\mathbf{V}' = v_\parallel \hat{\mathbf{z}} + \mathbf{V}'_\perp, \tag{6.90}$$
$$\mathbf{X}' = \mathbf{x} + v_\parallel \Delta t \, \hat{\mathbf{z}} + \boldsymbol{\rho}' - \boldsymbol{\rho}, \tag{6.91}$$

where $\Delta t = t' - t$. It is convenient to express the perpendicular variables $\mathbf{V}'_\perp$ and $\boldsymbol{\rho}'$ in terms of the rotation eigenvectors $\hat{\mathbf{e}}_\sigma$ defined in (4.20)-(4.21). We adopt the convention that a summation should be carried out on terms containing a repeated index $\sigma$, where $\sigma$ takes the values $+1, -1$. We then have

$$\mathbf{V}'_\perp = v_\perp \hat{\mathbf{e}}_\sigma e^{i\sigma(\Omega \Delta t + \zeta)}, \tag{6.92}$$
$$\boldsymbol{\rho}' = \frac{v_\perp}{i\sigma\Omega} \hat{\mathbf{e}}_\sigma e^{i\sigma(\Omega \Delta t + \zeta)}. \tag{6.93}$$

where $\zeta$ is the gyrophase angle for $\mathbf{v}$.

For the purpose of evaluating the conductivity tensor, we assume that the electric field has the form

$$\mathbf{E}(\mathbf{x}, \mathbf{v}, t) = \hat{\mathbf{E}}_k e^{i(\mathbf{k}\cdot\mathbf{x}-\omega t)},$$

where $\text{Im}(\omega) > 0$.

Substituting the above expressions into the equation for the perturbation, we find

$$\hat{f}_{\mathbf{k}}(\mathbf{v}) = -\frac{e}{m_s}\frac{f_0'(v)}{v}\int_{-\infty}^{t} dt'\, \mathbf{V}'\cdot\hat{\mathbf{E}}_k \exp[i(k_\parallel v_\parallel - \omega)\Delta t + i\mathbf{k}_\perp\cdot(\boldsymbol{\rho}'-\boldsymbol{\rho})].$$

In order to evaluate the integral we choose the coordinate system such that $\mathbf{k}$ lies in the $(\hat{\mathbf{x}}, \hat{\mathbf{z}})$ plane, so that

$$\mathbf{k}_\perp\cdot\boldsymbol{\rho} = \frac{k_\perp v_\perp}{\Omega}\sin(\zeta+\Omega\Delta t).$$

We then use the identity

$$e^{iz\sin\theta} = \sum_{n=-\infty}^{\infty} J_n(z)e^{in\theta},$$

where the $J_n$ are Bessel functions. There follows

$$\hat{f}_{\mathbf{k}}(\mathbf{v}) = \frac{e}{m_s}\frac{f_0'(v)}{v}\sum_{\ell,n=-\infty}^{\infty}\frac{J_n(\frac{k_\perp v_\perp}{\Omega})\mathbf{Q}_\ell(\frac{k_\perp v_\perp}{\Omega})\cdot\hat{\mathbf{E}}_k e^{i(n-\ell)\zeta}}{i(\omega - k_\parallel v_\parallel - \ell\Omega)}\qquad(6.94)$$

where

$$\mathbf{Q}_\ell(k_\perp v_\perp/\Omega) = v_\parallel J_\ell(k_\perp v_\perp/\Omega)\,\hat{\mathbf{z}} + \frac{1}{2}v_\perp J_{\ell+\sigma}(k_\perp v_\perp/\Omega)\,\hat{\mathbf{e}}_\sigma.$$

The two most important effects of gyration are evident in (6.94). First, the averaging of the acceleration by the perturbed field along the particle's gyration orbit results in the attenuation of the response for particles with Larmor radii larger than the perpendicular wavelength. This effect is described by the Bessel functions. Second, the wave-particle resonance is replaced by a sequence of resonances corresponding to velocities such that the wave frequency Doppler-shifted by the *parallel* velocity is a multiple of the Larmor frequency,

$$\omega - k_\parallel u = \ell\Omega.$$

The reason the perpendicular velocity does not appear is that since the particles are tied to the field line, they are unable to drift at the phase velocity of the wave in the perpendicular direction. They nevertheless exhibit a resonant response when the frequency of the perturbation in the rest frame of the guiding center matches the Larmor frequency. At this frequency the electrons (ions) remain in phase with a right (left) circularly polarized perturbation. If the perturbation varies along the gyration orbit ($k_\perp > 0$), the modulation the perturbation amplitude at the Larmor frequency causes the particle to experience resonance at harmonics of this frequency.

We may evaluate the conductivity tensor by calculating the perturbed current corresponding to (6.94). We find

$$\hat{\mathbf{J}}_{s\mathbf{k}} = e_s \int d^3 v \, \mathbf{v} \hat{f}_{\mathbf{k}} = \boldsymbol{\sigma}_s(\omega, \mathbf{k}) \hat{\mathbf{E}}_k,$$

where

$$\boldsymbol{\sigma}_s(\omega, \mathbf{k}) = \frac{e^2}{m_s} \int \int dv_\| dv_\perp^2 \, \frac{f_0'(v)}{2v} \sum_{\ell=-\infty}^{\infty} \frac{\mathbf{Q}_\ell^*(\frac{k_\perp v_\perp}{\Omega}) \mathbf{Q}_\ell(\frac{k_\perp v_\perp}{\Omega})}{i(\omega - k_\| v_\| - \ell\Omega)} \tag{6.95}$$

A surprising aspect of this result is that Landau damping requires that the wavevector have a nonvanishing parallel component. The reason for this is that all particles of a given species have the same Larmor frequency: thus phase-mixing is inoperative unless perturbed particles with different gyration phases are carried to the same point by their parallel drifts. The corresponding damping, called cyclotron damping, differs from Landau damping in that the wave acts on the perpendicular motion of the particles: this damping can occur even in the absence of a longitudinal electrostatic perturbation. In this case, phase mixing is caused by the superposition at one spatial point of particles with different *gyration* phases.

The mode equations for the propagation of waves in a magnetized plasma follow by summing the conductivity tensors (6.95) for all the species present and substituting the resulting total conductivity tensor into (4.8)-(4.9),

$$\mathbf{M}\hat{\mathbf{E}} = \left\{ (\hat{\mathbf{k}}\hat{\mathbf{k}} - \mathbf{I})n^2 + \mathbf{I} + \sum_s \frac{i\boldsymbol{\sigma}_s}{\omega\epsilon_0} \right\} \hat{\mathbf{E}} = 0.$$

We next describe the solution of the mode equations for the cases of perpendicular and parallel propagation.

## Perpendicular propagation

For $k_\| = 0$ the parallel and perpendicular components of the conductivity tensor decouple and the tensor becomes bloc-diagonal. There is thus a solution with the electric vector polarized along the equilibrium magnetic field ($\hat{\mathbf{E}}_\perp = 0$). The corresponding dispersion relation is $M_{zz} = 0$, or

$$1 - \frac{k^2 c^2}{\omega^2} - \frac{\pi}{\omega} \sum_{s,\ell} \frac{\omega_{ps}^2}{n_0} \int_0^\infty dv_\perp^2 \int_{-\infty}^\infty dv_\| \frac{J_n^2 f_{0s}(v)}{\omega - n\Omega_s} = 0, \tag{6.96}$$

For long wavelength modes, $k_\perp v_\perp \ll \Omega$, we may solve the dispersion relation by expanding the Bessel functions for small arguments. In this limit we recover $\omega^2 = \omega_{pe}^2 + k^2 c^2$, the dispersion relation for the ordinary electromagnetic mode.

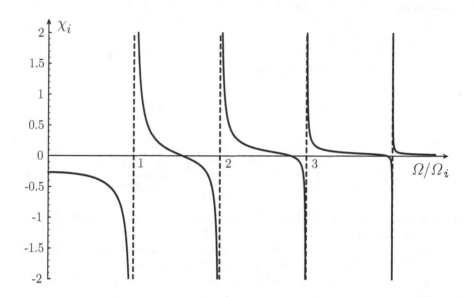

Figure 6.11: Long wavelength ion permittivity for $k_\perp \rho_i = \sqrt{2}$.

The roots of the dispersion relation (6.96) are thus called ordinary modes by extension, although unlike the ordinary mode in a cold plasma, they are affected by the presence of the magnetic field. In particular it is clear from (6.96) that it is generally possible to find roots of the dispersion relation close to the harmonics of the cyclotron frequencies, by virtue of the rapid variation of the response there. It follows that there are propagation bands near the harmonics of the cyclotron frequencies. These propagation band have no fluid equivalent.

The dispersion relation for waves with polarization perpendicular to the magnetic field can be solved in the limit $k^2 c^2 \gg \omega_{pe}^2$. In this limit $M_{xx} M_{yy} \gg M_{xy} M_{yx}$, and the mode equation is approximately diagonal. The corresponding families of modes are the extraordinary modes, electromagnetic modes with $E_x \ll E_y$, and the Bernstein modes which are electrostatic (longitudinal) modes.

We briefly describe the ion Bernstein waves. For the simple case of a Maxwellian distribution and $k_\parallel = 0$, the susceptibilities are given by

$$\chi_s = -\frac{2}{k^2 \lambda_{Ds}^2} \sum_{n=1}^{\infty} \frac{n^2 \Omega_s^2}{\omega^2 - n^2 \Omega_s^2} \Gamma_{ns}(b_s), \tag{6.97}$$

where $b_s = k_\perp^2 \rho_s^2 / 2$ and

$$\Gamma_{ns}(b_s) = I_n(b_s) e^{-b_s} = 2\pi \int_0^{\infty} dv_\perp \, v_\perp f_M'(v_\perp) J_n(k_\perp \rho_s)^2.$$

Here $I_n(b)$ is the modified Bessel function of the first kind. For the ion waves, we have $\omega \sim \Omega_i \ll \Omega_e$ and $k_\perp \rho_e \ll 1$. Hence the electron susceptibility takes

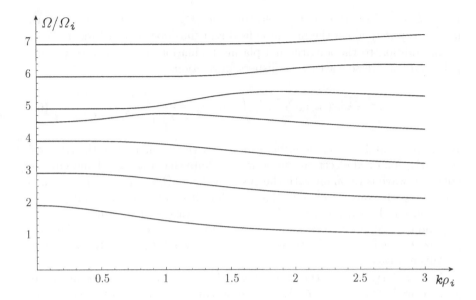

Figure 6.12: Dispersion curves for the Ion Bernstein Wave for a mass ratio $m_i/m_e = 40$.

the simplified form

$$\chi_e = \frac{\omega_{pe}^2}{\Omega_e^2}$$

The dispersion equation can then be written as

$$1 + \frac{\omega_{pe}^2}{\Omega_e^2} = -\chi_i(\omega)$$

The form of the dispersion relation $\omega = \omega(k_\perp)$ can be sketched by considering its limiting forms for $k_\perp \to \infty$ and $k \to 0$. First, for $k_\perp \to \infty$ we have $\Gamma_{ni}/k^2 \to 0$ so that

$$\lim_{k_\perp \to \infty} \omega^2 \to n^2 \Omega_i^2.$$

Second, for $k \to 0$ we note that $\Gamma_{ni} \sim k^{2n}$. The dependence of of $\chi_i$ on $\omega$ for $k_\perp \rho_i = \sqrt{2}$ is shown in Fig. 6.11. The dispersion relation is shown in Fig. 6.12. Its most important feature is that the frequency bands where ion-Bernstein waves propagate are separated by gaps. This makes it difficult to excite these waves.

## Parallel propagation

In the case of parallel propagation, $k_\perp = 0$, all the Bessel functions vanish except $J_0(0) = 1$. In this case the conductivity tensor is diagonal in the base $(\hat{e}_+, \hat{e}_-, \hat{z})$ describing circularly polarized waves. The parallel conductivity $\sigma_{zz}$ is identical to that for an unmagnetized plasma, reflecting the fact that for

$\mathbf{E_k} = E_\parallel \hat{\mathbf{z}} = E_\parallel \hat{\mathbf{k}}$, the mode is electrostatic ($\mathbf{B_k} = 0$) so that the perturbed motion is purely along the magnetic field and thus does not feel its effects.

In addition to the longitudinal plasma oscillation, there are two circularly polarized transverse waves with dispersion relations

$$\omega^2 = k^2 c^2 + \pi \omega \sum_s \omega_{ps}^2 \int dv_\perp dv_\parallel \, \frac{v_\perp^3}{v} \frac{f'_{0s}(v)}{k_\parallel v_\parallel \pm \Omega_s - \omega} \qquad (6.98)$$

Where the $+$ and $-$ signs describe respectively the left-hand and the right-hand circularly polarized electromagnetic waves. Note that the right-hand circularly polarized wave is resonant with the electrons ($\omega = |\Omega_e| + k_\parallel v_\parallel$). This resonance determines the rate of Landau damping for this wave. Reciprocally, Landau damping for the left-hand circularly polarized wave is determined by the corresponding resonance with the ions. At intermediate frequency, the right-hand polarized wave becomes the Whistler wave. At low frequency, both waves reduce to Alfvén waves.

In summary, we find that the Vlasov theory of waves in a hot magnetized plasma predicts many modes that have no equivalent in fluid theory. In some cases these modes propagate in frequency bands where fluid theory predicts that all signals should be cut off. These modes have many applications, ranging from diagnosing the conditions inside a magnetized plasma to heating the plasma and driving plasma currents.

## 6.7   Nonlinear solutions

An important nonlinear phenomenom is the trapping of particles by a finite-amplitude electrostatic potential. Implicit in the idea of trapping is that the rate of change of the electrostatic potential (in a frame moving with the wave) is slow compared to the frequency of oscillation of the particles in the well. If the evolution of $\phi$ is neglected altogether, the energy of individual particles becomes a constant of the motion. The solution of the Vlasov equation in the presence of a conserved quantity was given in Sec. 6.1. Throughout the present section we express all quantities in the rest frame of the wave. In this frame,

$$\frac{\partial f}{\partial t} = \frac{\partial \phi}{\partial t} = 0$$

The solution (6.9) is

$$f_s(x, v) = F_{s,\sigma}(E_s) \qquad (6.99)$$

where

$$E_s = m_s \frac{v^2}{2} + e_s \phi(x)$$

is the particle energy, $\sigma = v/|v|$ is the sign of the velocity and $s$ is the species index. At points $x$ such that $e_s\phi(x) = E_s$, particles with energy $E_s$ are reflected

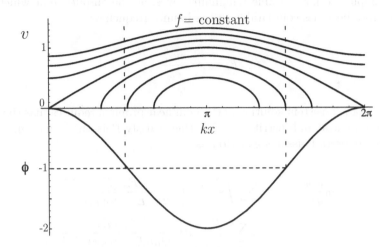

Figure 6.13: Sketch of the potential well and the corresponding contours of constant energy in phase space.

by the electrostatic potential. At these points the particle distribution must obey the so-called bounce condition,

$$F_{s,+}(E_s) = F_{s,-}(E_s).$$

where $F_{s,+}(E_s)$ and $F_{s,-}(E_s)$ are the densities of particles with energy $E_s$ and with positive and negative velocities respectively. The bounce condition expresses the fact that in steady state the number of particles flowing towards the potential barriers must equal the number of particles returning from the barrier after reflection. If two such "bounce" points are present, they give rise to trapped orbits.

The oscillation frequency for deeply trapped particles (particles near the bottom of a potential well) measures the characteristic time-scale for the motion of trapped particles in an electrostatic wave. For a sinusoidal wave, $E(x) = |E_k| \sin(kx)$, we may calculate this frequency by linearizing the equation of motion near the bottom of the potential well. For electrons, $m\ddot{x} = -ek|E_k|x$. The resulting frequency,

$$\omega_b = (ek|E_k|/m)^{1/2},$$

is called the bounce frequency. Note that particles near the the edge of the well, by contrast, take an infinite time to reach the peak so that their oscillation frequency vanishes. Hence there is a gradient in the rate of bouncing, which mixes the phases of particles at different energies. Perturbations of the distribution function will thus experience phase mixing with a characteristic time given by the bounce period $T_b = 2\pi/\omega_b$. That is, after a few bounce periods an initially nonstationary distribution will become constant on surfaces of constant energy, and the evolution will saturate. We may use this to estimate the typical saturation amplitude for unstable Langmuir waves, as the amplitude at which the growth rate becomes comparable to the bounce frequency:

$$E_{k,\text{sat}} \sim \frac{m\gamma_L^2}{ek}.$$

A general, consistent solution of the nonlinear problem requires that the electrostatic potential and distribution functions satisfy Poisson's equation. Using (6.99), we express Poisson's equation as

$$
\begin{aligned}
\epsilon_0 \frac{d^2\phi}{dx^2} \;=\; & e \sum_\sigma \int_{-e\phi(x)}^\infty dE \, \frac{F_{e\sigma}(E)}{[2m(E + e\phi(x))]^{1/2}} \\
& -e \sum_\sigma \int_{e\phi(x)}^\infty dE \, \frac{F_{i\sigma}(E)}{[2m(E - e\phi(x))]^{1/2}}
\end{aligned}
\tag{6.100}
$$

This can be used to construct exact solutions of the nonlinear Vlasov equation in one dimension.

There are two especially important applications for the solutions of the nonlinear Vlasov equation. The first concerns the problem of determining the electrostatic potential at the boundary of a plasma. Recall that quasi-neutrality is ordinarily violated in a narrow layer near the boundary, the Langmuir sheath. In Chapter 3 we analysed this sheath for the case of cold ions, $T_i \ll T_e$, using fluid theory. Here we apply the Vlasov theory to determine the effect of the thermal spread in the ion distribution function when $T_i \sim T_e$.

The second application concerns the determination of the energy distribution function required to generate a particular waveform of the electrostatic potential. The principal result is surprising: any desired $\phi(x)$ can be realized with the proper choice of the particle distribution functions. The corresponding solutions are called BGK waves after their discoverers Bernstein, Greene and Kruskal.[9]

We start by presenting the calculation of the Langmuir sheath at the surface of a plasma.

## Langmuir sheath

Poisson's equation (6.100) is solved by multiplying both sides by $d\phi/dx$ and integrating. There follows

$$\frac{\epsilon_0}{2} \left( \frac{d\phi}{dx} \right)^2 = V(\phi),
\tag{6.101}$$

where

$$V(\phi) = \int_{-\infty}^{x(\phi)} dx\, \rho(x) E(x)$$

$$= \sum_{\sigma} \int_{-e\phi}^{\infty} dE\, F_e(E, \sigma)[2(E + e\phi)/m]^{1/2}$$

$$+ \sum_{\sigma} \int_{e\phi}^{\infty} dE\, F_i(E, \sigma)[2(E - e\phi)/m]^{1/2} \qquad (6.102)$$

is the work required to establish the given charge distribution.

Since $V$ depends on $x$ only through $\phi$, it is possible to integrate (6.101),

$$x - x_0 = \int_{\phi_0}^{\phi} d\hat{\phi}\, \frac{1}{[2V(\hat{\phi})/\epsilon_0]^{1/2}}, \qquad (6.103)$$

where $\phi_0 = \phi(x_0)$. Equation (6.103) determines the sheath potential $\phi(x)$ implicitly, once $F_i$ and $F_e$ are known.

Recall that the trapped electrons must satisfy the bounce condition,

$$F_{e+}(E) = F_{e-}(E).$$

We saw that in the fluid case, all the electrons slower than $(m_i/m_e)^{1/2} v_t$ are trapped. The effect on $F_{e-}(E)$ of the capture of electrons by the wall may thus be neglected without compromising the accuracy of $V(\phi)$, and the electron distribution function may be assumed to be Maxwellian. The electron contribution to the work $V(\phi)$, the first term of (6.102), is then

$$V_e(\phi) = T_e \exp\left(\frac{-e\phi}{T_e}\right). \qquad (6.104)$$

The ion distribution function, by contrast, depends on the reflection and absorption properties of the wall as well as on collisional and ionization processes in the presheath. Here we will leave $F_i$ unspecified.

The equation for the sheath potential, (6.103), clearly requires that $V(x) > 0$. This leads to a kinetic version of the Langmuir sheath criterion, discussed in Chapter 3. In order for the electric field to vanish in the bulk plasma, we must impose the asymptotic condition $V(x) \to 0$ as $x \to -\infty$. A real solution for the electric field requires then that $dV/d\phi < 0$ for $|x| \gg \lambda_D$, since $\phi < 0$. In order for a well behaved solution to exist, it is thus necessary that

$$\left\langle \frac{1}{v^2} \right\rangle_i \leq \frac{m_i}{T_e}, \qquad (6.105)$$

according to (6.102) and (6.104). Here the brackets represent the average over the ion distribution function. Equation (6.105) is the kinetic Bohm criterion, due to Harrison and Thompson. The striking aspect of this criterion is the singular dependence on the distribution of slow ions: in fact one sees from (6.105) that the ion distribution function must vanish faster than $v$ for small $v$ in order for the criterion to be satisfied.

## BGK waves

The solution of the inverse problem, finding the distribution function know-ing the potential waveform, can be obtained by interpreting Poisson's equation(6.100) as an integral equation for the distribution of trapped parti-cles. We consider a region of space where the potential forms a well and express the term $d^2\phi/dx^2$ as a function of $\phi$,

$$\epsilon_0 d^2\phi/dx^2 = -en(e\phi).$$

Poisson's equation can then be expressed as

$$\int_{e\phi}^{e\phi_{\text{Max}}} dE \frac{F_i(E)}{[2m(E - e\phi)]^{1/2}} = n_{ti}(e\phi) \tag{6.106}$$

where we have used the bounce condition to write $F_{i+}(E) = F_{i-}(E) = F_i(E)$ for the trapped ions and where

$$\begin{aligned}
n_{ti}(e\phi) &= n(e\phi) + \sum_\sigma \int_{-e\phi(x)}^\infty dE \frac{F_{e\sigma}(E)}{[2m(E + e\phi(x))]^{1/2}} \\
&- \sum_\sigma \int_{e\phi_{\text{Max}}}^\infty dE \frac{F_{i\sigma}(E)}{[2m(E - e\phi(x))]^{1/2}} \tag{6.107}
\end{aligned}$$

is the density of trapped ions at the point $x$ where the potential takes the value $\phi$. Equation (6.106) is an Abel integral-equation for the trapped ion distribution $F_i$ which may be solved with Laplace transforms. The solution is

$$F_i(E) = \frac{\sqrt{2m}}{\pi} \int_E^{e\phi_{\text{Max}}} d\Phi \frac{dn_{ti}}{d\Phi} \frac{1}{[2m(e\Phi - E)]^{1/2}}, \qquad E < e\phi_{\text{Max}}. \tag{6.108}$$

An arbitrary waveform can now be constructed as follows. We subdivide the space into intervals separated by extrema of the potential and consider one such interval, assuming that the distribution function has already been calculated for points lying to the left of this interval. To fix ideas assume that $d\phi/dx > 0$ in the chosen interval, and let $\phi_{\text{min}}$ and $\phi_{\text{Max}}$ be the values of $\phi$ at the left and right end of the interval. All distributions but that of the trapped electrons are then determined by continuity from their value to the left of the interval. The trapped electron distribution may then be determined by inverting Abel's operator. This operation can be continued from interval to interval until the entire distribution function is determined.

It can be shown that BGK solutions reduce to CVK modes in the limit of small amplitudes. This provides a physical interpretation for the Dirac delta-function appearing in the CVK modes: the delta-function indicates the presence of a finite number of particles trapped in the infinitesimal well of a linear wave. The need for these delta-functions can be understood from the fact that the trapped particle distribution is determined by the free particle distribution and $\phi$, through the Abel inversion procedure.

# Additional reading

Landau's original paper, published in 1946, remains the best reference on Vlasov's equation and Landau damping.[48] The original paper by Vlasov is now mostly of historical interest. A physical interpretation of Landau damping based on the exchange of energy between particles and fields has been proposed by J. Dawson.[19] Van Kampen's book,[74] by contrast, stresses the phase-mixing aspect of Landau damping and shows how the damping rate can be calculated without pushing the integration contour below the real axis. The complete dielectric permittivity tensor resulting from Vlasov's equation in the presence of a magnetic field is given by Krall and Trivelpiece[43] The nonlinear saturation of the beam-plasma instability is described in a classic paper by O'Neil, Winfrey and Malmberg.[55]

# Problems

1. Give the form of the Vlasov equation that is appropriate for particles subject to dissipative (non-Hamiltonian) forces.

2. Compute the density and thus the plasma frequency $\omega_{pe}$ for a one-dimensional equilibrium distribution function $f_0(v)$ with the form of a top hat, $f_0(v) = F$ for $-v_0 < v < v_0$ and $f_0(v) = 0$ elsewhere. Express your answer in terms of $m$, $e$, $v_0$ and $F$. Calculate the dielectric and solve the dispersion relation for the eigen-frequencies of the system. Comment on the relevance of Landau damping to your result.

3. Can an equilibrium distribution function in a four-dimensional phase space be such that one of the surfaces of constant $f$ has the topology of a Klein bottle? (a Klein bottle is a closed continuous surface without an interior or exterior.)

4. The adjoint of an operator $A$ is defined by

$$\langle A^* f | g \rangle = \langle f | A g \rangle$$

Calculate the adjoint $V^*$ of the Vlasov operator, and show that the bra $\langle \hat{f}_{k,j} |$ in (6.84) is an eigen function of $V^*$.

5. Discuss the conditions that must be imposed on an initial perturbation in order that it satisfy the generalized entropy conservation property.

6. Consider the following one-dimensional equilibrium distribution describing two counter-streaming plasmas:

$$f_0(v) = F(1 + v^2)e^{-v^2/w^2}.$$

(a) Apply Gardner's theorem to find a sufficient condition for stability in terms of $F$ and $w$.

(b) Find the necessary *and* sufficient stability condition given by the Penrose criterion in terms of $F$ and $w$ and compare to Gardner's condition obtained in (a).

7. Calculate the density and the dispersion relation for the distribution function given by

$$f_0(v) = f_-(v) + f_+(v), \tag{6.109}$$

where

$$f_\pm(v) = \frac{F_0}{(v \pm v_0)^2 + v_T^2}. \tag{6.110}$$

Solve the dispersion relation, and state the stability condition. Compare your result to that given by perturbation theory, (6.42), and by the Penrose stability criterion.

8. The generalized entropy conservation property and the phase-space fluid evolution interpretation of the Vlasov equation suggests that the linearized Vlasov evolution problem be formulated in terms of the velocity of a phase space fluid element. In a two-dimensional space $(x, v)$, phase space volume conservation further implies that the phase space velocity $(\dot{x}, \dot{v})$ may be written in terms of a stream function. What is this stream function?

9. Calculate the conductivity for electrostatic perturbations in an anisotropic magnetized plasma described by the equilibrium distribution $f_0(v_\perp, v_\parallel)$. Compare your answer to (6.95).

# Chapter 7

# Binary Collisions

## 7.1 Correlations and collisions

### Ensemble average

We noted in Chapter 3 that the exact (microscopic) distribution $\mathcal{F}_s(\mathbf{x}, \mathbf{v}, t)$ for species $s$ satisfies a first order partial differential equation of deceptively simple appearance:

$$\frac{\partial \mathcal{F}_s}{\partial t} + \mathbf{v} \cdot \nabla \mathcal{F}_s + \mathbf{a}_s \cdot \frac{\partial \mathcal{F}_s}{\partial \mathbf{v}} = 0, \tag{7.1}$$

where

$$m_s \mathbf{a}_s = e_s (\mathbf{E} + \mathbf{v} \times \mathbf{B}) \tag{7.2}$$

is the Lorentz acceleration. We pointed out that (7.1) cannot be considered simple because its solution would require knowing the trajectories of every particle in the system. Furthermore, even if it could be computed, the exact tracing of all these trajectories could not be compared, directly, with finite-resolution experimental data. What is wanted is a much smoother distribution, $f_s \equiv \bar{\mathcal{F}}_s$, that has been averaged over an ensemble of similarly prepared plasma systems. One has in mind an ensemble whose subsystems differ in their microscopic initial conditions, but share the same macroscopic state—for example, the same density, energy and momentum.

The kinetic equation describing the evolution of $\bar{\mathcal{F}}_s$ differs from (7.1) because, as we also noted in Chapter 3, the ensemble average of the acceleration term is not the product of the ensemble averages of its two factors:

$$\langle \mathbf{a_s} \cdot \frac{\partial \mathcal{F}_s}{\partial \mathbf{v}} \rangle \neq \bar{\mathbf{a}}_s \cdot \frac{\partial \bar{\mathcal{F}}_s}{\partial \mathbf{v}} \tag{7.3}$$

Observe that equality would hold if the Lorentz acceleration were statistically independent of the particle distribution. Hence the difference between the two sides of (7.3) measures the effect of *correlations*—between particles and fields or, because the particle positions determine the fields, between the trajectories of individual particles. This difference is traditionally denoted by

$$C = \bar{\mathbf{a}}_s \cdot \frac{\partial \bar{\mathcal{F}}_s}{\partial \mathbf{v}} - \langle \mathbf{a}_s \cdot \frac{\partial \mathcal{F}_s}{\partial \mathbf{v}} \rangle \tag{7.4}$$

and, after it has been expressed as an operator on $f_s$, called the collision operator. In this Chapter we derive the form of the collision operator, study its relation to particle-field correlations, and display its most important properties.

## Particle discreteness

It is helpful to begin with a physical sense of the difference between $\mathcal{F}$ and $\bar{\mathcal{F}}$ (we suppress species subscripts when they are not needed). The exact distribution, measuring phase-space density on all scales down to the quantum limit, can be expressed as a combination of Dirac delta functions:

$$\mathcal{F}(\mathbf{x}, \mathbf{v}, t) = \sum_{i=1}^{N} \delta(\mathbf{x} - \mathbf{x}_i(t))\delta(\mathbf{v} - \mathbf{v}_i(t)) \tag{7.5}$$

where $(\mathbf{x}_i(t), \mathbf{v}_i(t))$ specifies the trajectory of the $ith$ particle. We can imagine measuring this so-called "microscopic" distribution by sitting at some phase-space point and counting particles. Most of the time we detect nothing: the appearance of a particle is rare and fleeting. The ensemble average replaces this pointillist distribution by a continuous *fluid* in phase-space. An observer of $\bar{\mathcal{F}}$ will detect some fraction of a particle almost anywhere, as the individual drops of plasma "rain" are replaced by an infinitely fine mist.

In order for the the mist to maintain in time its statistical relation to the rain, it must respond to more than the averaged electromagnetic field: that is the message of (7.3). It is true that the microscopic field seen by an individual particle may be dominated by the locally averaged field, due to external sources or collective effects; it is furthermore likely that the ensemble average will approximate this mean micro-field. However, it is also clear that any particle will see, at least occasionally, sharp departures from the mean field, especially when the trajectories of two particles bring them close together. Thus particle discreteness causes fluctuations about the mean field, and a particle subject to such random kicks will eventually depart, like a Brownian pollen grain, from the mean-field orbit. Note that the correlation time for such kicks can be vanishingly short; the essential correlation is between the field fluctuations and the particle distribution that is perturbed by them.

Of course discreteness is not the only possible source of electromagnetic fluctuations. Turbulent fluctuations, arising from plasma instabilities, are often more important; if the turbulence is sufficiently fine-scaled to conserve local energy and momentum, its statistical effects could be incorporated into a turbulent collision operator. We study the effects of turbulence in Chapter 9. By concentrating in this Chapter on the effects of discreteness alone, we study the *minimum* plasma noise level, present in even the most quiescent system.

## Modified binary collisions

Here and throughout this Chapter we assume that the plasma parameter is large,

$$\Lambda \gg 1$$

implying that each charged particle is affected simultaneously by the Coulomb fields of many surrounding particles. The superimposition of these random fields will vanish in the mean; it is the occasional imbalance caused by accidental close approach that has effect. Just as the near overlap of two trajectories is rare, so are interactions simultaneously involving three or more trajectories rarer still. In this sense, the dominant Coulomb-collisional process is binary, as in a neutral gas, despite the long range of the Coulomb interaction. Yet the two-body collision in a plasma is importantly affected by many-body physics; in particular, it is the Debye shielding due to many charged particles that makes the effective interaction range finite. Thus the collisions studied here are two-body events, modified by many-body effects; the desired operator should have the form anticipated in Chapter 3:

$$C_s = \sum_{s'} C_{ss'}, \tag{7.6}$$

at least in an approximate sense.

The Brownian response to a random sequence of impulses appears in the kinetic equation as velocity-space diffusion, a key part of the collision operator $C$. In fact a standard form for the collision operator consists of velocity diffusion combined with only one other term, the so-called *dynamical friction*, representing an average drag in response to many random impacts. Dynamical friction decelerates individual particles, pulling the distribution towards the velocity-space origin, while diffusion tends to spread the distribution over a larger velocity range. It is the compromise between these two competing tendencies that produces the Maxwellian distribution of collisional equilibrium.

## General features

When we speak of a *local* collision operator, we mean one that conserves particles, momentum and energy at each spatial point. As in Chapter 3, we express local conservation in terms of $C_{ss'}$, the operator describing collisional effects *on* species $s$, *due to* species $s'$. The conservation laws then have the form

$$\int d^3v\, C_{ss'} = 0, \tag{7.7}$$

$$\int d^3v\, m_s \mathbf{v} C_{ss'} = -\int d^3v\, m_{s'} \mathbf{v} C_{s's}, \tag{7.8}$$

$$\int d^3v\, \frac{1}{2} m_s v^2 C_{ss'} = -\int d^3v\, \frac{1}{2} m_{s'} v^2 C_{s's} \tag{7.9}$$

expressing respectively the conservation of particles, momentum and energy. We demand that our collision operator satisfy these laws identically, for any particle distributions: they constrain the form of the operator, not that of the distribution.

There are other properties that one expects a physically reasonable collision operator to have. It should have appropriate geometrical symmetries; for example, if there is no equilibrium field establishing a preferred direction in space, then $C$ should be rotationally symmetric. It should transform properly when viewed in a moving frame, being symmetric under Lorentz or, in the more common, non-relativistic case, Galilean transformations. Perhaps most importantly, the collision operator should satisfy some version of Boltzmann's H-theorem, making entropy increase and vanishing when acting on Maxwellian distributions.

Let us consider the last property in more detail. The conventional expression for entropy,

$$s_s = - \int d^3v\, f_s \log f_s \tag{7.10}$$

in which we use the Chapter 3 notation $f \equiv \bar{\mathcal{F}}$, agrees with (3.53) in the Maxwellian case. Simple differentiation yields the collisional rate of entropy change,

$$\left(\frac{d}{dt}\right)_c s = - \int d^3v\, [C + (\log f)C].$$

Here the species subscript is suppressed; the notation $(d/dt)_c$ refers to change due exclusively to collisions, ignoring flow, macroscopic forces and so on. Particle conservation implies that the first term on the right-hand side vanishes, so we have

$$\left(\frac{d}{dt}\right)_c s = - \int d^3v\, (\log f)C. \tag{7.11}$$

For the Boltzmann operator, one can show that this quantity is positive, vanishing only when $f_s$ for each species is Maxwellian. In particular one expects a physically sensible operator to satisfy

$$C_{ss'}(f_{Ms}, f_{Ms'}) = 0. \tag{7.12}$$

Notice that collisional effects vanish only when *both* species are Maxwellian at the same temperature and velocity: scattering by an arbitrarily distributed species need not push $f_s$ toward Maxwellian form. For example, if particles of species $s$ suffer Coulomb collisions with those of an externally driven beam, the operator $C(f_{Ms}, f_{beam})$ would not vanish, but rather drive $f_s$ away from an initially Maxwellian state.

## 7.2   Fokker-Planck scattering

The typical depiction of Brownian motion shows a jagged path, with abrupt, large-angle changes in direction. The collisional motion of charged particles in a plasma with large parameter, $\Lambda \gg 1$, is very different. Because of the long-range of the (shielded) Coulomb interaction, small-angle deflections dominate:

most trajectories are smooth, with marked changes in direction occurring only after many scatterings accumulate. A random walk in which mild changes predominate is called a *Fokker-Planck* process.

It is easy to determine, without detailed dynamics, the form of the collision operator associated with Fokker-Planck scattering. Suppose that the random force acting on a particle changes on a time scale $t_c$, its correlation time (or autocorrelation time): after an interval $t_c$ the force can be assumed statistically independent from its previous value. Then, for time intervals $\Delta t$ larger than $t_c$, the effect of collisions on the particle is described by a probability distribution, $P(\mathbf{v}; \Delta\mathbf{v}, \Delta t)$, giving the probability that a particle with velocity $\mathbf{v}$ will change its velocity to $\mathbf{v} + \Delta\mathbf{v}$ in the time interval $\Delta t$. Note that the shortness of $t_c$ allows the probability distribution to be specified without regard to the particle's past history.

We use $P$ to express the effect of collisions on the *particle* distribution, $f$, in terms of the integral

$$f(\mathbf{v}, t) = \int d^3\Delta v \, f(\mathbf{v} - \Delta\mathbf{v}, t - \Delta t) P(\mathbf{v} - \Delta\mathbf{v}; \Delta\mathbf{v}, \Delta t), \qquad (7.13)$$

stating simply that particles having velocity $\mathbf{v}$ at time $t$ had some other velocity, with designated probability, at a time $\Delta t$ earlier.

Equation (7.13) describes any (stationary) random walk. In the Fokker-Planck case we know in addition that $P(\mathbf{v}; \Delta\mathbf{v}, \Delta t)$ is a localized function of $\Delta\mathbf{v}$, rapidly decreasing as $|\Delta\mathbf{v}|$ increases. This property allows us to expand the distribution function in powers of $\Delta\mathbf{v}$. Similarly, the shortness of the correlation time allows us to choose $\Delta t$ to be small on the time scale of changes in $f$, so that a Taylor's series expansion in $\Delta t$ is also permitted. In other words we can write

$$f(\mathbf{v}, t) = \int d^3\Delta v \, [f(\mathbf{v}, t) P(\mathbf{v}; \Delta\mathbf{v}, \Delta t) - \frac{\partial f}{\partial t} \Delta t P(\mathbf{v}; \Delta\mathbf{v}, \Delta t)$$
$$- \frac{\partial P f}{\partial \mathbf{v}} \cdot \Delta\mathbf{v} + \frac{1}{2} \frac{\partial^2 P f}{\partial \mathbf{v} \partial \mathbf{v}} : \Delta\mathbf{v}\Delta\mathbf{v} - ...] \qquad (7.14)$$

Here the first term on the right cancels with the left-hand side, since

$$\int d^3\Delta\mathbf{v} \, P(\mathbf{v}; \Delta\mathbf{v}, \Delta t) = 1;$$

the partial derivative with respect to time is appropriately replaced by the collision operator, since (7.13) includes only collisional effects:

$$\frac{\partial f}{\partial t} \to C(f);$$

and, in the remaining terms, $f(\mathbf{v}, t)$ can be removed from the integral. Hence we have

$$C(f) = -\frac{\partial}{\partial \mathbf{v}} \cdot \left( \frac{\langle \Delta\mathbf{v} \rangle}{\Delta t} f \right) + \frac{1}{2} \frac{\partial^2}{\partial \mathbf{v} \partial \mathbf{v}} : \left( \frac{\langle \Delta\mathbf{v}\Delta\mathbf{v} \rangle}{\Delta t} f \right) \qquad (7.15)$$

where the angle-brackets denote expectation values relative to the distribution $P$.

The acceleration vector $\langle \Delta \mathbf{v} \rangle / \Delta t$ evidently measures the average force due to random impacts, and can be identified with the dynamical friction anticipated in the previous section. We denote this velocity-dependent force by $\mathbf{R}'$:

$$\mathbf{R}' = m \frac{\langle \Delta \mathbf{v} \rangle}{\Delta t} \tag{7.16}$$

where $m$ is the particle mass. The dynamical friction should not be confused with the friction force $\mathbf{F}$ defined by (3.28); the latter, as a moment of the collision operator, is necessarily independent of velocity. The second-rank tensor

$$\mathbf{D} = \frac{1}{2} \frac{\langle \Delta \mathbf{v} \Delta \mathbf{v} \rangle}{\Delta t} \tag{7.17}$$

is called the diffusion tensor, for obvious reasons. It is characteristic of a random walk that $\langle \Delta \mathbf{v} \Delta \mathbf{v} \rangle$ grows linearly in time. Hence (noting that $\Delta t$ is large compared to $t_c$) we expect $\mathbf{D}$ to be, like $\mathbf{R}'$, independent of $\Delta t$.

An issue of notation deserves mention here. Often the diffusion and dynamical friction coefficients are defined by the form

$$C(f) = -\frac{1}{m} \frac{\partial}{\partial \mathbf{v}} \cdot (\mathbf{R}f) + \frac{\partial}{\partial \mathbf{v}} \cdot \mathbf{D} \cdot \frac{\partial f}{\partial \mathbf{v}} \tag{7.18}$$

which differs from (7.15) in that the diffusion tensor has been extracted from one velocity derivative. A simple calculation shows that the two forms coincide if we identify

$$\mathbf{R} = \mathbf{R}' + m \frac{\partial}{\partial \mathbf{v}} \cdot \mathbf{D}$$

Equation (7.18) or (7.15) has the form anticipated in the previous section: the effect of collisions on the distribution consists of diffusion plus drag. The operator of primary interest in plasma physics is slightly more complicated in its bilinear dependence on the distribution functions of two colliding species. Thus the quantities $\mathbf{D}$ and $\mathbf{R}$ occurring in $C_{ss'}$ depend upon $f_{s'}$:

$$C_{ss'} = -\frac{1}{m_s} \frac{\partial}{\partial \mathbf{v}} \cdot [\mathbf{R}_{ss'}(f_{s'}) f_s] + \frac{\partial}{\partial \mathbf{v}} \cdot \mathbf{D}_{ss'}(f_{s'}) \cdot \frac{\partial f_s}{\partial \mathbf{v}}. \tag{7.19}$$

The collisional conservation laws constrain the coefficients of the Fokker-Planck operator. To examine these constraints, we first note that $C$ can be written as a velocity-space divergence:

$$C = -\frac{\partial}{\partial \mathbf{v}} \cdot \mathbf{\Gamma} \tag{7.20}$$

with

$$\mathbf{\Gamma} = -\mathbf{D} \cdot \frac{\partial f}{\partial \mathbf{v}} + \frac{1}{m} \mathbf{R}f.$$

The divergence form, together with Gauss' theorem in velocity-space, makes particle conservation obvious. (All these arguments assume that the distribution, along with its velocity gradient, vanishes as $v \to \infty$.) The minus sign in (7.20) appears because $C$ is on the right-hand side of the kinetic equation.

The vector $\boldsymbol{\Gamma}$ represents the velocity-space flow associated with collisions. The total force associated with $m\boldsymbol{\Gamma}$,

$$\int d^3v\, m\boldsymbol{\Gamma} = \int d^3v\, m\mathbf{v}C$$

simply measures the momentum exchange, or friction force, defined by (3.28). After making $\boldsymbol{\Gamma}$ explicit and integrating by parts, we find that momentum conservation can be expressed as

$$\int d^3v\, f_s \left( m_s \frac{\partial}{\partial \mathbf{v}} \cdot \mathbf{D}_{ss'} + \mathbf{R}_{ss'} \right) + \int d^3v\, f_{s'} \left( m_{s'} \frac{\partial}{\partial \mathbf{v}} \cdot \mathbf{D}_{s's} + \mathbf{R}_{s's} \right) = 0$$
(7.21)

This result does not require the parenthesized terms to vanish; recall that $\mathbf{D}_{ss'}$ and $\mathbf{R}_{ss'}$ depend on $f_{s'}$.

Similarly we find that energy conservation requires

$$\int d^3v\, f_s \left( \mathbf{v} \cdot \mathbf{R}_{ss'} + m_s \frac{\partial}{\partial \mathbf{v}} \cdot \mathbf{v} \cdot \mathbf{D}_{ss'} \right)$$
$$+ \int d^3v\, f_{s'} \left( \mathbf{v} \cdot \mathbf{R}_{s's} + m_{s'} \frac{\partial}{\partial \mathbf{v}} \cdot \mathbf{v} \cdot \mathbf{D}_{s's} \right) = 0. \qquad (7.22)$$

Finally we consider the requirement that collisions between two Maxwellian species, at the same temperature, have no effect:

$$C\left( f_{Ms}, f_{Ms'} \right) = 0, \qquad (7.23)$$

for $T_s = T_{s'}$. Using

$$\frac{\partial f_M}{\partial \mathbf{v}} = -\frac{2\mathbf{v}}{v_t^2} f_M$$

we find that (7.19) allows (7.23) to be expressed as

$$C\left( f_{Ms}, f_{Ms'} \right) = \frac{1}{T} f_{Ms} \left( \mathbf{v} - \frac{T}{m_s} \frac{\partial}{\partial \mathbf{v}} \right) \cdot \left( \mathbf{R}_{Mss'} + \frac{m_s^2}{T} \mathbf{v} \cdot \mathbf{D}_{Mss'} \right) = 0,$$

where the $M$-subscripts on $\mathbf{R}$ and $\mathbf{D}$ indicate their evaluation on the Maxwellian $f_{Ms'}$ and $T$ is the common temperature of the two species. This first order differential equation has a nontrivial solution which, however, grows exponentially with $v^2$. The only physically acceptable solution satisfies

$$\mathbf{R}_{Mss'} + \frac{m_s^2}{T} \mathbf{v} \cdot \mathbf{D}_{Mss'} = 0, \qquad (7.24)$$

a useful characterization of the Maxwellian coefficients. Its physical content is that "radial" diffusion—diffusion away from the velocity-space origin—must balance drag in any collisional equilibrium.

Our remaining task is to compute **R** and **D** explicitly, using the Coulomb force law and appropriate statistics. A direct evaluation of (7.16) (7.17), using Newton's law in the screened potential field, is straightforward[61]. However, we find the following approach, based on a quasilinear treatment of the kinetic equation, more illuminating.

## 7.3    Balescu-Lenard operator

We refer to the difference between some physical quantity and its ensemble average as the "fluctuation" in that quantity, and distinguish it with a tilde. Thus the fluctuation in the distribution function is

$$\tilde{\mathcal{F}} \equiv \mathcal{F} - \bar{\mathcal{F}} = \mathcal{F} - f, \tag{7.25}$$

In a turbulent plasma, fluctuations—perhaps corresponding to turbulent eddies or vortices—are driven by plasma instability. When such processes are sufficiently stochastic and fine-scaled, they can be treated statistically, by coarse graining. The nonlinear result is usually irreversible and dissipative. However, the discussion of the previous Section shows that even an entirely stable plasma must have fluctuations: particle discreteness forces $\mathcal{F}$ to differ from is ensemble average. This source of fluctuation alone, together with quasilinear theory, yields a collision operator first found independently by Balescu and Lenard. The Balescu-Lenard operator is sufficiently complicated to be rarely used in its exact form. But its derivation is instructive, and approximate versions of the operator are widely used.

Before plunging into the analysis we emphasize its basis in discreteness: all Balescu-Lenard fluctuations are rooted in the difference between the microscopic distribution of (7.5) and its smoothed average—that is, in the difference between rain and mist. The plasma is presumed stable. Thus we study the irreducible collisional effects present in even a fully quiescent plasma. (A more detailed version of the following derivation may be found in Appendix A.)

The acceleration is decomposed, like $\mathcal{F}$, into averaged and fluctuating pieces,

$$\mathbf{a} = \bar{\mathbf{a}} + \tilde{\mathbf{a}}$$

to describe the electromagnetic response to discreteness. Inserting these representations into the definition of the collision operator, (7.4), and noting that any fluctuation has vanishing mean, we see that

$$C = -\left\langle \tilde{\mathbf{a}} \cdot \frac{\partial \tilde{\mathcal{F}}}{\partial \mathbf{v}} \right\rangle. \tag{7.26}$$

Our task—the traditional goal of kinetic theory—is to express this quantity as an operator on $\bar{\mathcal{F}}$.

Consistent with the discussion of the previous Section, we assume that the fluctuating force is small in the sense that

$$\tilde{\mathbf{a}} \cdot \frac{\partial \mathcal{F}}{\partial \mathbf{v}} \ll \mathbf{v} \cdot \nabla \mathcal{F},$$

or

$$\frac{|\tilde{a}|}{k v_t^2} \ll 1,$$

where $k$ is a characteristic wavenumber (inverse wavelength) of the fluctuations. Only in this case can we expect to find an operator of Fokker-Planck form. Secondly we assume spatial and temporal scale separation: $\tilde{\mathcal{F}}(\mathbf{x}, \mathbf{v}, t)$ and $\tilde{a}(\mathbf{x}, t)$ are supposed to vary much more sharply with respect to $\mathbf{x}$ and $t$ than does $\bar{\mathcal{F}}$. Separation of spatial scales is intrinsic to the construction of a local collision operator. Thirdly, and only to control the size of the calculation, we assume that the averaged fields vanish,

$$\bar{\mathbf{a}} = 0.$$

There is no fundamental problem in allowing for nonfluctuating fields, but they severely complicate the analysis. We discuss later whether the $\bar{\mathbf{a}} = 0$ result can be applied to a magnetized plasma.

With these assumptions it is consistent to compute $\tilde{\mathcal{F}}$ and the fluctuating (Coulomb) field $\tilde{E} = -\nabla\Phi$ from the coupled, linearized equations

$$\frac{\partial \tilde{\mathcal{F}}_s}{\partial t} + \mathbf{v} \cdot \nabla \tilde{\mathcal{F}}_s = \frac{e_s}{m_s} \nabla\Phi \cdot \frac{\partial \tilde{\mathcal{F}}_s}{\partial \mathbf{v}}, \tag{7.27}$$

$$-\nabla^2 \Phi = \sum_s \frac{e_s}{\epsilon_0} \int d^3 \mathbf{v} \tilde{\mathcal{F}}_s \tag{7.28}$$

We assume the plasma to be stable, and therefore solve this linear system as an *initial value* problem, in which the perturbations $\tilde{\mathcal{F}}$ and $\Phi$ are driven by the initial condition, $\tilde{\mathcal{F}}(t = 0)$.

All that matters about the initial fluctuation is its statistics. We assume that at $t = 0$, before the discreteness-induced field has acted, the particles are uncorrelated. Then (7.5) implies that

$$\langle \tilde{\mathcal{F}}_s(\mathbf{x}, \mathbf{v}, 0) \tilde{\mathcal{F}}_{s'}(\mathbf{x}', \mathbf{v}', 0) \rangle = \delta_{ss'} \delta(\mathbf{x} - \mathbf{x}') \delta(\mathbf{v} - \mathbf{v}') \bar{\mathcal{F}}_s(\mathbf{v}). \tag{7.29}$$

Collisions are not included in the system (7.27)–(7.28). Of course accounting for collisional effects is awkward at a point in the calculation where C has not been computed, but there is also a physical excuse: the fluctuation time scale, essentially $t_c$, is small compared to the collisional time scale, on which many weak scatterings accumulate. This argument characterizes weak turbulence theory, which uses (7.27)–(7.28). Strong turbulence theory includes collisional effects in the calculation of fluctuations; as will be noted in Chapter 9, the demarcation between weak and strong turbulence occurs at surprisingly low field levels.

We solve (7.27)–(7.28) by Fourier transforms, as in Chapter 6, using the notation

$$\tilde{\mathcal{F}}(\mathbf{x}, \mathbf{v}, t) = \frac{1}{(2\pi)^4} \int d^3k \, d\omega \, e^{-i(\omega t - \mathbf{k} \cdot \mathbf{x})} \mathcal{G}(\mathbf{k}, \mathbf{v}, \omega) \tag{7.30}$$

$$\Phi(\mathbf{x}, t) = \frac{1}{(2\pi)^4} \int d^3k \, d\omega \, e^{-i(\omega t - \mathbf{k} \cdot \mathbf{x})} \varphi(\mathbf{k}, \omega). \tag{7.31}$$

One finds that

$$\mathcal{G}_s = \frac{i\mathcal{G}_{0s}}{\omega - \mathbf{k} \cdot \mathbf{v}} - \frac{e_s}{m_s} \frac{\mathbf{k} \cdot \partial \bar{\mathcal{F}}_s / \partial \mathbf{v}}{\omega - \mathbf{k} \cdot \mathbf{v}} \varphi \qquad (7.32)$$

with

$$\varphi = i \sum_s \frac{e_s}{k^2 \epsilon_0 \kappa(\mathbf{k}, \omega)} \int d^3 v \, \frac{\mathcal{G}_{0s}}{\omega - \mathbf{k} \cdot \mathbf{v}}. \qquad (7.33)$$

Here $\mathcal{G}_{0s}$ is the transformed initial value,

$$\mathcal{G}_{0s} = \int d^3 x \, e^{-i\mathbf{k} \cdot \mathbf{x}} \tilde{\mathcal{F}}(\mathbf{x}, \mathbf{v}, t = 0).$$

and $\kappa(\mathbf{k}, \omega)$ is the plasma dielectric discussed in Chapter 6, (6.72).

The content of these results is transparent. The fluctuation in $\mathcal{F}$ includes the ballistically propagated initial value, given by the first term in (7.32), plus a correction measuring acceleration by the fluctuating field. Equation (7.33) on the other hand displays the *shielded* electrostatic response to the discreteness-induced charge perturbation. Note that ballistic propagation—use of straight-line orbits—reflects key simplifications: the $\bar{\mathbf{a}} = 0$ assumption as well as the neglect of collisional effects on $\tilde{\mathcal{F}}$.

Substitution of the inverse transforms of these functions into (7.26) is straightforward, but one physical issue deserves comment: we take advantage of time-scale separation and plasma stability to simplify the inverse transform in time. First, since the fluctuations relax quickly on the time scale of $\bar{\mathcal{F}}$, it is appropriate to insert the time-asymptotic solutions,

$$\tilde{\mathcal{F}} \to \lim_{t \to \infty} \tilde{\mathcal{F}}(t) \qquad (7.34)$$

into (7.26). In general the inverse transform would depend upon the zeroes of the dielectric, $\kappa$, in addition to the ballistic pole at $\omega = \mathbf{k} \cdot \mathbf{v}$. However in a stable plasma all contributions from the former will decay in time, so that (7.34) allows us to keep only the residue from the ballistic pole. Second, we apply similar reasoning to the final form of $C$, which involves the factor

$$\frac{e^{ikut}}{ku} = e^{ikut} \left[ P \left( \frac{1}{ku} \right) - i\pi \delta(ku) \right] \qquad (7.35)$$

where $ku \equiv \mathbf{k} \cdot (\mathbf{v} - \mathbf{v}')$, $P$ refers to the principal value and we have noted that $ku$ should be interpreted as slightly above the real axis, for proper behavior as $t \to \infty$. Note, by the Riemann-Lebesgue lemma, that the principal value term decays on a time-scale measured by $\mathbf{k} \cdot \mathbf{v}$—very rapid compared to the scale of $\bar{\mathcal{F}}$. Hence in the spirit of (7.34) we retain only the delta-function term. Both of these exploitations of time-scale separation are referred to as the 'Bogoliubov' or adiabatic hypothesis.

Since both $\Phi$ and $\tilde{\mathcal{F}}$ are proportional to $\mathcal{G}_{0s}$, statistics enter the quasilinear operator through the quantity

$$\langle \mathcal{G}_{0s}^* \mathcal{G}_{0s'}' \rangle \equiv \langle \mathcal{G}_{0s}^*(\mathbf{k}, \mathbf{v}) \mathcal{G}_{0s'}(\mathbf{k}', \mathbf{v}') \rangle.$$

Here the asterisk denotes complex conjugation: as usual, it is convenient to express $\Phi$ in terms of the complex conjugate of its transform. The reader can verify that (7.29) yields

$$\langle \mathcal{G}_{0s}^* \mathcal{G}_{0s'}' \rangle = (2\pi)^3 \delta_{ss'} \delta(\mathbf{k} - \mathbf{k}') \delta(\mathbf{v} - \mathbf{v}') \bar{\mathcal{F}}_s. \qquad (7.36)$$

Finally we mention a point of mathematical neatness. The two terms of (7.32) yield two corresponding terms in the collision operator, one quadratic in $\varphi$, and therefore having $|\kappa|^2$ in its denominator, and one proportional to $\varphi g_0$, having a single power of $\kappa^*$ in its denominator. (In fact the two terms contribute diffusion and friction, respectively, to the final operator.) This awkwardness suggests the substitution

$$\frac{1}{\kappa^*} = \frac{\kappa_r - i\kappa_i}{|\kappa|^2}$$

where $\kappa_r$ and $\kappa_i$ are the real and imaginary parts, respectively, of the dielectric. We computed these functions in Chapter 6. It can be seen that the former is even in $\mathbf{k}$ and therefore does not contribute to $C$, while the latter contributes a term involving the velocity derivative of $\bar{f}(\mathbf{v}')$.

With these observations, it is not hard to derive from (7.26), (7.32) and (7.33) the Balescu-Lenard collision operator:

$$C_s = -\frac{\partial}{\partial \mathbf{v}} \cdot \frac{1}{m_s} \sum_{s'} \int d^3 \mathbf{v}' \, \mathbf{K}_{ss'}(\mathbf{v}, \mathbf{v}') \cdot \left[ f_s \frac{1}{m_{s'}} \frac{\partial f_{s'}'}{\partial \mathbf{v}'} - f_{s'}' \frac{1}{m_s} \frac{\partial f_s}{\partial \mathbf{v}} \right] \qquad (7.37)$$

Here we have reverted to the notation of Chapter 3,

$$f_s \equiv \bar{\mathcal{F}}$$

and introduced the tensor

$$\mathbf{K}_{ss'}(\mathbf{v}, \mathbf{v}') \equiv \frac{e_s^2 e_{s'}^2}{8\pi^2 \epsilon_0^2} \int d^3 k \, \delta(\mathbf{k} \cdot \mathbf{v} - \mathbf{k} \cdot \mathbf{v}') \frac{\mathbf{k}\mathbf{k}}{k^4 |\kappa(\mathbf{k}, \mathbf{k} \cdot \mathbf{v})|^2}. \qquad (7.38)$$

We also use the Boltzmann abbreviation, $f' \equiv f(\mathbf{v}')$. Notice that

$$\mathbf{K}_{ss'}(\mathbf{v}, \mathbf{v}') = \mathbf{K}_{s's}(\mathbf{v}', \mathbf{v}). \qquad (7.39)$$

A striking feature of the Balescu-Lenard operator is its compromise between binary and many-particle interaction. Thus the square-bracketed factor in (7.37) has the form, bilinear in $f_s$ and $f_{s'}'$, associated with two-particle collisions; it is in fact the weak-scattering version of a similar factor in the Boltzmann operator. On the other hand, the quantity $\mathbf{K}$ depends nonlinearly upon the distributions of all species present, through the dielectric $\kappa$. As we have noted, the presence of the dielectric reflects many-particle shielding of the Coulomb interaction. Despite this departure from bilinearity, we show presently that the operator is well-approximated, in most circumstances, by a nearly bilinear version.

By comparison with (7.19) we infer the corresponding Fokker-Planck coefficients

$$\mathbf{D}_{ss'} = \frac{1}{m_s^2} \int d^3v' \, \mathbf{K}_{ss'} f'_{s'}, \tag{7.40}$$

$$\mathbf{R}_{ss'} = \frac{1}{m_{s'}} \int d^3v' \, \mathbf{K}_{ss'} \cdot \frac{\partial f'_{s'}}{\partial \mathbf{v}'} \tag{7.41}$$

After substituting these forms into (7.21) and (7.22) it is straightforward to verify that the Balescu-Lenard operator conserves momentum and energy. For example, momentum conservation reduces to the requirement that the quantity

$$\int d^3v \, d^3v' \, f_s f'_{s'} \left( \frac{1}{m_s} \frac{\partial}{\partial \mathbf{v}} \cdot \mathbf{K}_{ss'} - \frac{1}{m_{s'}} \frac{\partial}{\partial \mathbf{v}'} \cdot \mathbf{K}_{ss'} \right)$$

change sign under the interchange of $s$ and $s'$; that it does follows from (7.39).

A conventional argument shows that the Balescu-Lenard operator satisfies an H-theorem, forcing entropy to increase. Recall that the entropy change is given by (7.11); considering for simplicity the like-species case ($s = s'$) and suppressing the $s$-subscript we have

$$\left( \frac{ds}{dt} \right)_c = -\frac{1}{m^2} \int d^3v \, d^3v' \, \log f \frac{\partial}{\partial \mathbf{v}} \cdot \left[ f f' \mathbf{K}(\mathbf{v}, \mathbf{v}') \cdot \left( \frac{\partial \log f}{\partial \mathbf{v}} - \frac{\partial \log f'}{\partial \mathbf{v}'} \right) \right]$$

or, after partial integration,

$$= \frac{1}{m^2} \int d^3v \, d^3v' \, f f' \frac{\partial \log f}{\partial \mathbf{v}} \left[ \mathbf{K}(\mathbf{v}, \mathbf{v}') \cdot \left( \frac{\partial \log f}{\partial \mathbf{v}} - \frac{\partial \log f'}{\partial \mathbf{v}'} \right) \right]. \tag{7.42}$$

We next symmetrize the integrand with respect to $\mathbf{v}$ and $\mathbf{v}'$; that is, we replace (7.42) by half of its sum with the interchanged, $\mathbf{v} \leftrightarrow \mathbf{v}'$, form. Because of the symmetry (7.39), the result is simply

$$\left( \frac{ds}{dt} \right)_c = \frac{1}{2m^2} \int d^3v \, d^3v' \, f f' \mathbf{g} \cdot \mathbf{K} \cdot \mathbf{g}$$

where we use the abbreviation

$$\mathbf{g} \equiv \frac{\partial \log f}{\partial \mathbf{v}} - \frac{\partial \log f'}{\partial \mathbf{v}'}.$$

But the definition of $\mathbf{K}$ shows that its diagonal elements are positive in any coordinate system, so

$$\mathbf{g} \cdot \mathbf{K} \cdot \mathbf{g} \geq 0 \tag{7.43}$$

and the result,

$$\left( \frac{ds}{dt} \right)_c \geq 0, \tag{7.44}$$

follows.

An often admired virtue of the Balescu-Lenard operator is that it requires no long-range cut-off. Recall that the "bare" Coulomb force has infinite range: the total (Rutherford) cross-section derived from Coulomb scattering diverges because of the unbounded effect of very weak, large-impact-parameter events. Indeed, if $\kappa$ in (7.37) were replaced by unity, the integral would evidently diverge at small $k$, corresponding to large impact parameter, and a cut-off would be required. (Such a cut-off is required, for precisely this reason, in the Boltzmann operator for Coulomb collisions.) However

$$\kappa \sim k^{-2}$$

for $k \to 0$, so that the $\kappa$ factors remove the need for a small-$k$ cut-off in (7.37). In other words, Debye shielding makes the effective range of the Coulomb interaction finite, and the Balescu-Lenard operator takes this shielding into account.

Unfortunately the operator remains divergent because of its inaccurate treatment of close encounters: it requires a cut-off at large $k$. It is not surprising that the quasilinear treatment, like the Fokker-Planck development, breaks down for hard collisions. What is unfortunate is that the large-$k$ contribution to the collision integral ultimately dominates, giving a cut-off dependent collision frequency. On the other hand the divergence is merely logarithmic, indicating that a more careful analysis, allowing for a distinctive analysis at large $k$, would have minor effect. These points are made explicit in the next section.

# 7.4 Landau-Boltzmann operator

Landau derived the operator described in this Section by expanding the Boltzmann collision operator for small momentum transfer; his result is know as the Landau operator or Landau-Boltzmann operator. We consider it here as an approximation to the Balescu-Lenard operator. (We do not propose a theory for the occurrence of binary names on each of the three basic forms of the operator, Fokker-Planck, Balescu-Lenard and Landau-Boltzmann. None of the pairs represent collaboration.) Our argument is simplified by considering only the electron contribution to $\kappa$, which can be written as

$$k^2 \kappa(\mathbf{k}, \mathbf{k} \cdot \mathbf{v}) = k^2 + k_D^2 \alpha(\mathbf{v}, \hat{\mathbf{k}})$$

where $k_D \equiv 1/\lambda_D$ is the inverse Debye length, $\hat{\mathbf{k}} = \mathbf{k}/k$ is the unit vector and

$$\alpha(\mathbf{v}, \hat{\mathbf{k}}) \equiv \frac{v_t^2}{2} \int d^3v' \, \frac{\mathbf{k} \cdot \partial \bar{F}/\partial \mathbf{v}'}{\mathbf{k} \cdot \mathbf{v}' - \mathbf{k} \cdot \mathbf{v}}$$

Here $\bar{F}$ is the average distribution normalized to unity,

$$\int d^3v \, \bar{F} = 1, \tag{7.45}$$

rather than the density, n. For reasonably smooth $\bar{F}$, the magnitude of $\alpha$ will be of order unity; the key observation is that it is independent of the magnitude of $\mathbf{k}$.

Expressing (7.38) in terms of $\alpha$ we find

$$\mathbf{K} = \frac{e_s^2 e_{s'}^2}{8\pi^2 \epsilon_0^2} \int d^3k\, \delta(\mathbf{k} \cdot \mathbf{v}' - \mathbf{k} \cdot \mathbf{v}) \frac{\mathbf{kk}}{|k^2 + k_D^2 \alpha|^2} \qquad (7.46)$$

We see again how shielding prevents an infrared disaster: if $\alpha$ were zero (that is, if $\kappa$ were unity), the integral would be logarithmically divergent at small $k$. We can also see the problem at large $k$: when $k \gg k_D$, the integral becomes

$$\int dk\, k^2 \delta(\mathbf{k} \cdot \mathbf{v}' - \mathbf{k} \cdot \mathbf{v}) \frac{\mathbf{kk}}{k^4}$$

Since $\delta(\mathbf{k} \cdot \mathbf{v}' - \mathbf{k} \cdot \mathbf{v}) = (1/k)\delta(\hat{\mathbf{k}} \cdot \mathbf{v}' - \hat{\mathbf{k}} \cdot \mathbf{v})$ this expression is logarithmically divergent.

We denote the required cut-off by $k_{max}$. It is determined by the physical requirement that interactions with $k < k_{max}$ are correctly treated by the Balescu-Lenard analysis, which rules out, in particular, hard collisions. If the physics requires $k_{max} \leq k_D$, then the value of the integral depends upon the functional form of $\alpha$ and no simple approximation is available. In the opposite case, however, the integral is dominated by the large-$k$ region and the form of $\alpha$ is unimportant. In other words, when

$$\Lambda_c \equiv k_{max}\lambda_D \gg 1, \qquad (7.47)$$

variation of $\alpha$ is irrelevant and a simple, asymptotic evaluation of the integral is possible; this is the essential approximation needed to derive the Landau-Boltzmann operator from its Balescu-Lenard ancestor. It is clear that (7.47) requires many "soft" collisions to occur at separations less than the Debye length—a requirement that is plausibly satisfied when there are many particles inside a Debye sphere.

Supposing (7.47) to hold, we see that $\mathbf{K}$ has the form of the tensor

$$\mathbf{T} = \int d^3k\, \delta(\mathbf{k} \cdot \mathbf{u}) S(k) \mathbf{kk}$$

where $S$ is an arbitrary function of $|\mathbf{k}|$ alone. The reader can verify that $\mathbf{T}$ is given by

$$\mathbf{T} = A\,(\mathbf{I} - \hat{\mathbf{u}}\hat{\mathbf{u}})\,. \qquad (7.48)$$

Here $A = T_{11}$ in the frame where $\mathbf{u} = (0, 0, u)$ and $\hat{\mathbf{u}} = \mathbf{u}/u$. Thus we need only compute the component

$$K_{11} = \frac{e_s^2 e_{s'}^2}{8\pi^2 \epsilon_0^2} \frac{1}{|\mathbf{v} - \mathbf{v}'|} \int dk_1 dk_2 dk_3\, \delta(k_3) \frac{k_1 k_1}{|k^2 + k_D^2 \alpha|^2}$$

Using polar coordinates in $(k_1, k_2)$-space, one quickly finds

$$K_{11} = \frac{e_s^2 e_{s'}^2}{8\pi \epsilon_0^2} \frac{1}{|\mathbf{v} - \mathbf{v}'|} \int \frac{dk}{k} \frac{k^4}{|k^2 + k_D^2 \alpha|^2}$$

Here again the importance of Debye shielding for $k < k_D$ is apparent. In the case of (7.47), the dominant contribution to $K_{11}$ is

$$K_{11} = \frac{e_s^2 e_{s'}^2}{8\pi\epsilon_0^2} \log \Lambda_c \frac{1}{|\mathbf{v} - \mathbf{v}'|}$$

and we have

$$\mathbf{K} = \frac{e_s^2 e_{s'}^2 \log \Lambda_c}{8\pi\epsilon_0^2} \mathbf{U} \tag{7.49}$$

where the tensor $\mathbf{U}$ has components

$$\mathbf{U} = \frac{1}{u^3}(u^2\mathbf{I} - \mathbf{uu}), \tag{7.50}$$

where $\mathbf{I}$ is the unit tensor and $\mathbf{u} = \mathbf{v} - \mathbf{v}'$.

The quantity $\log \Lambda_c$ is called the *Coulomb logarithm*. A plausible choice for the cut-off is that corresponding to the classical distance of closest approach, $r_c$, where $4\pi\epsilon_0 r_c = 2e^2/mv_t^2$:

$$k_{max} \sim 1/r_c.$$

Because

$$\frac{1}{r_c} = 4\pi n\lambda_D^2,$$

we find that $\Lambda_c$ approximates the plasma parameter $\Lambda$, which has already been assumed large. At sufficiently high temperature, the quantum uncertainty in the electron's position $\hbar/m_e v_{te}$ becomes comparable to the distance of closest approach and must replace the latter in $\log \Lambda_c$. Spitzer states the resulting formula as

$$\log \Lambda_c = \begin{cases} 23.4 - 1.15 \log n + 3.45 \log T_e, & \text{for } T_e < 50\text{eV} , \\ 25.3 - 1.15 \log n + 2.3 \log T_e. & \text{for } T_e > 50\text{eV} , \end{cases}$$

where $T_e$ is measured in electron volts and n is measured in $cm^{-3}$. Because of the logarithm, this quantity is rarely very different from $\log \Lambda$.

After substituting (7.49) into (7.37) we obtain the Landau-Boltzmann collision operator given by

$$C_{ss'}(f_s, f_{s'}) = \frac{\gamma_{ss'}}{m_s} \frac{\partial}{\partial \mathbf{v}} \cdot \int d^3 v' \, \mathbf{U} \cdot \left( \frac{f_{s'}'}{m_s} \frac{\partial f_s}{\partial \mathbf{v}} - \frac{f_s}{m_{s'}} \frac{\partial f_{s'}'}{\partial \mathbf{v}'} \right), \tag{7.51}$$

where

$$\gamma_{ss'} = \frac{e_s^2 e_{s'}^2 \log \Lambda_c}{8\pi\epsilon_0^2} \tag{7.52}$$

and $\mathbf{U}$ is given by (7.50).

The Landau-Boltzmann operator is a large-$\Lambda_c$ approximation to the result of Balescu-Lenard. In so far as $\Lambda_c$ is comparable to $\Lambda$, no major new assumption is needed: $\Lambda$ was already assumed large in Balescu-Lenard theory. On the other hand the relative simplification, which removes the $k$-integral from (7.37) by performing it, is enormous. Note in particular that just one artifact of many-body interaction has survived: the Debye shielding implicit in (7.47). Because this shielding enters only logarithmically, we have reduced the Balescu-Lenard operator to one that is very nearly bilinear. In fact the Landau-Boltzmann operator is simple and tractable enough to have become an important tool in plasma physics research.

Application of the Landau-Boltzmann operator to distributions with steep velocity gradients—such as high-intensity beam plasmas—requires some care. For example, the function $\alpha$ might become large enough to affect the Coulomb logarithm. The most important limitation of the operator, however, is inherited from Balescu-Lenard: neglecting nonfluctuating fields. One wonders in particular about the application of either operator to a magnetized plasma, where the ensemble-averaged force is large. The problem is that (7.32) and (7.33) use straight, rather than helical, trajectories.

We have found that for large $\Lambda_c$ the dominant collisional process occurs at interparticle distances less than the Debye length, $\lambda_D$. It is therefore clear that if the gyroradius, $\rho$, of interacting particles exceeds $\lambda_D$, the assumption of straight-line trajectories remains valid. In the opposite case,

$$\rho \ll \lambda_D, \tag{7.53}$$

however, the collision operator must be modified to treat helical orbits. One finds that (7.53) can indeed pertain to *electron* collisions in some magnetized plasmas.

The collision operator that includes helical trajectories, allowing $\rho < \lambda_D$, has been calculated. It is extremely complicated, involving in particular anisotropy between the $\parallel$ and $\perp$ directions in velocity space, and will not be presented here. It is noteworthy, however, that for many processes the effect of magnetic fields is surprisingly simple: one has only to replace $\lambda_D$ by $\rho$ in the Coulomb logarithm.

## 7.5  Calculus of collisions

### Symmetry and conservation laws

Correlations should appear the same to observers in relative (uniform) motion: the collision operator should not favor any particular rest frame. That the Landau-Boltzmann operator satisfies this requirement is evident from (7.51): note that $C$ depends upon $\mathbf{v}$ only through the tensor $\mathbf{U}$ (aside from its dependence on $f_s$ and $f_{s'}$, which are Galilean scalars). Now if $\mathbf{v}$ is subject to a Galilean transformation,

$$\mathbf{v} \to \mathbf{v} - \mathbf{V}$$

we can clearly shift the integration variable, $\mathbf{v}'$, by the same amount, leaving $\mathbf{U}$ and therefore $C$ unchanged. Rotational symmetry of the Landau-Boltzmann operator can be demonstrated similarly, using the tensor character of $\mathbf{U}$.

The Fokker-Planck coefficients associated with (7.51) can be found by substituting (7.49) for $\mathbf{K}$ into the Balescu-Lenard versions, (7.40) and (7.41). Thus we have the diffusion tensor

$$\mathbf{D}_{ss'} = \frac{\gamma_{ss'}}{m_s^2} \int d^3v' \, \mathbf{U} f'_{s'}, \tag{7.54}$$

and dynamical friction,

$$\mathbf{R}_{ss'} = \frac{\gamma_{ss'}}{m_{s'}} \int d^3v' \, \mathbf{U} \cdot \frac{\partial f'_{s'}}{\partial \mathbf{v}'}. \tag{7.55}$$

Note that

$$\frac{\partial}{\partial \mathbf{v}} \cdot \mathbf{U} = -\frac{\partial}{\partial \mathbf{v}'} \cdot \mathbf{U}. \tag{7.56}$$

This identity and partial integration can be used to show that

$$\mathbf{R}_{ss'} = \frac{m_s^2}{m_{s'}} \frac{\partial}{\partial \mathbf{v}} \cdot \mathbf{D}_{ss'}, \tag{7.57}$$

a relation that does *not* pertain to the general Balescu-Lenard operator. Similarly the symmetry (7.39) is much stronger in the Landau-Boltzmann case, where $\mathbf{K}_{ss'}(\mathbf{v}, \mathbf{v}')$ is separately symmetric under exchange of either $s \leftrightarrow s'$ or $\mathbf{v} \leftrightarrow \mathbf{v}'$.

To verify that the Landau-Boltzmann operator conserves momentum, we begin with the general Fokker-Planck expression of momentum conservation, (7.21), and use (7.57) to find

$$\int d^3v \, m_s \mathbf{v} C_{ss'} = \int d^3v \, f_s m_s^2 \left( \frac{1}{m_s} + \frac{1}{m_{s'}} \right) \frac{\partial}{\partial \mathbf{v}} \cdot \mathbf{D}_{ss'}, \tag{7.58}$$

or, in view of (7.54),

$$\int d^3v \, m_s \mathbf{v} C_{ss'} = \gamma_{ss'} \left( \frac{1}{m_s} + \frac{1}{m_{s'}} \right) \int d^3v d^3v' \, f_s f'_{s'} \frac{\partial}{\partial \mathbf{v}} \cdot \mathbf{U}$$

Momentum conservation requires this quantity to change sign if $s$ and $s'$ are interchanged; we can simultaneously interchange the integration variables $\mathbf{v}$ and $\mathbf{v}'$ and observe that all factors are symmetric under this double interchange except the divergence, which changes sign in view of (7.56). The demonstration of energy conservation, starting from (7.22), is only slightly more complicated.

## Rosenbluth potentials

For several applications it is convenient to write the collision operator in terms of certain integrals of the distribution, called Rosenbluth potentials. Thus one finds that

$$C_{ss'} = -\frac{\gamma_{ss'}}{m_s m_{s'}} \frac{\partial}{\partial \mathbf{v}} \cdot \left( 2 \frac{\partial H_{s'}}{\partial \mathbf{v}} f_s - \frac{m_{s'}}{m_s} \frac{\partial^2 G_{s'}}{\partial \mathbf{v} \partial \mathbf{v}} \cdot \frac{\partial f_s}{\partial \mathbf{v}} \right), \tag{7.59}$$

where

$$G_{s'}(\mathbf{v}) \equiv \int d^3v'\, f'_{s'}u, \qquad (7.60)$$

$$H_{s'}(\mathbf{v}) \equiv \int d^3v'\, f'_{s'}\frac{1}{u}. \qquad (7.61)$$

To verify that (7.59) is the Boltzmann-Landau operator, we take note of the identities

$$U_{\alpha\beta} = \frac{\partial^2 u}{\partial u_\alpha \partial u_\beta}, \qquad (7.62)$$

$$\frac{\partial U_{\alpha\beta}}{\partial u_\beta} = 2\frac{\partial}{\partial u_\alpha}\frac{1}{u}, \qquad (7.63)$$

$$\frac{\partial^2 U_{\alpha\beta}}{\partial u_\alpha \partial u_\beta} = -8\pi\delta(\mathbf{u}). \qquad (7.64)$$

It follows from (7.62) that the diffusion tensor can be expressed as

$$D_{ss'\alpha\beta} = \frac{\gamma_{ss'}}{m_s^2}\frac{\partial^2 G_{s'}}{\partial v_\alpha \partial v_\beta}. \qquad (7.65)$$

Similarly the dynamical friction becomes

$$R_{ss'\alpha} = \frac{2\gamma_{ss'}}{m_{s'}}\frac{\partial H_{s'}}{\partial v_\alpha}. \qquad (7.66)$$

Substituting these expressions into (7.19) we obtain the collision operator of (7.59).

The functions $G_s$ and $H_s$ are called potentials because of their formal similarity to the electrostatic potential. That is, since $u$ is a distance in velocity space, $1/u$ corresponds to the potential at $\mathbf{v}$ due to a point "charge" at $\mathbf{v}'$. In this sense $H_s$ corresponds to the Coulomb potential resulting from the charge density $f_s$: the identity

$$\nabla_v^2 H_s \equiv \frac{\partial^2 H_s}{\partial v_\alpha \partial v_\alpha} = -4\pi f_s,$$

which follows from (7.64), is analogous to Poisson's equation. One similarly finds that $2H_s$ acts as a velocity-space charge density for $G_s$:

$$\nabla_v^2 G_s = 2H_s. \qquad (7.67)$$

The analogy with electrostatics, first noted by Rosenbluth, MacDonald and Judd, brings some of the methods and insights of potential theory to bear on collisional physics. For example, Rosenbluth, MacDonald and Judd used expansion formulae from multipole electrostatics to express $C$ in terms of non-Cartesian velocity-space coordinates. The potentials are used more generally as a convenient framework for collisional analysis, such as approximation of $C_{ss'}$ for small mass-ratio $m_s/m_{s'}$.

As an example we use the potential formalism to study the collision operator in the case of Maxwellian distributions. We compute first

$$H_{sM} \equiv \int d^3v'\, f'_{s'M}(v') \frac{1}{|\mathbf{v} - \mathbf{v}'|}$$

by using polar coordinates $(v, \theta, \phi)$, where $\theta$ is the angle between $\mathbf{v}$ and $\mathbf{v}'$. We recall the expansion,

$$\frac{1}{|\mathbf{v} - \mathbf{v}'|} = \frac{1}{v_>} \sum_{k=0}^{\infty} \left(\frac{v_<}{v_>}\right)^k P_k(\cos\theta)$$

where $v_>$ and $v_<$ are respectively the larger and smaller members of the pair $(v, v')$ and the $P_k$ are Legendre polynomials. Since $f_M$ is independent of angle the angular integrals are trivial and eliminate all but the first term in the Legendre expansion; the remaining $v'$-integral is straightforward and yields

$$H_{s'M} = \frac{n_{s'}}{v_{ts'}x}\,\mathrm{erf}\,(x), \tag{7.68}$$

where erf denotes the error function and

$$x \equiv v/v_{ts'}.$$

A similar calculation shows that

$$G_{s'M} = \frac{n_s/v_{ts'}}{2x}\left[\frac{d\,\mathrm{erf}\,(x)}{dx} + (1 + 2x^2)\mathrm{erf}\,(x)\right].$$

Differentiation of these functions provides the diffusion tensor,

$$\mathbf{D}_{ss'M} = \frac{\gamma_{ss'}n_{s'}T_{s'}}{m_s^2 m_{s'}}\frac{1}{v^3}\left[\mathbf{I}F_1(x) + 3\frac{\mathbf{vv}}{v^2}F_2(x)\right], \tag{7.69}$$

and dynamical friction,

$$\mathbf{R}_{ss'M} = \frac{2\gamma_{ss'}n_{s'}}{m_{s'}}\frac{\mathbf{v}}{v^3}\left[\mathrm{erf}\,(x) - x\frac{d\,\mathrm{erf}\,(x)}{dx}\right]. \tag{7.70}$$

Here we use the abbreviations

$$F_1(x) = x\frac{d\,\mathrm{erf}\,(x)}{dx} + (2x^2 - 1)\,\mathrm{erf}\,(x), \tag{7.71}$$

$$F_2(x) = \left(1 - \frac{2}{3}x^2\right)\mathrm{erf}\,(x) - x\frac{d\,\mathrm{erf}\,(x)}{dx}. \tag{7.72}$$

Finally we combine these results to compute the collision operator when both species are Maxwellian:

$$C_{ss'M} = -\frac{2\gamma_{ss'}n_{s'}}{m_s m_{s'}v_{ts'}v_{ts}^2}\left(\frac{T_s - T_{s'}}{T_s}\right)\left[\frac{\mathrm{erf}}{x} - \left(1 + \frac{m_{s'}T_{s'}}{m_s T_s}\right)\frac{d\,\mathrm{erf}}{dx}\right]f_{sM}. \tag{7.73}$$

Note in particular that this quantity vanishes, as it must, when $T_s = T_{s'}$.

The strong energy dependence in (7.73) and its predecessors shows that the effects of Coulomb collisions are far from uniform in velocity space. In particular we see from (7.69) and (7.70) that collisional effects become weak at large speed. This circumstance is best known for its relation to the phenomenon called *runaway*, discussed in Chapter 8, but in fact it has a pervasive influence. To mention one example we recall the collisionality parameter introduced in Chapter 3: $\lambda_{\mathrm{mfp}}/L$, where $\lambda_{\mathrm{mfp}} = v_t/\nu$ is the mean-free path and $L$ a gradient scale-length. It is clear that the *speed-dependent* ratio, $v/L\nu(v)$, more reliably estimates the importance of collisions, and that supra-thermal particles can become effectively collisionless in a plasma, unless $\lambda_{\mathrm{mfp}}/L$ is very small indeed. The corresponding experimental observation is the formation of extended high-energy tails in all but the most collisional distribution functions.

## Collisional moments

The moment equations studied in Chapter 3 involve the collisional rates of change of momentum and energy, given respectively by

$$\mathbf{F}_{ss'} = \int d^3v\, m_s \mathbf{v} C_{ss'}(f_s, f_{s'}), \tag{7.74}$$

$$W_{ss'} = \int d^3v\, \frac{1}{2} m_s (\mathbf{v} - \mathbf{V}_s)^2 C_{ss'}(f_s, f_{s'}). \tag{7.75}$$

These quantities depend upon the distributions and are therefore specific to the particular plasma system being studied. But we have noted that approximately Maxwellian distributions occur under a wide range of circumstances, even in nonequilibrium and long mean-free-path systems. Furthermore the non-Maxwellian corrections to the distribution often affect $\mathbf{F}$ and $W$ very little. Hence it makes sense to compute the collisional moments for the Maxwellian case.

Of course if both species are Maxwellian at the same temperature and flow, the calculation is trivial: $C = \mathbf{F} = W = 0$. We obtain non-trivial results by allowing the Maxwellians of the two species to differ. In the case of the friction force $\mathbf{F}$, we suppose different flow velocities, and calculate the momentum change suffered by species $s$ in the rest frame of species $s'$. Thus we consider the distributions $f_{s'} = f_{s'M}(\mathbf{v})$ and

$$f_s = f_{sM}(\mathbf{v} - \mathbf{V}_s) \cong f_{sM}(\mathbf{v}) \left[ 1 + \frac{2\mathbf{v} \cdot \mathbf{V}_s}{v_{ts}^2} \right], \tag{7.76}$$

where $\mathbf{V}_s$ is the relative velocity, and we have assumed $|\mathbf{V}_s| \ll v_{ts}$. Next we combine (7.57) and (7.58) to write

$$\mathbf{F}_{ss'} = \left( 1 + \frac{m_{s'}}{m_s} \right) \int d^3v\, f_s \mathbf{R}_{ss'} \tag{7.77}$$

or, in view of (7.66),

$$\mathbf{F}_{ss'} = 2\gamma_{ss'} \left( \frac{1}{m_s} + \frac{1}{m_{s'}} \right) \int d^3v \, f_s \frac{\partial H_{s'}}{\partial \mathbf{v}}. \tag{7.78}$$

This form can be used to compute the Landau-Boltzmann friction for any distributions; in the present case, $H_{s'}$ can be replaced by $H_{s'M}$, given by (7.68), and $f_s$ is given by (7.76), in which only the second term contributes. After straightforward manipulation we find

$$\mathbf{F}_{ss'} = -\frac{8\gamma_{ss'}n_s n_{s'} v_{ts'}^2}{3v_{ts}^5} \left( \frac{1}{m_s} + \frac{1}{m_{s'}} \right) \mathbf{V}\mathcal{I}$$

where

$$\mathcal{I} \equiv \frac{2}{\sqrt{\pi}} \int_0^\infty dx \, x \exp(-v_{ts'}^2 x^2/v_{ts}^2) \left[ \mathrm{erf} \ (x) - x\frac{d\,\mathrm{erf} \ (x)}{dx} \right].$$

The integral can be evaluated through integration by parts, with the result

$$\mathbf{F}_{ss'} = -\frac{m_s n_s}{\tau_{ss'}} \mathbf{V}_s, \tag{7.79}$$

where the momentum exchange time $\tau_{ss'}$ is given by

$$\tau_{ss'} \equiv \frac{3\sqrt{\pi}}{8} \frac{m_{s'}}{m_s + m_{s'}} \frac{m_s^2 \left( v_{ts}^2 + v_{ts'}^2 \right)^{3/2}}{\gamma_{ss'} n_{s'}}. \tag{7.80}$$

From the equation of motion we see that $\tau_{ss'}$ measures the time-scale for collisional loss of momentum—the decay time (e-folding time) for $m_s n_s \mathbf{V}_s$. At a deeper level, $\tau_{ss'}$ is the time required for a substantial change in *direction* of a ($s$-species) particle's trajectory, due to accumulated collisions with particles of species $s'$. For that reason it is sometimes called the "90° scattering time," or simply the collision time. The quantity $\nu$, which we have used to represent the size of the collision operator, can be taken to represent the inverse of the collision time $\tau$. Similarly we can define the mean-free path for $s - s'$ collisions by

$$\lambda_{\mathrm{mfp,ss'}} \equiv v_{ts} \tau_{ss'} \tag{7.81}$$

Two important special cases of our formula are the ion-ion collision time

$$\tau_{ii} = \frac{3\sqrt{2\pi}}{8} \frac{m_i^2 v_{ti}^3}{\gamma_{ii} n_i}, \tag{7.82}$$

and the electron-ion collision time

$$\tau_{ei} = \frac{3\sqrt{\pi}}{8} \frac{m_e^2 v_{te}^3}{\gamma_{ei} n_i}. \tag{7.83}$$

This formula approximates (7.80) for small $m_e/m_i$, assuming $T_i \sim T_e$.

Note that the collision times increase sharply with increasing temperature,

$$\tau \propto T^{3/2}$$

reflecting the reduced interaction time for more rapidly moving particles. That hotter plasmas are less collisional is a ruling consideration in many contexts.

The calculation of the energy moment $W_{ss'}$ from (7.73) is similar to, but somewhat easier than, the calculation of $\mathbf{F}_{ss'}$. We present only the result, supposing both species to be at rest:

$$W_{ss'} = 3\frac{m_s}{m_{s'}} \left( \frac{m_s}{m_s + m_{s'}} \right) \frac{n_s}{\tau_{ss'}} (T_{s'} - T_s). \tag{7.84}$$

This expression is thermodynamically reasonable in that species $s$ gains energy collisionally only if it is cooler than species $s'$.

In the case of disparate masses the factor $m_s/m_{s'}$ distinguishes the time scale for energy exchange from that for momentum exchange. Thus electrons exchange heat with ions at the rate

$$W_{ei} = 3\frac{m_e}{m_i}\frac{n_e}{\tau_{ei}} (T_i - T_e) \tag{7.85}$$

thousands of times slower than the momentum exchange rate. The point is that energy exchange becomes relatively inefficient at small mass ratio—as shown by the elastic (but momentum reversing) bounce of a basketball against the floor.

## Linearized collision operator

Most applications of the collision operator, including the transport issues we examine in Chapter 8, involve small departures from thermodynamic equilibrium. In this case the distribution functions for all species are nearly Maxwellian,

$$f_s = f_{sM} + f_{s1},$$

where the correction term is relatively small:

$$\frac{f_1}{f_M} \sim \epsilon \ll 1.$$

Any bilinear collision operator then has the form

$$C_{ss'}(f_s, f_{s'}) = C_{ss'}(f_{sM}, f_{s'M}) + C_{ss'}(f_{sM}, f_{s'1}) + C_{ss'}(f_{s1}, f_{s'M}) + O(\epsilon^2).$$

The *linearized operator*, consisting of the $O(\epsilon)$ terms in this expression, is abbreviated by the notation

$$C_{ss'}^l \equiv C_{ss'}(f_{sM}, f_{s'1}) + C_{ss'}(f_{s1}, f_{s'M}). \tag{7.86}$$

To make the linearized operator explicit, it is convenient to use the notation

$$\hat{f} \equiv \frac{f_1}{f_M}.$$

Thus we substitute $f = f_M(1 + \hat{f})$ into our Landau-Boltzmann formula, (7.51), and find

$$C_{ss'}^l = \frac{\gamma_{ss'}}{m_s} \frac{\partial}{\partial \mathbf{v}} \cdot \int d^3v' \mathbf{U} \cdot$$
$$\left[ \frac{1}{m_s} \frac{\partial \hat{f}_s}{\partial \mathbf{v}} - \frac{1}{m_{s'}} \frac{\partial \hat{f}'_{s'}}{\partial \mathbf{v}'} + \left( \frac{\mathbf{v}'}{T_{s'}} - \frac{\mathbf{v}}{T_s} \right) (\hat{f}_s + \hat{f}'_s) \right]. \qquad (7.87)$$

The linearized operator inherits essential properties of the full collision operator, since these must hold to each order in $\epsilon$. In particular it satisfies the same conservations laws and a very similar, quadratic $H$-theorem. The corresponding (quadratic) entropy production is examined in Chapter 8.

## 7.6 Approximations based on mass ratio

### Collision times

The Landau-Boltzmann operator is an integro-differential operator in three velocity dimensions, coupling all plasma species. In the case of like species collisions $C_{ss}$ the mathematical monster is difficult to evade; but for unlike species of disparate mass, such as electrons colliding with ions, much simpler versions of the operator can be justified. The simplifications we consider assume comparable species temperature, so that

$$v_{tl} \gg v_{th}$$

where the $l$ and $h$ subscripts denote "light" and "heavy" respectively. We also assume the distributions are smooth, and characterized by a thermal speed such that

$$\frac{\partial f_s}{\partial v} \sim \frac{f_s}{v_{ts}}.$$

Certain distributions, such as those dominated by cold beams, might disobey this estimate. Additional approximations based on large $Z_s = e_s/e$ are occasionally useful, but we prefer to take the various species charges (as well as their densities) to be comparable.

The first simplifying observation is that collisional scattering of a heavy species by a lighter one is weaker than the effect of like-species collisions. For (7.80) shows that

$$\frac{\tau_{hh}}{\tau_{hl}} \sim \left( \frac{m_h}{m_l} \right)^{1/2}$$

In particular, since the ion collision frequency $\nu_i$ is estimated well by $1/\tau_{ii}$, the ion mean-free path is set by ions alone. The fact that

$$C_{hh} \gg C_{hl}$$

does not imply that the $C_{hl}$ terms are negligible; in Chapter 8 we will find that their correction to $C_{hh}$ can have considerable importance.

On the other hand an analogous estimate shows that

$$\frac{\tau_{ll}}{\tau_{lh}} \sim 1.$$

Thus electrons are scattered by ions at least as much as by self-collisions: $\nu_e \sim \nu_{ee} \sim \nu_{ei}$. (Indeed, when the ionic charge $Z_i$ is considered large, scattering by ions dominates.) In this sense the collisional coupling between ions and electrons is not reciprocal.

Finally, although these quantities do not appear in the same kinetic equation, it is useful to compare the collision rates of the two species :

$$\frac{\nu_h}{\nu_l} \sim \left(\frac{m_l}{m_h}\right)^{1/2}.$$

It follows that for comparable temperatures the ion and electron mean-free paths are comparable,

$$\lambda_{\mathrm{mfp,i}} \sim \lambda_{\mathrm{mfp,e}};$$

if one species is collision-dominated, so is the other.

## Lorentz operator

Although collisions with a heavier species cannot be neglected in the kinetics of a light species, the collision operator $C_{lh}$ can be approximated by something much simpler than its full Landau-Boltzmann form. The strongest and most popular approximation treats the heavier scatterers as infinitely massive, with $v_{th} \to 0$. This so-called Lorentz gas is rather like a Pachinko game, in which the light particles are scattered in velocity-space angle, without change in energy. The corresponding $C_{lh}$ is a second-order differential operator, without integral terms, describing diffusion in $\mathbf{v}/v$.

To construct the Lorentz operator, we observe from (7.57) that $R_{lh}$ is negligible, and compute $D_{lh}$ from (7.54). Asssuming first that the heavy species is at rest $(\mathbf{V}_h = 0)$, we have $f_h(\mathbf{v}') = n_h \delta(\mathbf{v}')$ and

$$\mathbf{D}_{lh} = \frac{\gamma_{lh} n_h}{m_l^2} \mathbf{U}(\mathbf{v}),$$

where $\mathbf{U}(\mathbf{v}) = (v^2 \mathbf{I} - \mathbf{v}\mathbf{v})/v^3$ is the usual tensor evaluated at $v' = 0$. The same result can be obtained from the $v_{ts'} \to 0$ limit of (7.69). This form yields the operator

$$C_{lh}^0(f_l) = \frac{\gamma_{lh} n_h}{m_l^2} \frac{\partial}{\partial \mathbf{v}} \cdot \mathbf{U}(\mathbf{v}) \cdot \frac{\partial f_l}{\partial \mathbf{v}}. \qquad (7.88)$$

Here the 0-superscript reminds us that the heavy species is assumed stationary.

When the heavy species is in motion, we must use $\mathbf{U}(\mathbf{v} - \mathbf{V})$ instead of (7.50). Equivalently, because of the Galilean invariance of $C$, we can express $f_l$

in the rest frame of the scatterers. In the most important case $V_h \ll v_{tl}$, either point of view can be used to derive the linearized operator:

$$C_{lh}^l(f_l) = \frac{\gamma_{lh} n_h}{m_l^2} \frac{\partial}{\partial \mathbf{v}} \cdot \left[ \mathbf{U}(\mathbf{v}) \cdot \frac{\partial f_l}{\partial \mathbf{v}} - 4 \frac{\mathbf{V}_h}{v_{tl}^2 v} \right]. \tag{7.89}$$

The simplicity of $C_{lh}^0$ is most apparent in spherical coordinates. Hence, choosing an arbitrary $z$-axis (such as the direction of an imposed magnetic field), we write

$$\mathbf{v} = \hat{\mathbf{z}} v \xi + v \sqrt{1 - \xi^2} (\hat{\mathbf{x}} \cos \phi + \hat{\mathbf{y}} \sin \phi)$$

Then we express

$$L(f) \equiv \frac{\partial}{\partial \mathbf{v}} \cdot \mathbf{U}(\mathbf{v}) \cdot \frac{\partial f}{\partial \mathbf{v}} \tag{7.90}$$

in terms of $(q^1, q^2, q^3) = (v, \xi, \phi)$, using the standard formula for the divergence in spherical coordinates:

$$L = \frac{1}{v} \frac{\partial}{\partial q^\alpha} v U^{\alpha \beta} \frac{\partial f}{\partial q^\beta}.$$

Here

$$U^{\alpha \beta} \equiv \frac{\partial q^\alpha}{\partial \mathbf{v}} \cdot \mathbf{U} \cdot \frac{\partial q^\beta}{\partial \mathbf{v}}.$$

The key observation is that $\partial q^1 / \partial \mathbf{v} = \mathbf{v}/v$, while the form of $\mathbf{U}$ shows that

$$\mathbf{v} \cdot \mathbf{U}(\mathbf{v}) = \mathbf{U}(\mathbf{v}) \cdot \mathbf{v} = 0.$$

Hence all the "radial" components of $U^{\alpha \beta}$ vanish, $U^{1\beta} = 0 = U^{\alpha 1}$, and derivatives with respect to $v$ do not appear in the collision operator: as we have remarked, only angular scattering occurs.

Next we compute

$$\frac{\partial \xi}{\partial \mathbf{v}} = \frac{\hat{\mathbf{z}}}{v} - \frac{\mathbf{v} v_z}{v^2}$$

and

$$\frac{\partial \phi}{\partial \mathbf{v}} = \frac{v_x \hat{\mathbf{y}} - v_y \hat{\mathbf{x}}}{v^2 (1 - \xi^2)^2}$$

to make $L$ explicit and express the Lorentz operator as

$$C_{lh}^0(f_l) = \frac{\gamma_{lh} n_h}{m_l^2} \frac{1}{v^3} \left[ \frac{\partial}{\partial \xi} (1 - \xi^2) \frac{\partial f}{\partial \xi} + \frac{1}{(1 - \xi^2)} \frac{\partial^2 f}{\partial \phi^2} \right]. \tag{7.91}$$

The corresponding form of the linearized operator that allows for finite $\mathbf{V}_h$ is

$$C_{lh}^l = C_{lh}^0(f_l) + \frac{\gamma_{lh} n_h}{m_l^2} \frac{1}{v^3} \frac{4\mathbf{V} \cdot \mathbf{v}}{v_{tl}^2} f_{lM}. \tag{7.92}$$

## Effective charge

A felicity of (7.91) is its simple dependence on the parameters of the heavy species, which appear only through the coefficient $\gamma_{lh}n_h$. Hence, when several heavy species are present, the collision term for $f_l$ has the form

$$C_l = C_{ll} + \left( \sum_h \gamma_{lh}n_h \right) \frac{1}{m_l^2} L(f_l)$$

accounting for the various heavy species by a simple additive constant. (There is no such simplification when the colliding species have comparable masses.) In particular, for electron scattering we have, from the definition (7.52),

$$\gamma_{ei} = \gamma_{ee}Z_i^2$$

whence

$$C_e = C_{ee} + \frac{\gamma_{ee}n_e}{m_e^2} Z_{\text{eff}} L(f_e). \tag{7.93}$$

where

$$Z_{\text{eff}} \equiv \frac{\sum_i n_i Z_i^2}{n_e} \tag{7.94}$$

is the *effective* $Z$ of the plasma. We see that electron scattering is affected by the charge states of the ion species present only through $Z_{\text{eff}}$. Its definition is natural (scaling with $Z$ rather than $Z^2$) in a quasineutral plasma, where

$$n_e = \sum_i Z_i n_i.$$

Note however that the simplification of $C_e$ works only if all the ion species are at rest in a common frame; otherwise the velocities $\mathbf{V}_i$ will enter through $U$ and the operator $L$ will depend on species. Fortunately a common ion flow velocity is usually enforced by ion-ion collisional friction.

Since the coefficient $\gamma_{ee}n_e/m_e^2$ also appears in $C_{ee}$, $Z_{\text{eff}}$ measures the relative importance of ion collisions on electron kinetic evolution. In a pure hydrogen plasma, where $Z_{\text{eff}} = 1$, $C_{ee}$ and $C_{ei}$ are of roughly equal importance; but when $Z_{\text{eff}} \gg 1$ (as commonly occurs, for example, in plasma processing contexts) electron-electron collisions have little effect and are generally ignored. Thus the electron kinetic equation is especially tractable in a heavy-ion plasma.

Equation (7.93) suggests measuring the electron collision time by the quantity

$$\tau_e \equiv \frac{3\sqrt{\pi}}{8} \frac{m_e^2 v_{te}^3}{\gamma_{ee}Z_{\text{eff}}n_e}. \tag{7.95}$$

We occasionally use this definition, although it does not describe the full dependence of the collision operator on effective charge. For example, the balance between energy scattering, present in the first term of (7.93), and the Lorentz scattering in the remaining term, depends upon $Z_{\text{eff}}$ in a matter not summarized by $\tau_e$.

## Scattering of a heavier species

We have noted that like-particle collisions dominate in the scattering of a heavy species $h$. Nonetheless for some applications the correction term, $C_{hl}$, is important. We compute the linearized form of this operator in the rest frame of the heavy species, assuming that the flow velocity $\mathbf{V}_l$ in this frame is suitably small. Recall that the Maxwellian contribution $C_{hl}(f_{hM}, f_{lM})$ does not generally vanish, being responsible in particular for energy exchange; however, because the energy exchange is proportional to $m_l/m_h$ the Maxwellian contribution to $C_{hl}$ often neglected.

To begin we observe from (7.19) and (7.57) that the linearized operator $C_{hl}^l$ is dominated by dynamical friction, $\mathbf{R}_{hl}$; the contribution from diffusion is smaller in the mass-ratio. If we further assume

$$f_{l1}/f_{lM} \sim f_{h1}/f_{hM}$$

then we find, estimating velocity derivatives in the usual way, that the heavy-species perturbation makes a relatively small contribution and we have

$$C_{hl}^l = -\frac{1}{m_h}\mathbf{R}_{hl}(f_{l1}) \cdot \frac{\partial f_{hM}}{\partial \mathbf{v}} + O\left(\sqrt{\frac{m_e}{m_i}}\right).$$

To compute the dynamical friction we use the formula (7.77) for the mean friction force, neglecting an obviously small term:

$$\mathbf{F}_{hl} = \int d^3 v f_{hM}\mathbf{R}_{hl}(f_{1l}) + O\left(\frac{m_e}{m_i}\right)$$

To lowest order in $v_{ti}/v_{te}$, the heavy distribution acts like a delta-function (for $T_i \sim T_e$) so we have

$$\mathbf{R}_{hl} = \frac{\mathbf{F}_{hl}}{n_h} + O\left(\sqrt{\frac{m_e}{m_i}}\right)$$

and the desired operator is

$$C_{hl}^l = -\frac{\mathbf{F}_{hl}}{m_h n_h} \cdot \frac{\partial f_{hM}}{\partial \mathbf{v}}. \tag{7.96}$$

Thus, to lowest order in the mass ratio, the collisional effect of a light species on a heavy one is to replace the electric field in the kinetic equation by an effective field,

$$\mathbf{E}_* \equiv \mathbf{E} + \frac{\mathbf{F}_{hl}}{e_h n_h}. \tag{7.97}$$

As in the case of wind resistance, the heavier particles experience a simple drag force.

## 7.7   Collisions with neutral particles

### Charge trajectory

Many plasmas, including those prevalent in interstellar space, as well as the cold plasmas used in industry, coexist with a population of neutral atoms or molecules. Sometimes it is the ionization of this neutral background which maintains the plasma density, even when the neutral density is relatively small. In any case the exchange of particles, momentum and energy between charged and uncharged particles can affect or even dominate the plasma dynamics. Ionization, which produces plasma from neutrals, and its inverse process recombination are key contributions to the source terms $I$ treated in Chapter 3. A closely related mechanism called *charge exchange* is studied here because it most resembles a collisional process.

Charge exchange (or "charge transfer") occurs when a neutral atom loses an electron to a nearby ion, usually neutralizing it. Note that the number of plasma particles (and neutral particles) is unchanged: charge exchange conserves particles locally. In the most important case of "resonant" exchange, the electron's initial and final energy states are the same, so that the total kinetic energy, as well as the total momentum, of the two particles is unchanged. But charge exchange is not energy conserving for either species alone, and often represents an important loss mechanism for plasma energy and momentum.

Figure 7.1 shows that the charge exchange process can be viewed in two ways. One can say that a charged particle occasionally loses its charge and moves as a neutral, until suffering a second exchange. This description is most useful in understanding the transport induced by charge exchange. Alternatively one can follow the path of charge itself and view the process as an ordinary (charge-conserving) collision. For example the exchange between a hot ion and a cold neutral can be legitimately interpreted as a transfer of ion *energy* to the neutral species, without charge transfer.

### Charge exchange operator

Figure 7.1 also suggests a crude model for the effect of charge exchange on the ion distribution function $f_i$. Ions at phase point $(\mathbf{x}, \mathbf{v})$ are lost (through gaining electrons) at the rate $\nu_x f_i$ proportional to their number and to the charge exchange frequency $\nu_x$. They are simultaneously replenished from the neutral distribution $f_n$ at a rate $\nu'_x f_n$ proportional to the number of neutrals at that phase point. In other words the time rate of change of $f_i$ due to charge exchange should have the approximate form

$$\left(\frac{d}{dt}\right)_{cx} f_i \equiv C_x(f_i, f_n) = -\nu_x f_i + \nu'_x f_n.$$

While particles are not conserved at each velocity (the newly-born ion will in general have a different velocity from the neutralized ion), the total number of

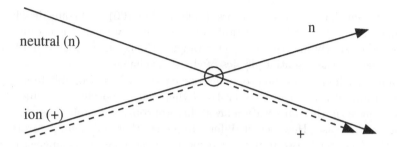

Figure 7.1: Charge exchange. The paths of two atoms are indicated by solid lines; the (+) label refers to an ionized atom, while (n) labels a neutral atom. Both paths are straight: the atoms do not suffer a mechanical collision. However, because of the charge-exchange event occurring at the intersection, the trajectory of *charge* changes direction, as indicated by the dashed lines. The result can be seen to resemble a mechanical collision, without charge exchange, between an ion and a neutral.

ions cannot change. Therefore we must impose

$$\int d^3v \, C_x = 0,$$

whence $\nu'_x = (n_i/n_n)\nu_x$ and we have

$$C_x(f_i, f_n) = -\nu_x \left( f_i - \frac{n_i}{n_n} f_n \right). \qquad (7.98)$$

In fact this simple form is often used in applications. The *charge-exchange frequency* $\nu_x$ is used to define a charge exchange mean-free path,

$$\lambda_{xi} \equiv v_{ti}/\nu_x, \qquad (7.99)$$

closely analogous to the $\lambda_{\text{mfp}}$ for Coulomb collisions.

A more accurate version of $C_x$ takes into account the dependence of $\nu_x$ on the speeds of the exchanging particles. The microscopic reaction rate is proportional to the charge-exchange cross-section $\sigma_x$, and to the rate $|\mathbf{v} - \mathbf{v}'|$ at which atoms with velocities $\mathbf{v}$ and $\mathbf{v}'$ can intersect. It must also be proportional to the number of available neutrals. Thus the frequency for ion loss depends upon velocity according to

$$\nu_x(\mathbf{v}; f_n) = \int d^3v' \, \sigma_x(\mathbf{v}, \mathbf{v}')|\mathbf{v} - \mathbf{v}'|f_n(\mathbf{v}').$$

Using an analogous expression for the gain frequency $\nu'_x$ we obtain the charge-exchange operator

$$C_x(f_i, f_n) = -\int d^3v' \, \sigma_x(\mathbf{v}, \mathbf{v}')|\mathbf{v} - \mathbf{v}'| \left[ f_n(\mathbf{v}')f_i(\mathbf{v}) - f_n(\mathbf{v})f_i(\mathbf{v}') \right] \qquad (7.100)$$

It can be seen that the two operators, (7.98) and (7.100), will coincide when the product $\sigma_x(\mathbf{v}, \mathbf{v}')|\mathbf{v} - \mathbf{v}'|$ is constant. In fact $\sigma_x|\mathbf{v} - \mathbf{v}'|$ is not constant, but it varies rather slowly in most parameter ranges. Hence the simple form of (7.98) often preserves the essential physics of charge exchange.

Charge exchange has the properties one expects of a reasonable binary colli- sion operator: it is bilinear, conservative (when neutral species are included with plasma species in the conservation laws), has appropriate symmetries and causes entropy to increase. However, it differs from the other collision operators stud- ied in this Chapter in two important respects. First, charge exchange scattering is not soft, in the sense of producing mainly small velocity changes. Recall that the trajectory of charge, after the "collision," simply extends the initial trajectory of the neutral particle. Since the initial velocities of the exchang- ing particles are uncorrelated, large-angle charge-exchange scattering events are common. In other words charge exchange is not a Fokker-Planck process, and we cannot approximate (7.100) by a differential operator.

The second important difference is that charge exchange does not necessarily drive either distribution towards Maxwellian form. Instead we see that the fully relaxed distributions are related in the charge-exchange case by

$$f_i(\mathbf{x}, \mathbf{v}) = \frac{\nu_x(\mathbf{v}; f_i)}{\nu_x(\mathbf{v}; f_n)} f_n(\mathbf{x}, \mathbf{v}). \tag{7.101}$$

Here the slow variation of $\sigma_x|\mathbf{v} - \mathbf{v}'|$ implies that $f_i$ and $f_n$ attain nearly the same velocity dependence in charge-exchange equilibrium—as predicted by the simple form (7.98). But neither relaxed distribution need be Maxwellian.

## Ion-neutral coupling

Charge exchange, like unlike-species Coulomb collisions, represents a coupling process—in this case between ions and a neutral species. Here we examine the coupling from the viewpoint of moment equations.

Moments of the charge-exchange operator, corresponding to momentum and energy exchange are defined precisely as in Chapter 3. The charge-exchange mo- ments add to the collisional moments and satisfy analogous laws. For example,

$$\mathbf{F}_{x,sn} \equiv \int d^3v \, m_s \mathbf{v} C_x(f_s, f_n),$$

is the "friction force" experienced by plasma-species $s$ due to charge exchange with neutral species $n$; the neutral species is subject to the corresponding force $\mathbf{F}_{x,ns} = -\mathbf{F}_{x,sn}$. Thus we can account for charge exchange in the species-specific moment equations of Chapter 3, (3.51) and (3.52), by simply assuming the collisional moments to include both Coulomb and charge-exchange scattering.

The *neutral* moment equations are obviously given by the $e \to 0$ limit of the plasma moments. In a charge-exchange environment, elastic neutral collisions are usually negligible in comparison with $C_x$; if we also ignore sources (such as recombination) for simplicity, the equation of motion for a neutral species $n$ exchanging charge with a single ion species $i$ is

$$m_n n_n \frac{d\mathbf{V}_n}{dt} + \nabla p_n + \nabla \cdot \boldsymbol{\pi}_n = \mathbf{F}_{x,ni}.$$

Observe that a nearly Maxwellian, unaccelerated neutral fluid must balance charge-exchange friction against its pressure gradient: $\mathbf{F}_{x,in} = -\nabla p_n$. In that (typical) case the charge-exchange coupling to ions is very simple: the ion equation of motion

$$m_i n_i \frac{d\mathbf{V}_i}{dt} + \nabla p_i + \nabla \cdot \boldsymbol{\pi}_i - e_i n_i(\mathbf{E} + \mathbf{V}_i \times \mathbf{B}) = \mathbf{F}_i + \mathbf{F}_{x,in}$$

becomes

$$m_i n_i \frac{d\mathbf{V}_i}{dt} + \nabla (p_i + p_n) + \nabla \cdot \boldsymbol{\pi}_i - e_i n_i(\mathbf{E} + \mathbf{V}_i \times \mathbf{B}) = \mathbf{F}_i. \qquad (7.102)$$

That is, because of charge exchange, the ion fluid moves as if subject to the combined pressure $p_i + p_n$. Since charge-exchange enters (7.102) in no other way, we infer that the force of charge-exchange in a plasma with neutrals is measured by the relative pressure $p_n/p_i$.

# Additional reading

Our derivation of the Balescu-Lenard operator is strongly influenced by the very clear discussion of Fried[27]. An alternative treatment, covering several topics we have been forced to omit, may be found in the thorough discussion of Balescu[5]; indeed Balescu develops much of the material in Chapters 7 and 8 of the present book.

Properties of the Landau-Boltzmann operator are derived in many books, including Balescu, above, and most of the texts cited at the end Chapter 1; the discussion by Braginskii[13] is perhaps especially useful. Spitzer's discussion of the Coulomb logarithm[67] is authoritative and clear.

Interaction between plasmas and neutral gases is usefully reviewed by Tendler and Heifetz[72].

# Problems

1. This and the next two problems concern the detailed derivation of the Balescu-Lenard collision operator.

   First, by performing the inverse Fourier transform of (7.32), show that the perturbed distribution is given by

   $$\mathcal{F}_s(\mathbf{x}, \mathbf{v}, t) = \mathcal{F}_s(\mathbf{x} - \mathbf{v}t, \mathbf{v}, 0) - \frac{e_s}{m_s} \int_0^\infty d\tau \tilde{\mathbf{E}}(\mathbf{x} - \mathbf{v}\tau, t - \tau) \cdot \frac{\partial \bar{\mathcal{F}}_s}{\partial \mathbf{v}} \quad (7.103)$$

   Obtain the same result by solving the linearized kinetic equation by integration along the characteristics.

2. Use the result of the previous exercise, (7.103), to show that the diffusion tensor can be expressed as

   $$\mathbf{D}_s = \frac{e_s^2}{m_s^2} \int_0^\infty d\tau \langle \tilde{\mathbf{E}}(\mathbf{x}, t) \tilde{\mathbf{E}}(\mathbf{x} - \mathbf{v}\tau, t - \tau) \rangle. \quad (7.104)$$

   Thus collisional diffusion is proportional to the time-integrated correlation function of the fluctuating electric field.

3. For the relevant case of spatially and temporally homogeneous statistics, the correlation function in (7.104) depends only on the difference between its two arguments; thus we have

   $$\mathbf{D}_s = \frac{e_s^2}{m_s^2} \int_0^\infty d\tau C_{\alpha\beta}^E$$

   where

   $$C_{\alpha\beta}^E = \langle \tilde{\mathbf{E}}(\mathbf{v}\tau, \tau) \tilde{\mathbf{E}}(0, 0) \rangle.$$

   Use (7.33) and (7.36) to show that

   $$C_{\alpha\beta}^E = \frac{1}{(2\pi)^3} \sum_s e_s^2 \int d^3k d^3v' \frac{\mathbf{k}\mathbf{k}\bar{\mathcal{F}}_s}{k^4 |\epsilon(\mathbf{k}, \mathbf{k} \cdot \mathbf{v}')|^2} e^{iku\tau}$$

   where, as in the text, $ku \equiv \mathbf{k} \cdot (\mathbf{v} - \mathbf{v}')$ and we have retained only the ballistic pole contributions to the integral. When this expression is substituted into (7.104) and the integral is evaluated by means of (7.35), the Balescu-Lenard diffusion tensor of (7.38) is reproduced.

4. Show explicitly that the small mass-ratio approximation implies $\tau_{ee} \sim \tau_{ei}$, while $\tau_{ii} \ll \tau_{ie}$.

5. Suppose that the linearized operator $C_{ss'}^l(\hat{f}_s, \hat{f}_{s'})$ vanishes. What can you conclude about the distributions $\hat{f}_s$ and $\hat{f}_{s'}$?

6. Suppose that the distribution of neutral particles in a plasma is known, that charge exchange is the dominant dissipative process, and that the plasma-neutral system has come to equilibrium. Can one then determine the ion distribution? Discuss in the context of (7.101).

# Chapter 8

# Collisional Transport

## 8.1 Physics of diffusion

### Fick's law

In the previous chapter we found it useful to distinguish two kinds of force that affect charged particle motion in a plasma: the mean or macroscopic force, such as an externally imposed electric field, and the random impulse associated with collisions. Thus one pictures a particle orbit as random collisional motion imposed on the mean-field trajectory. The local flow of the plasma as a whole—the quantity we have denoted by $n\mathbf{V}$—will reflect both types of motion. A crucial observation is that this flow need not vanish in the absence of the macroscopic force; the bulk fluid can move even when its constituent particles experience only random impacts.

A simple way to visualize the situation is to imagine a large collection of sparklers—fireworks that emit sparks from a wire coated with incendiary flakes. Sparklers display a nearly symmetrical pattern of emitted sparks about the wire axis, with each spark having a finite range. Imagine the sparklers arranged in several parallel rows, where the row spacing approximates the sparkler range; two rows of such an array are depicted in Figure 8.1. For simplicity the sparklers are uniformly distributed on each row, but the row populations are allowed to vary. This arrangement corresponds to a spark density that depends on a single coordinate, denoted by $x$ in the Figure. Also indicated is a window or screen placed between two of the rows; the screen should be small compared to the entire array, but still large enough to extend over many sparklers on each side (not as represented in the Figure). By counting sparks as they cross the screen from either side, we can measure their velocity distribution $f(x, v_x)$ at the location $x$ of the screen.

It is clear that when the sparkler density increases with increasing $x$, a screen placed between any two rows will see an anisotropic distribution $f$, with net flow in the negative-$x$ direction. Indeed, the spark flow will be proportional to the

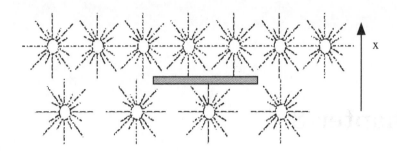

Figure 8.1:  An aerial view of two rows of sparklers. Each sparkler emits isotropically, yet the spark distribution observed by the indicated screen (grey rectangle) is anisotropic, corresponding to a net downward flow.

sparkler density gradient,

$$\text{flow} \propto -\nabla(\text{density}).$$

This is the sparkler version of an archetypal "transport" relation, called Fick's law. It is an essentially geometrical result, pertaining despite the fact that each sparkler emits isotropically: the total number of upward-moving sparks equals the number of downward moving sparks.

Because of the short range of an individual spark, the spark-flow between any two rows depends only upon the density in those two rows; that is, the flow is determined locally. For larger spark range the screen would detect sparks from several rows, making the flow depend non-locally upon total number of sparklers within range—upon some global spark-density profile. (Of course the flow will continue to depend upon density inhomogeneity, vanishing for a uniform array.) In that case no local transport relation like Fick's law can pertain.

The same geometrical facts apply to flow in a plasma (or neutral gas). The role of the sparklers is replaced by $90°$ collisions: the isotropy of spark emission corresponds to rotational invariance of Coulomb collisions, and the spark range corresponds to the collisional mean-free path, $\lambda_{\text{mfp}}$. The analogy works because, after a time $\nu^{-1}$, where $\nu$ is the collision frequency, each particle in the plasma has forgotten its previous velocity. Therefore a typical particle trajectory is one that was enforced at a location

$$\lambda_{\text{mfp}} \equiv v_t/\nu \tag{8.1}$$

distant. Thus it might as well have come from a sparkler at that location, and, as in the sparkler case, geometry will induce a net plasma flow in the direction opposite to the density gradient. When the mean-free path is short compared to the gradient scale-length,

$$\frac{\lambda_{\text{mfp}}}{L} \ll 1, \tag{8.2}$$

this flow is described by the local transport relation,

$$n\mathbf{V} = -D\nabla n, \tag{8.3}$$

which is Fick's law with diffusion coefficient $D$.

The sparkler analogy captures the essence of Fick's law, and gives insight into the significance of the mean-free path. But it cannot model many other transport properties. In particular, it fails to represent the fact that the resulting flow can affect the density profile and cause fluid evolution.

Generally we expect the flow to reduce the gradient that drives it, and thus to increase the fluid entropy. But the density is affected only when the flow has a divergence. Specializing to the simplest case of one-dimensional diffusion with constant coefficient, we have $nV = -D\partial n/\partial x$ and particle conservation implies

$$\frac{\partial n}{\partial t} = D\frac{\partial^2 n}{\partial x^2}. \tag{8.4}$$

A uniform density gradient is not eroded (because each slice in $x$ passes on the fluid it receives from its neighbor), but diffusion dissipates any density peaks or troughs, on a time scale

$$\tau \sim \frac{(\Delta L)^2}{D}$$

where $\Delta L$ measures the size of the density perturbation. This dissipative character, which increases fluid entropy by removing spatial order, is a distinctive feature of diffusive flow. Heat conduction, which carries heat down the fluid temperature gradient, is similarly dissipative; but some other types of flow, even those proportional to gradients, have no effect on entropy. An example is the diamagnetic flow derived in Chapter 4, which is proportional to the density or temperature gradient, but has the wrong direction to dissipate order.

A second distinguishing feature of diffusion concerns propagation. By solving the diffusion equation (8.4) one finds that an initial density-perturbation spike diffuses according to

$$\Delta n \propto t^{-1/2}e^{-x^2/4Dt}. \tag{8.5}$$

That is, the spread in $x$ increases roughly (aside from a logarithmic correction) as the square-root of the time. Note that if the same perturbation were convected by bulk fluid motion, the distance traveled would be proportional to $t$, at least over distances shorter than the scale-length for change in the fluid velocity.

To avoid confusion we note that that Fick's law requires mild gradients and does not apply to a delta-function density. Hence (8.5) is physical only at sufficiently large $t$; explicitly one finds that $t$ must far exceed the collision time of the underlying random process.

## Diffusion coefficient

In Chapter 7 we studied diffusion in velocity space due to (Coulomb) collisions between particles. We assumed the correlation time for changes in particle motion was short, and used a random walk argument to obtain the coefficient given by (7.17). Precisely the same argument in (one-dimensional) coordinate space yields the spatial diffusion coefficient

$$D = \frac{\langle(\Delta x)^2\rangle}{\Delta t}. \tag{8.6}$$

Here we can associate $\Delta t$ with the inverse collision frequency for $90°$ scattering:

$$\Delta t \sim 1/\nu. \tag{8.7}$$

Similarly $\Delta x$ measures the size of such excursions—the step size of the random walk. As remarked in Chapter 7, the time-independence of $D$ is characteristic of a random walk, in which $\Delta x \propto t^{1/2}$.

The formula (8.7) can be used to estimate diffusion coefficients in various circumstances. In an unmagnetized plasma or neutral gas, for example, particles move a distance $\Delta x \sim \lambda_{\mathrm{mfp}}$, before embarking on a new, uncorrelated step. According to (8.7) and (8.1), the corresponding diffusion coefficient is estimated by

$$D \sim \nu(v_t/\nu)^2 = v_t^2/\nu. \tag{8.8}$$

Therefore Fick's law prescribes the flow speed of diffusion to be roughly

$$V \sim \frac{v_t^2}{\nu L} = v_t \frac{\lambda_{\mathrm{mfp}}}{L}$$

where $L$ is the gradient scale-length.

Note the characteristic *inverse* dependence on collision frequency: collisions impede entropy increase by slowing particle motion across the gradient. On the other hand, since collisions provide the essential randomizing agent, we cannot believe (8.8) as $\nu \to 0$. The point, already emphasized, is that the entire local approach makes sense only in the short mean-free-path case of (8.2). It follows in particular that diffusive velocities are always much slower than the thermal speed.

It is instructive to express Fick's law as

$$nV \sim v_t[n(x + \lambda_{\mathrm{mfp}}) - n(x)]. \tag{8.9}$$

When the bracketed difference is expanded in a Taylor's series, the first term can be seen to reproduce (8.3) and (8.8). Of course this formula is reliable only in when $\lambda_{\mathrm{mfp}}$ is sufficiently small. But extrapolating nonetheless to longer mean-free path, we see (correctly) that the non-local transport process at finite $\lambda_{\mathrm{mfp}}/L$ involves all derivatives of the density. At the same time the flow driven by density variation becomes more rapid, approaching the thermal speed as $\lambda_{\mathrm{mfp}}$ approaches the system scale-length. Thus two key characteristics of transport at short mean-free path—*weak* flows depending on *local* gradients—are inextricably linked.

## 8.2   Local transport theory

### General features

Transport theory describes the behavior of a plasma that is close to equilibrium. It predicts the plasma response to so-called *thermodynamic forces*, which characterize small departures from equilibrium. In thermodynamic equilibrium, the plasma temperature, for example, must be uniform in space. Warming one boundary of such a plasma drives the system slightly away from its equilibrium state. The corresponding thermodynamic force in this case is the temperature gradient, which we require to be weak in an appropriate sense. A collisional plasma responds to this driving force in an especially simple way: by exhibiting a heat flux $\mathbf{q}$ that is linearly related to the *local* temperature gradient. Thus we have the typical transport relation

$$\mathbf{q}(\mathbf{x}) = -\kappa(\mathbf{x})\nabla T(\mathbf{x}), \qquad (8.10)$$

in terms of the thermal conductivity, $\kappa$, which is computed by solving an appropriately simplified kinetic equation. Fick's law describes the response to a density gradient (or to a gradient in the chemical potential). More generally the departure from equilibrium is characterized by a set of thermodynamic forces, $A_i(\mathbf{x})$, $i = 1, \ldots, n$, analogous to $\nabla T$. Then, again assuming the forces to be sufficiently weak, the plasma responds through a corresponding set of flows, $\Gamma_i(\mathbf{x})$, with transport relations

$$\Gamma_i = -\sum_j L_{ij} A_j. \qquad (8.11)$$

Here the quantities $L_{ij}$ are called transport coefficients; the entire set $L_{ij}$ is called the *transport matrix*. The object of transport theory is to identify the flows corresponding to specified thermodynamic forces, and then to compute, from a kinetic equation, the transport matrix.

In the previous section we noted that collisions, by restricting free-particle motion to distances shorter than the gradient scale-length, localize the fluid response, yielding transport laws of the form of (8.11). It is in this context that the formalism of (8.11) was developed, by Chapman and Cowling and others, as a linearized extension of equilibrium thermodynamics. Transport theory for an unmagnetized plasma follows a closely analogous development, and depends equally upon short mean-free path: there is no linear transport matrix for an unmagnetized, low-collisionality plasma.

The case of a magnetized plasma is more interesting: magnetized plasma transport is strongly anisotropic. Along the direction of the magnetic field, behavior is qualitatively similar to the unmagnetized case, and local only in the case of short mean-free path. But perpendicular to the magnetic field, localization can be provided by the field itself: even for arbitrarily weak collisions, their small gyroradii inhibits particle excursions across the density gradient. Hence a local description of the perpendicular response of a magnetized plasma, following (8.11), is possible in any collisionality regime.

We will find that the form of the transport coefficients is correspondingly anisotropic in a magnetized plasma. Along the magnetic field, collisions impede free particle motion, so that such quantities as the thermal conductivity $\kappa$ vary inversely with collision frequency. Dissipative flow across the magnetic field, on the other hand, requires collisions; thus perpendicular transport coefficients are proportional to the collision frequency.

In summary, the local description of (8.11) is always approximate, and not always possible: it requires an appropriate small parameter. In the case of an unmagnetized plasma, or a magnetized plasma responding to parallel forces, the only useful localizing parameter is the mean-free-path ratio, $\Delta = \lambda_{\mathrm{mfp}}/L$; a local description of parallel transport is possible only when $\Delta \ll 1$. In the case of a magnetized plasma, the small parameter $\delta = \rho/L$ can take the role of $\Delta$, allowing a local description of the response to perpendicular forces. Note that the weakness of the thermodynamic forces, mentioned at the start of this section, is measured by the smallness of $\Delta$ or $\delta$.

Just as it requires weak gradients, conventional transport theory also requires slow evolution in time. Thus parallel transport theory, like neutral fluid theory, uses the assumption,

$$\partial/\partial t \sim \nu \Delta^2, \tag{8.12}$$

where $\nu$ is the collision frequency. Of course one similarly assumes

$$\partial/\partial t \sim \nu \delta^2, \tag{8.13}$$

in studying perpendicular transport of a magnetized plasma. We will refer to either form of the slowness assumption as the *transport ordering*. Departures from the transport ordering are not always fatal. For example, an oscillatory time dependence can be incorporated into the theory, even when the oscillation frequency, $\omega$, is comparable to $\nu$; the transport coefficients derived in this case, sometimes called "AC coefficients," depend upon $\omega$. However the derivation of a transport matrix requires that the form of any rapid time dependence be specified *a priori*.

## Transport closure

Although it may be generally useful to know, for example, that the heat flow in a collisional plasma is proportional to the local temperature gradient, the key application of transport theory is in providing a closure for plasma moment equations. In Chapter 3 we derived exact fluid (moment) equations for the evolution of density, temperature and flow velocity of each plasma species. These five scalar equations, however, necessarily involve higher-order moments: they do not provide a closed system. When the mean free path is short ($\Delta \ll 1$), transport theory can be used to evaluate the higher moments and close the system. This program, the most prominent example of a rigorously asymptotic fluid closure, is called Chapman-Enskog theory.

There is no corresponding transport closure at long mean-free path. In the magnetized case, schemes based on the small gyroradius parameter $\delta$ are easily constructed, but intrinsically incomplete: they leave the parallel response undetermined. That is why the purely fluid closures presented in Chapter 4, MHD and the drift model, are not rigorously asymptotic. Both rely on small-$\delta$ alone, but neither closes the parallel dynamics in a rigorous way; for example the parallel heat flows and the stress anisotropy are neglected without physical justification. In both cases the rigorous versions (including, *inter alia*, anisotropy and heat) require reaching outside the fluid context, and incorporating a kinetic description of the parallel response. A model for this sort of fluid-kinetic hybrid closure is "kinetic MHD," discussed in Chapter 4.

Even in the case of short mean-free path

$$\Delta \ll 1$$

which is assumed for the remainder of this chapter, misunderstandings of the closure issue are possible. Thus we consider the matter more concretely, beginning with the exact moments derived in Chapter 3. Omitting source terms (which, being specified, do not affect closure) for simplicity, we have the familiar five conservation laws:

$$\frac{dn_s}{dt} + n_s \nabla \cdot \mathbf{V}_s = 0, \tag{8.14}$$

$$m_s n_s \frac{d\mathbf{V}_s}{dt} + \nabla p_s + \nabla \cdot \boldsymbol{\pi}_s - e_s n_s (\mathbf{E} + \mathbf{V}_s \times \mathbf{B}) = \mathbf{F}_s, \tag{8.15}$$

$$\frac{3}{2}\frac{dp_s}{dt} + \frac{5}{2}p_s \nabla \cdot \mathbf{V}_s + \boldsymbol{\pi}_s : \nabla \mathbf{V}_s + \nabla \cdot \mathbf{q}_s = W_s. \tag{8.16}$$

The first step in understanding any closure is to identify its basic variables—the quantities that the system is supposed to advance in time. For the present we choose these variables to be the densities, temperatures and flows velocities of each plasma species: five (scalar) variables per species. Then (since the electromagnetic field variables can be presumed known from Maxwell's equations), our five conservation laws will close the system once they are expressed in terms of the electromagnetic field and these variables alone. In other words, expressing the quantities

$$\mathbf{q}_s, \boldsymbol{\pi}_s, \mathbf{F}_s, W_s \tag{8.17}$$

in terms of

$$n_s, p_s, \mathbf{V}_s, \tag{8.18}$$

will constitute a closure of the system. It is a *transport closure* if these expressions have the form of (8.11): that is, we identify the moments (8.17) as fluxes, and use kinetic theory to derive transport equations in which the forces $A_i$ are expressed in terms of the basic variables (8.18). This scheme is the essence of Chapman-Enskog theory.

This choice of variables and fluxes is not unique. For example, one could instead use only the particle and energy conservation laws, as a basic dynamical system for the variables $n_s$ and $p_s$. The flow velocities would in this case join the list of moments, (8.17), that are to be determined from kinetic theory.

What is potentially confusing in the Chapman-Enskog development is the role of kinetic theory. Since closure requires solving the kinetic equation, why is it useful to consider moment equations at all? Once the distributions $f_s$ are known, what else is there to determine? These questions have sensible answers only because the parameter $\Delta$ is small; they depend in particular upon the transport ordering. We see first of all that the closed fluid system describes dynamics on the slow time-scale of (8.12). Hence we would have to include all

$O(\Delta^2)$-terms in an equivalent kinetic description. But we shall find that the calculation of the fluxes demands less accurate kinetic theory: only the $O(\Delta)$ terms are needed. Thus Chapman-Enskog theory never calculates the distributions to an accuracy that displays time-dependence: the $\Delta$-order corresponding to time evolution is described exclusively by moment equations.

In other words small-$\Delta$ allows the use of a crude and therefore tractable kinetic equation, without time variation. Yet the results of this simplified kinetic analysis can be consistently inserted into the higher-order moment equations that describe evolution. As a scheme for squeezing the essential second-order information out of a first-order kinetic description, Chapman-Enskog theory is clearly analogous to the calculations of perpendicular motion, based on small gyroradius, of Chapter 4. In both cases moment equations are manipulated to reduce the burden on kinetic theory. Both procedures stem from the recognition that phase space has more dimensions than coordinate space.

## 8.3  Kinetic description of a magnetized plasma

### Guiding center distribution

The guiding center distribution characterizes a magnetized plasma—one in which the ratio of gyroradius to scale-length is small:

$$\delta \equiv \frac{\rho}{L} \ll 1.$$

Here we study this distribution as a tool for understanding plasma transport.

In chapter 2 we learned that the energy,

$$\mathcal{E}(\mathbf{x}, t) = \frac{1}{2}mv^2 + e\Phi(\mathbf{x}, t)$$

and the lowest-order magnetic moment,

$$\mu(\mathbf{x}, \mathbf{v}, t) = \frac{mv_\perp^2}{2B(\mathbf{x}, t)}$$

are approximate constants of the motion in a magnetized plasma. For $\mu$ is the lowest order approximation to the adiabatically invariant magnetic moment, while $\mathcal{E}$ changes only because of electromagnetic work,

$$\frac{d\mathcal{E}}{dt} = -e\mathbf{v} \cdot \frac{\partial \mathbf{A}}{\partial t}, \tag{8.19}$$

which is small in the case of the transport ordering. (The magnitude of electromagnetic work is considered in more detail below.) It follows that a distribution function $f$ depending on position and velocity only through $\mathcal{E}$ and $\mu$ would vary slowly, being constant on the time scale of the transit frequency, $\omega_t$:

$$\frac{df(\mathcal{E}, \mu, t)}{dt} = O(\delta\omega_t). \tag{8.20}$$

The spatial dependence of such a distribution is severely constrained; to describe the collisional dissipation of plasma fluid gradients, more general spatial dependence must be permitted. The point is that the slow time variation of the transport ordering reflects a *balance* between gradient-induced flow and collisions, rather than the collisionless equilibrium described by (8.20). Thus we consider distributions that depend upon the guiding center position, $\mathbf{x}_{gc}(\mathbf{x}, \mathbf{v}, t)$, as well as $\mathcal{E}$ and $\mu$. Here $\mathbf{x}_{gc}$ is the solution to the guiding center equation,

$$\frac{d\mathbf{x}_{gc}}{dt} = \mathbf{v}_{gc} = \mathbf{b}v_{\parallel} + \mathbf{v}_D$$

derived in Chapter 2—the instantaneous center of gyration of a particle trajectory. It can be expressed in terms of the particle position, $\mathbf{x}$, and the vector gyroradius

$$
\begin{aligned}
\boldsymbol{\rho}(\mathbf{x}, \mathbf{v}, t) &= \frac{\mathbf{b}(\mathbf{x}, t) \times \mathbf{v}}{\Omega(\mathbf{x}, t)} \\
&= \frac{v_{\perp}}{\Omega}(-\mathbf{e}_1 \cos\gamma + \mathbf{e}_2 \sin\gamma)
\end{aligned}
\tag{8.21}
$$

as

$$\mathbf{x}_{gc}(\mathbf{x}, \mathbf{v}, t) = \mathbf{x} - \boldsymbol{\rho}(\mathbf{x}, \mathbf{v}, t) \tag{8.22}$$

Recall here that $\gamma$ is the gyrophase angle and $\mathbf{e}_1$ and $\mathbf{e}_2$ are unit vectors normal to $\mathbf{b}$.

Because the guiding center position does not change on the time scale of the gyrofrequency $\Omega$, a function that depends on $\mathbf{x}$ and $\mathbf{v}$ only through $\mathbf{x}_{gc}$, $\mathcal{E}$, and $\mu$ will satisfy

$$\frac{dF(\mathbf{x}_{gc}, \mathcal{E}, \mu, t)}{dt} = O(\omega_t). \tag{8.23}$$

Here the arbitrary function $F$ must be suitably smooth; in particular we require

$$\frac{\partial F}{\partial \mathbf{x}_{gc}} \sim \frac{F}{L}. \tag{8.24}$$

The relatively fast variation allowed by (8.23) (compare (8.20)) shows the essential role of collisions in the transport ordering.

Such distributions are called *guiding center* distributions and denoted by

$$f_{gc}(\mathbf{x}, \mathbf{v}, t) = F(\mathbf{x}_{gc}, \mathcal{E}, \mu, t). \tag{8.25}$$

The guiding center distribution evidently resides in a five-dimensional phase-space, where gyrophase information is ignored: $F$ is independent of $\gamma$ at fixed $\mathbf{x}_{gc}$. Note however that at fixed $(\mathbf{x}, \mathbf{v})$, $f_{gc}$ depends upon gyrophase through (8.22). Since moments of any distribution are performed at fixed $\mathbf{x}$, not fixed $\mathbf{x}_{gc}$, this dependence is important.

It is convenient to write

$$f_{gc}(\mathbf{x}, \mathcal{E}, \mu, \gamma) = \bar{f}(\mathbf{x}, \mathcal{E}, \mu) + \tilde{f}(\mathbf{x}, \mathcal{E}, \mu, \gamma),$$

where
$$\bar{f}(\mathbf{x}, \mathcal{E}, \mu) = \langle f_{gc} \rangle_\gamma \equiv \oint \frac{d\gamma}{2\pi} f_{gc}.$$

We will need the form of $\bar{f}$ to derive perpendicular transport coefficients in Section 8.6. To calculate it, we write the kinetic equation in terms of the variables $(\mathbf{x}, \mathcal{E}, \mu, \gamma)$

$$\frac{\partial f}{\partial t} + \mathbf{v} \cdot \nabla f + \frac{d\mathcal{E}}{dt} \frac{\partial f}{\partial \mathcal{E}} + \frac{d\gamma}{dt} \frac{\partial f}{\partial \gamma} = C(f).$$

In a magnetized plasma, the $\gamma$-derivative term in this equation dominates all the others. For the equations of motion imply that (to an accuracy that suffices here)

$$\frac{d\gamma}{dt} = \Omega,$$

which is by definition the largest frequency characterizing a magnetized plasma. It is therefore useful to introduce the operator

$$\mathcal{L} = \frac{\partial}{\partial t} + \mathbf{v} \cdot \nabla + \frac{d\mathcal{E}}{dt} \frac{\partial}{\partial \mathcal{E}},$$

and to write the kinetic equation in the form

$$\mathcal{L}f - \Omega \frac{\partial \tilde{f}}{\partial \gamma} = C(f). \tag{8.26}$$

Here we have noted that the large term acts only on $\tilde{f}$; it follows in particular that

$$\tilde{f} \sim \delta \bar{f}. \tag{8.27}$$

Note that (8.27) allows the (lowest order) identification

$$\bar{f}(\mathbf{x}, \mathcal{E}, \mu) = F(\mathbf{x}, \mathcal{E}, \mu, t),$$

so we can replace $\bar{f}$ by $F$, implicitly evaluating the latter at $\mathbf{x}$ instead of $\mathbf{x}_{gc}$.

Since the average annihilates the $\gamma$-derivative, we see that

$$\langle (\mathcal{L} - C)f \rangle_\gamma = 0. \tag{8.28}$$

This expression is an abstract form of the *drift kinetic equation*, studied in the next subsection. Here we combine it with (8.26) to write

$$\Omega \frac{\partial \tilde{f}}{\partial \gamma} = (\mathcal{L} - C)f - \langle (\mathcal{L} - C)f \rangle_\gamma$$

or, after neglecting $O(\delta^2)$ terms,

$$\Omega \frac{\partial \tilde{f}}{\partial \gamma} = -\tilde{\mathcal{L}} F. \tag{8.29}$$

Next we recall

$$\mathbf{v} = \mathbf{b}v_\| + \mathbf{v}_\perp,$$

with

$$\mathbf{v}_\perp = v_\perp(\mathbf{e}_1 \sin\gamma + \mathbf{e}_2 \cos\gamma),$$

to infer

$$\bar{\mathbf{v}} = \mathbf{b}v_\|,$$
$$\tilde{\mathbf{v}} = \mathbf{v}_\perp.$$

Therefore (8.19) allows us to write

$$\tilde{\mathcal{L}} = \mathbf{v}_\perp \cdot \nabla - ev_\perp \cdot \frac{\partial \mathbf{A}}{\partial t} \frac{\partial}{\partial \mathcal{E}}$$

and straightforwardly solve (8.29). The result is conveniently expressed as

$$\tilde{f} = -\boldsymbol{\rho} \cdot \nabla F + m\mathbf{v} \cdot \left(\frac{\mathbf{b}}{B} \times \frac{\partial \mathbf{A}}{\partial t}\right) \frac{\partial F}{\partial \mathcal{E}}. \tag{8.30}$$

The first term in this expression results from a Taylor series expansion,

$$F(\mathbf{x} - \boldsymbol{\rho}, \mathbf{v}, t) \approx F(\mathbf{x}, \mathcal{E}, \mu) - \boldsymbol{\rho} \cdot \nabla F(\mathbf{x}, \mathcal{E}, \mu). \tag{8.31}$$

which is justified by (8.24). The second term reflects the perturbation in the kinetic energy due to the electromagnetic $\mathbf{E} \times \mathbf{B}$ drift.

## Drift-kinetic equation

To determine the form of the function $F$ we must solve the kinetic equation (8.28). To study parallel transport processes, only the lowest-order terms are required. Thus we consider the limit of zero gyroradius, in which $\langle \tilde{\mathcal{L}}\tilde{f} \rangle$ is neglected and we have simply

$$\bar{\mathcal{L}}F = C(F).$$

or

$$v_\| \nabla_\| F + ev_\| E_I \frac{\partial F}{\partial \mathcal{E}} = C(F). \tag{8.32}$$

Here we use a convenient abbreviation

$$E_I \equiv -\frac{\partial A_\|}{\partial t}$$

for the induced, parallel electric field.

That the electromagnetic work should be included in this order is not obvious. It might be expected to vanish at zero gyroradius, because of the transport ordering (and despite its factor of electric charge). Of course the size of the induced electric field is determined by the physical situation, not by some theoretical ordering. It might well be negligible, or it might dominate all other terms, preventing a quasistatic state—a situation we consider presently. Here we retain it in order to allow an "Ohmic" balance between collisional friction and the electric force:

$$F_{\|ei} \sim enE_I$$

The reader can verify, using (7.79), that this balance is consistent with the transport ordering and with a plausible estimate for the size of the electron parallel flow,

$$\frac{V_{\|e}}{v_{te}} \sim \Delta_e. \tag{8.33}$$

Hence we treat the electric-field term in (8.32) as comparable to the parallel-gradient term; both are measured by the transit frequency $\omega_t$:

$$v_\| \nabla_\| \sim ev_\| E_I \frac{\partial}{\partial \mathcal{E}} \sim \omega_t \tag{8.34}$$

Equation (8.32) is the lowest-order form of the *drift-kinetic equation*, which determines the guiding-center distribution function. It has the form

$$\frac{\partial f}{\partial t} + \mathbf{v}_{gc} \cdot \nabla f + \dot{\mathcal{E}}_{gc} \frac{\partial f}{\partial \mathcal{E}} = C \tag{8.35}$$

where $\mathbf{v}_{gc}$ is the guiding-center velocity studied in Chapter 2 and

$$\dot{\mathcal{E}} \equiv -e \left( \frac{\partial \phi}{\partial t} + \mathbf{v}_{gc} \cdot \frac{\partial \mathbf{A}}{\partial t} \right) - \mu \frac{\partial B}{\partial t}$$

is the guiding-center rate of energy change. The detailed derivation of (8.35) closely parallels the guiding-center orbit derivation of Chapter 2, so we omit it. It is straightforward to verify, using the transport ordering, that (8.35) reduces to (8.32) in lowest order.

## Short mean-free path

We have already noted the very different physics underlying parallel and perpendicular transport. The ways in which these two processes are analyzed are also dissimilar. Indeed, we shall find that the perpendicular transport problem is essentially solved by (8.30). Computing the parallel fluxes, on the other hand, requires solving the drift-kinetic equation. We consider the latter problem here, displaying the Chapman-Enskog approach to (8.32).

The *derivation* of (8.32) used only the magnetization parameter, $\delta$, to disentangle parallel from perpendicular physics; its *solution*, on the other hand, requires use of the collisionality parameter, $\Delta$. Recalling that the parallel transport problem is ill-posed unless the mean-free path is short, we now assume

$$\Delta \ll 1$$

and expand the distribution function in powers of $\Delta$:

$$F = F_0 + \Delta F_1 + O\left(\Delta^2\right)$$

Since the right-hand side of (8.32) is measured by $\nu F$, while (8.34) estimates the two terms on the left as $\Delta \nu F$, the decomposition of our drift-kinetic equation is straightforward. In lowest order we have

$$C(F_0) = 0, \tag{8.36}$$

and in first order,

$$v_{\parallel} \nabla_{\parallel} F_0 + e v_{\parallel} E_I \frac{\partial F_0}{\partial \mathcal{E}} = C^l(F_1), \tag{8.37}$$

where $C^l$ is the linearized operator discussed in Chapter 7; recall (7.86).

The lowest order equation, (8.36), is easy to solve: as we found in Chapter 7, the collision operator vanishes only when acting on Maxwellian distributions, in which both Maxwellians have the same temperature and flow. Specifically we recall that velocity diffusion and dynamical friction combine to produce Maxwellian distributions in each species, while energy exchange $W_L$, and moment exchange, $\mathbf{F}$, lead respectively to equal temperatures and flows of the colliding species.

The equal temperature requirement is realistic when the two colliding species have comparable masses. It can be relaxed, however, in the case of collisions between species of disparate mass, $m_l \ll m_h$. The point is that temperature equilibration, depending upon collisional terms that are small in the mass-ratio, is often slower than the transport processes under consideration. In this regard, note that the relevant mass-ratio approximation neglects not $W_{lh}$ but rather the conserved quantity

$$W_{lhL} = W_{lh} + \mathbf{V}_l \cdot \mathbf{F}_{lh} = O(m_l/m_h). \tag{8.38}$$

The distinction is important because $\mathbf{V}_l \cdot \mathbf{F}_{lh}$ is not small.

Regarding the equal flow requirement, we assume for simplicity that the lowest order flow velocity of each species vanishes:

$$V_{s0} = 0. \tag{8.39}$$

Then the lowest order distributions are

$$F_{s0} = f_{sM},$$

and our first-order equation then becomes

$$C^l(F_{s1}) = v_{\parallel} f_{sM} \left( \nabla_{\parallel} \log f_{sM} - \frac{e_s E_I}{T_s} \right), \tag{8.40}$$

a linear equation for the non-Maxwellian correction to the distribution function of species $s$. (Other species enter implicitly through $C^l$.) We will call an equation of this form, in which the unknown function appears as an argument of the linearized collision operator, a *Spitzer* problem, because of the pioneering work on its solution by Lyman Spitzer and his students. Its solution allows calculating the first-order parallel flows, $nV_\parallel$ and $q_\parallel$, thereby closing the moment equations for density and pressure evolution in the zero-gyroradius limit. Thus solving the Spitzer problem is the central task of parallel, linear transport theory.

The Spitzer problem is challenging because of the complicated form of the collision operator on its left-hand side; we attack this feature in the following two Sections. Here it is convenient to examine the relatively simple *right-hand* side. From

$$\log f_M = \log n - \frac{3}{2} \log T_s - (\mathcal{E} - e\Phi)/T_s + constant,$$

we find that

$$\nabla_\parallel \log f_{sM} - \frac{e_s E_I}{T_s} = A_{s1} + \left(x^2 - \frac{5}{2}\right) A_{s2}. \tag{8.41}$$

Here we used the abbreviation

$$x = v/v_t,$$

as in Chapter 7, and introduced the parallel forces,

$$A_{s1} \equiv \nabla_\parallel \log p_s - \frac{e_s E_\parallel}{T_s}, \tag{8.42}$$

$$A_{s2} \equiv \nabla_\parallel \log T_s, \tag{8.43}$$

which are to be considered thermodynamic forces in the sense of (8.11). The induced field $E_I$ has been replaced by the total parallel electric field,

$$E_\parallel = E_I - \nabla_\parallel \Phi,$$

essentially because the gradient in (8.41) is performed at fixed $\mathcal{E}$. Note also that we have chosen to use the pressure gradient rather than the density gradient in the force $A_{s1}$; this explains why 5/2 rather than 3/2 appears in (8.41).

## 8.4   The Spitzer problem

### Runaway

A skydiver's fall to the earth attains constant velocity because the drag due to air friction, $F_{drag}(v)$, increases with velocity. Thus there is an unaccelerated equilibrium at terminal velocity, $v_{term}$:

$$F_{drag}(v_{term}) = mg,$$

where $mg$ is the gravitational force. Clearly the Ohmic steady state, in which an applied electric field drives a steady current (rather than accelerating charged particles indefinitely) represents an analogous balance of forces. However the collisional balance in a plasma is more delicate: as emphasized in Chapter 7, collisional drag in a plasma is a non-monotonic function of particle speed that *decreases* with $v$ in the $v \gg v_t$ limit.

In Chapter 7 we computed the dynamical friction for the (typical) Maxwellian case:

$$\mathbf{R}_{ss'M} = \frac{2\gamma_{ss'}n_{s'}}{m_{s'}} \frac{\mathbf{v}}{v^3} \left[ erf(x) - x\frac{derf(x)}{dx} \right],$$

where $x \equiv v/v_{ts'}$. Recall that the drag on a moving particle comes from both dynamical friction and diffusion; when the "dragged" particle is lighter than the background particles, diffusion is more important:

$$\mathbf{F}_{ss'drag} \sim m_s \frac{\partial \mathbf{D}_{ss'}}{\partial \mathbf{v}} = \frac{m_{s'}}{m_s}\mathbf{R}_{ss'}. \tag{8.44}$$

However, since the two contributions are proportional, our discussion applies to both.

The magnitude of $\mathbf{R}_{ss'}$ is plotted in Figure 8.2. Recalling that the error function approaches unity for large $x$, we see that $\mathbf{R}_{ss'} \to \mathbf{v}/v^3$, ruling out any terminal velocity for supra-thermal particles. Instead an imposed electric field will accelerate the fast particles to relativistic speeds.

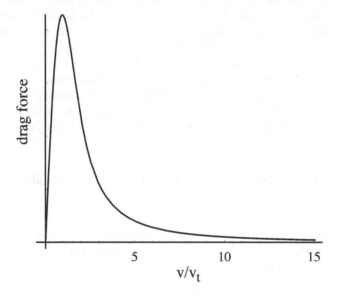

Figure 8.2: The collision frequency, as expressed by the dynamical friction, as a function of $x = v/v_t$. The peak occurs somewhat above the thermal speed where $x = 1$; the large-$x$ behavior is proportional to $x^{-3}$.

Such particles are described as "runaways." Runaway electrons are commonly observed in laboratory experiments, where they end up striking the walls of the vessel and creating X-rays. They also seem to play a role in various astrophysical phenomena.

Figure 8.2 shows that the drag force increases with $v$ up to some speed above the thermal speed. Hence collisional balance is accessible to sub-thermal and thermal particles if the driving field is less than the maximum drag: such particles, starting to the left of the hump, will not be accelerated sufficiently to surmount it. Therefore the maximum of the drag force, measured by

$$\frac{m_{s'}}{m_s} |\mathbf{R}_{ss'}| \sim \frac{2\gamma_{ss'} n_{s'}}{m_s v_{ts}^2},$$

corresponds to a limiting electric field:

$$E_c \sim \frac{2\gamma_{ss'} n_{s'}}{e_s m_s v_{ts}^2}.$$

If the actual driving field is much less than $E_c$, only an exponentially small tail of supra-thermal particles will run away; in typical experiments the driving field must exceed about a tenth of $E_c$ before runaway effects are observed. But if $E_I$ increases to an appreciable fraction of $E_c$, the bulk of the particles cannot reach "terminal" equilibrium and quasistatic evolution of the distribution function, as in the transport ordering, is ruled out.

Note that only the induced electric field, omitting electrostatic contributions, enters this discussion. The point is that particles in an electrostatic field will eventually collect in potential wells; they cannot run away. (Recall our discussion in Chapter 4 of the adiabatic electron response.)

It is clear that the smallest limiting field occurs for the lightest particles: runaway is mainly a problem for electrons. The electron version of $E_c$ is usually called the *Dreicer* field (in view of Harold Dreicer's early investigations), denoted by $E_D$, and expressed in terms of the electron collision frequency, $\nu_e \sim 1/\tau_{ei}$. Recalling (7.83) and ignoring numerical factors of order unity, we see that

$$E_D = m_e v_{te} \nu_e / e. \tag{8.45}$$

In summary, electron runaway is avoided if the Dreicer condition,

$$E_I \ll E_D, \tag{8.46}$$

is satisfied. It is interesting to check the consistency between this condition and our previous estimate, (8.34), for the size of the parallel field. Substituting from (8.34) and (8.45) we find that

$$\frac{E_I}{E_D} \sim \frac{\omega_{te}}{\nu_e}.$$

Thus, happily, the Dreicer condition precisely coincides with the requirement of short (electron) mean-free path.

## Parallel kinetics of a light species

The Spitzer problem can be solved analytically, albeit approximately, when the desired distribution is mainly affected by collisions with a much heavier species.

In that case the collision operator is approximated by the simple Lorentz-gas form, with trivial linearization. Because the Lorentz-gas approximation is most accurate for electron scattering by ions, we consider that case for concreteness. However the same manipulations apply to ion scattering by a heavier ion species, when the ion masses can be considered disparate.

Recall that like-particle collisions—which require a much more complicated operator—can be comparable in effect to scattering by the heavier species; the latter dominates only when $Z_{eff} \gg 1$. Therefore the present analysis considers electron transport in a high-$Z$ plasma. For simplicity we assume a single ion species, which is taken to be at rest.

With these simplifications we need only recall (7.91) to express the electron version of (8.40), as

$$\frac{\nu_{ei}}{x^3} \frac{\partial}{\partial \xi} (1 - \xi^2) \frac{\partial \hat{f}}{\partial \xi} = v_{te} x \xi \left[ A_{e1} + \left( x^2 - \frac{5}{2} \right) A_{e2} \right]. \tag{8.47}$$

Here

$$\nu_{ei} \equiv \frac{\gamma_{ei} n_i}{m_e^2 v_{te}^3} = \frac{3\sqrt{\pi}}{8} \frac{1}{\tau_{ei}},$$

the driving forces $A_{e1}$ and $A_{e2}$ are given by (8.42) and (8.43), $\xi$ measures the parallel velocity,

$$v_\| \equiv \xi v,$$

and we have introduced the abbreviation

$$\hat{f} \equiv F_{e1}/f_{eM}. \tag{8.48}$$

We integrate (8.47) once with respect to $\xi$ to find

$$\frac{\partial \hat{f}}{\partial \xi} = \frac{v_{te}}{2\nu_{ei}} x^4 \left[ A_{e1} + \left( x^2 - \frac{5}{2} \right) A_{e2} \right] \frac{\xi^2}{1 - \xi^2} + \frac{C}{1 - \xi^2}$$

where $C$ is an integration constant (a function of $x$ but not $\xi$). After choosing $C$ to enforce regularity of the distribution at $|\xi| = 1$ we find that

$$\frac{\partial \hat{f}}{\partial \xi} = -\frac{v_{te}}{2\nu_{ei}} x^4 \left[ A_{e1} + \left( x^2 - \frac{5}{2} \right) A_{e2} \right].$$

The desired distribution is now obtained from another integration:

$$\hat{f} = -\frac{v_{te}}{2\nu_{ei}} x^4 \xi \left[ A_{e1} + \left( x^2 - \frac{5}{2} \right) A_{e2} \right]. \tag{8.49}$$

Here we should include a new integration constant, adding to $\hat{f}$ an arbitrary function of speed alone. It is obvious that the Lorentz operator cannot determine a $\xi$-independent piece of $\hat{f}$; even the exact collision operator (including like and unlike species collisions) vanishes when acting on a Maxwellian, and therefore would allow homogeneous solutions corresponding to corrections to the lowest order density. Fortunately we can ignore such terms because, being even in $\xi$, they cannot contribute to the parallel heat transport and flow processes under consideration.

Equation (8.49) gives the simplest example (small $m_e/m_i$, large $Z$) of a Spitzer function. Its main features are common to all such functions. Thus it is explicitly first order in the collisionality parameter $\Delta$, for obvious reasons; it has the same $\xi$-dependence as the driving term in (8.40) because of rotational symmetry of the collision operator; and it is a linear combination of the two driving forces in (8.40) because of the linearity of that operator. In other words we could have written, from inspection of (8.47),

$$\hat{f} = \frac{v_{te}}{\nu_{ei}} \xi \left[ A_{e1} h_1(x) + A_{e2} h_2(x) \right]. \tag{8.50}$$

Analysis was required only to determine the form of the dimensionless functions $h_i$.

We next use our Spitzer function to compute the desired parallel flows of particles and heat:

$$nV_{e\parallel} = \int d^3 v f_{eM} \hat{f} v_{\parallel} \tag{8.51}$$

$$\frac{q_{e\parallel}}{T_e} = \int d^3 v f_{eM} \hat{f} v_{\parallel} \left( x^2 - \frac{5}{2} \right). \tag{8.52}$$

Thus the particle flow is

$$nV_{e\parallel} = -\frac{nv_{te}^2}{2\nu_{ei}} \int_{-1}^{1} d\xi \xi^2 \frac{2}{\sqrt{\pi}} \int dx e^{-x^2} x^7 \left[ A_{e1} + \left( x^2 - \frac{5}{2} \right) A_{e2} \right]$$

or

$$nV_{e\parallel} = -\frac{32}{3\pi} \frac{nT_e \tau_{ei}}{m_e} \left( A_{e1} + \frac{3}{2} A_{e2} \right). \tag{8.53}$$

A similar calculation yields the heat flow

$$\frac{q_{e\parallel}}{T_e} = -\frac{200}{3\pi} \frac{nT_e \tau_{ei}}{m_e} \left( \frac{6}{25} A_{e1} + A_{e2} \right) \tag{8.54}$$

The two parallel flows have similar form; we restrict our comments to $nV_{e\parallel}$. By considering the case of constant temperature and vanishing electric field, we recover Fick's law from (8.53). The diffusion coefficient is evidently

$$D_{e\parallel} = \frac{32}{3\pi} \frac{T_e \tau_{ei}}{m_e},$$

in agreement with our previous estimate, (8.8).

On the other hand we can allow all the gradients to vanish, and examine the effect of the electric field alone. In this case, since the ions are assumed to be at rest, the parallel current is simply

$$J_{\parallel} = -enV_{e\parallel},$$

or, according to (8.53),

$$J_{\parallel} = \frac{32}{3\pi} \frac{e^2 n \tau_{ei}}{m_e} E_{\parallel}.$$

Thus the ($Z = \infty$ version of) Spitzer conductivity is

$$\sigma_{\parallel} = \frac{32}{3\pi} \frac{e^2 n \tau_{ei}}{m_e}. \tag{8.55}$$

Finally note that, aside from the expected linear dependence on $\nabla_{\parallel} n$ and $E_{\parallel}$, the parallel flow can be driven by a parallel temperature gradient. Similarly, (8.54) shows that a density gradient can drive heat flow. In other words the transport matrix of (8.11) includes the "thermoelectric" terms $L_{12}$ and $L_{21}$. Notably, both thermoelectric coefficients are the same:

$$L_{12} = L_{21} = \frac{16}{\pi} \frac{n T_e \tau_{ei}}{m_e}. \tag{8.56}$$

This is an example of a symmetry, called *Onsager symmetry*, which we study more systematically in Section 8.5.

The thermoelectric flows result from the velocity dependence of the Coulomb cross-section. Thus suppose the temperature increases with increasing coordinate $z$, and consider the flux across a surface, with normal in the $z$-direction, at $z = 0$. We use a $-$ ($+$) subscript for particles crossing from $z < 0$ ($z > 0$), thus defining the typical speeds $v_{\pm}$, densities $n_{\pm}$, collision frequencies $\nu_{\pm}$ and pressures $p_{\pm}$. Then the rate at which particles cross the surface from $z < 0$ will be proportional to the number of particles within a mean-free path $v_-/\nu_-$ of the surface, and to their speed: $n_- v_{t-}^2/\nu_- \propto p_-/\nu_-$; the rate for the opposing flow is similarly proportional to $p_+/\nu_+$. If the collision frequency were independent of energy, then the only particle flux through the surface would reflect the pressure difference: $nV_z \propto p_- - p_+ \propto \nabla p$. However, because the collision frequencies differ, a thermoelectric flow $nV_z \propto \nu_- - \nu_+$ occurs even in the absence of a pressure gradient.

## Parallel kinetics of a heavy species

Collisional transport of a heavy species necessarily involves like-particle collisions, and the corresponding operator is too complicated for analytic solution in closed form. In practice the heavy species problem is attacked by a combination of numerical and variational methods we will consider presently—or by crude approximation. Our purpose here is simply to point out a useful simplification of the pertinent kinetic equation.

Again for concreteness we consider a two species plasma of electrons and ions, viewed from a frame in which the latter are at rest. The relevant unlike-species collision operator, $C_{ie}$, is given by (7.96):

$$C_{ie}^l = -\frac{\mathbf{F}_{ie}}{m_i n_i} \cdot \frac{\partial f_{iM}}{\partial \mathbf{v}}.$$

Only the gyroaverage of this operator enters the ion Spitzer problem; performing the average and making the velocity derivative explicit we obtain the operator

$$C_{ie}^l = \frac{v_\|}{p_i} F_{ie\|} f_{iM}.$$  (8.57)

and the ion Spitzer equation,

$$C_{ii}^l(F_{i1}) = v_\| f_{iM} \left[ \nabla_\| \log p_i + \left( x^2 - \frac{5}{2} \right) \nabla_\| \log T_i - \frac{eE_{*\|}}{T_i} \right],$$  (8.58)

where

$$E_{*\|} \equiv E_\| + \frac{F_{ie\|}}{en_i}.$$

is the effective electric field introduced in Chapter 7.

The distinctive feature of the ion Spitzer problem is that most of the driving terms on the right-hand side of (8.58) cancel; only the temperature gradient survives. The point is that ion parallel force balance requires, in lowest order,

$$\nabla_\| p_i - enE_\| = F_{ie\|},$$  (8.59)

or

$$\frac{eE_{*\|}}{T_i} = \frac{\nabla_\| p_i}{p_i}.$$

Therefore the ion Spitzer problem reduces to

$$C_{ii}^l(\hat{f}_i f_{iM}) = v_\| \frac{\nabla_\| T_i}{T_i} \left( x^2 - \frac{5}{2} \right),$$  (8.60)

where $\hat{f}_i f_{iM} = F_{i1}$ as before. Thus there is no Fick's law for ions in lowest order; a light breeze of electrons has distorted the sparkler symmetry just enough to offset diffusion.

It is evident that the ion Spitzer function has the form

$$\hat{f}_i = \frac{v_{ti}}{\nu_{ii}} A_{2i} h_i(x)$$

analogous to (8.50). Unfortunately the complicated integro-differential nature of $C_{ii}$ precludes any simple calculation of $h_i(x)$.

We have noted that the corresponding complication of $C_{ee}$ prevents analytical solution of the electron Spitzer problem for $Z_{eff} \sim 1$. Nonetheless there is an analytical method for accurately computing all the ion and electron transport coefficients, for any $Z_{eff}$. This method is the topic of the following Section.

## 8.5   Variational transport theory

### Entropy production

We have encountered entropy production on two earlier occasions. A fluid representation of entropy production was introduced in Chapter 3, where we found, in (3.57), that the entropy density $s_s = n_s \log(T_s^{3/2}/n_s)$ increases at the rate

$$\Theta_s \equiv \frac{W_{Ls}}{T_s} - \frac{\pi_s : \nabla \mathbf{V}_s}{T_s} - \frac{\mathbf{V}_s \cdot \mathbf{F}_s}{T_s} - \frac{\mathbf{q}_s}{T_s} \cdot \frac{\nabla T_s}{T_s}. \tag{8.61}$$

On the other hand in Chapter 7 we considered the statistical entropy,

$$s_s = -\int d^3 v \, f_s \log f_s, \tag{8.62}$$

and found that collisions caused it to increase at the rate

$$\Theta_{sK} \equiv -\int d^3 v \, (\log f_s) C_s; \tag{8.63}$$

recall (7.11). Here we consider the relation between these two expressions, in the context of the Spitzer problem.

To begin we transform (8.61), in three steps.

1. Eliminate the friction term, $\mathbf{V} \cdot \mathbf{F}$, using the dot product of $\mathbf{V}$ with the equation of motion, (3.51).

2. Approximate for small $\delta$ by retaining only the parallel flows of particles and heat. This approximation corresponds to the neglect of gyromotion and perpendicular drifts in the lowest order drift-kinetic equation, (8.32).

3. Approximate for small $\Delta$: use the transport ordering and the weak flow assumption, (8.39), to remove acceleration and viscosity terms. Also use the small mass ratio to neglect $W_L$.

Here our two approximations are precisely those used in deriving the Spitzer problem. The reader can verify that they yield the following entropy production rate:

$$\Theta_s = -nV_{s\|} \left( \frac{\nabla_\| p_s}{p_s} - \frac{e_s E_\|}{T_s} \right) - \frac{q_{s\|}}{T_s} \frac{\nabla_\| T_s}{T_s}. \tag{8.64}$$

We see that each term on the right contains one of the driving forces, $A_{s1}$ or $A_{s2}$, of the Spitzer problem, multiplied by the corresponding flow: $nV_{s\|}$ with $A_{s1}$ and $q_{s\|}$ with $A_{s2}$. It is therefore natural to enumerate the flows according to

$$\Gamma_{s1} = nV_{s\|} \tag{8.65}$$

$$\Gamma_{s2} = \frac{q_{s\|}}{T_s}, \tag{8.66}$$

and thus to express the entropy production rate in the form

$$\Theta_s = -\sum_j \Gamma_{sj} A_{sj}. \tag{8.67}$$

Note that the choice of flows and forces is not unique; for example, we might have defined $A_{s1}$ in terms of the density gradient rather than the pressure gradient, or we might have used the energy flow, $Q$, in place of the heat flow, $q$. Some of these choices will complicate the form of $\Theta_s$; choices that reproduce the simple form of (8.67)—where each term involves only a single flow and its corresponding force—are called *canonical*.

Turning our attention to the kinetic entropy production of (8.63), we first use (8.48) to express the distributions as

$$f_s = f_{sM}(1 + \hat{f}_s)$$

where $\hat{f} = O(\Delta)$. It is then easy to show that the first order, $O(\Delta)$, contribution to $\Theta_{sK}$ is proportional to the energy exchange, $W_L$, and that the second order contribution is

$$\Theta_{sK}^{(2)} = - \int d^3 v \hat{f}_s C_s^l(f_{sM} \hat{f}_s). \tag{8.68}$$

We next evaluate the quadratic entropy production on the right-hand side for a magnetized, short mean-free path plasma, using our approximate kinetic equation, (8.40). Thus we write

$$\Theta_{sK}^{(2)} = - \int d^3 v \hat{f}_s v_\| f_{sM} \left[ A_{s1} + \left( x^2 - \frac{5}{2} \right) A_{s2} \right]$$

Since the $A's$ are constant each term integrates trivially and we have

$$\Theta_{sK}^{(2)} = -n V_{s\|} A_{s1} - \frac{q_{s\|}}{T_s} A_{s2},$$

in agreement with the canonical expression derived above:

$$\Theta_{sK}^{(2)} = \Theta_s.$$

This unsurprising agreement reflects the assumption, in the kinetic derivation, that the lowest-order distribution is Maxwellian at each point. Thus the lowest-order statistical entropy coincides with the fluid entropy of (8.62). The linear correction to the local Maxwellian, $\hat{f}$, accounts for relaxation of the weak gradients that measure departure from full thermodynamic equilibrium.

## Self-adjointness

Equation (8.68) directs attention to the bilinear form,

$$\Psi[\hat{g}_1, \hat{g}_2] \equiv - \int d^3 v \hat{g}_1 C^l(\hat{g}_2 f_M). \tag{8.69}$$

where the $\hat{g}_n$ are *any* Maxwellian-normalized distribution functions. (That is, we require $\hat{g}_n F_M$ to have finite velocity-moments. Species subscripts are omitted for simplicity.) We call $\Psi$ the *collisional bilinear form*, or simply the collisional form.

Evaluated on arbitrary normalized distributions, $\Psi$ lacks physical significance, but when the solution to the Spitzer problem is substituted for both arguments, the collisional form becomes the entropy production. We reserve the notation $\hat{f}$ for the solution to the Spitzer problem, and therefore write

$$\Psi[\hat{f}, \hat{f}] = \Theta. \tag{8.70}$$

The collisional bilinear form provides the crucial tool for calculating transport coefficients when the mathematical complexity of like-species collisions cannot be evaded. Its power results from the self-adjointness of the linearized, like-species Landau-Boltzmann collision operator; that is,

$$\Psi[\hat{g}_1, \hat{g}_2] = \Psi[\hat{g}_2, \hat{g}_1]. \tag{8.71}$$

The demonstration of (8.71) follows straightforwardly from the expression (7.87) for $C^l$ given in Chapter 7, and is left as an exercise for the reader.

The unlike-species operator is exactly self-adjoint only when the two species have the same temperature, but it is usually treated as self-adjoint in any case. The point is that when the masses of the two species are comparable, energy exchange is sufficiently large to enforce temperature equilibration. On the other hand, for disparate masses we can approximate the operators, as in the previous section, and find that $C_{lh}$ is self-adjoint to lowest order in the mass ratio, while $C_{hl}$ combines harmlessly with other macroscopic forces. [Thus only $C_{hh}$ enters the heavy-species Spitzer problem, as in (8.60).]

Before using the collisional form to solve the Spitzer problem, we first deduce from it the Onsager symmetry of (8.56). (Onsager proved a deeper result: that the symmetry reflects time-translation invariance of the underlying collisional dynamics.) Although our argument uses the specific example of electron parallel transport for concreteness, it will be clear that the result is general, requiring only linearity (in order to define a transport matrix) and symmetry. In particular, there may be any number of driving forces: even the electron Spitzer problem acquires a larger transport matrix when zeroth-order flow velocity is allowed.

Because of the linearity of the operator $C^l$, (8.50) allows a decomposition of (8.47) into two Spitzer problems:

$$C^l(\xi h_1 f_M) = \nu \xi x f_M, \tag{8.72}$$

$$C^l(\xi h_2 f_M) = \nu \xi x f_M \left(x^2 - \frac{5}{2}\right), \tag{8.73}$$

where species-subscripts are suppressed. We have already solved these equations in the special case of Lorentz scattering, without like-particle collisions. But Onsager symmetry does not require such simplification, so we do not use our Lorentz results in the present demonstration. In other words the functions $h_n$ should be considered as defined by (8.72) and (8.73) but otherwise unspecified.

Now consider the particle flux,

$$\Gamma_1 = \int d^3 v v_\parallel f_M \frac{v_t}{\nu} \xi (A_1 h_1 + A_2 h_2),$$

as in (8.53). Recalling (8.11) we see that

$$L_{12} = \frac{v_t^2}{\nu} \int d^3v f_M x \xi^2 h_2. \tag{8.74}$$

We similarly examine the heat flux in (8.54) and conclude that

$$L_{21} = \frac{v_t^2}{\nu} \int d^3v f_M x \left( x^2 - \frac{5}{2} \right) \xi^2 h_2. \tag{8.75}$$

There is no obvious reason for the equality of these two coefficients; in particular the functions $h_1$ and $h_2$ do not in general differ by a simple factor of $x^2 - 5/2$. (The Lorentz example is exceptional in this respect.) But (8.72) allows us to write $L_{12}$ as

$$L_{12} = \frac{v_t^2}{\nu^2} \int d^3v \xi h_2 C^l(\xi h_1 f_M),$$

or, in view of the self-adjointness property,

$$L_{12} = \frac{v_t^2}{\nu^2} \int d^3v \xi h_1 C^l(\xi h_2 f_M).$$

After using (8.73) to eliminate the collision operator from this expression, we find that its right-hand side coincides with that of (8.75), completing the demonstration.

Parallel transport is an exceptionally simple context for Onsager symmetry because the magnetic field does not enter. When there is a magnetic field, the relation between $L_{mn}$ and $L_{nm}$ is more complicated, and, depending about parity with respect to velocity reversals, may include a minus sign. We do not pursue these complications.

As a practical tool in computing the transport matrix, Onsager symmetry obviously helps in reducing the number of unknowns. Its power is summarized by the following recipe: if, by whatever method, one manages to express the *entropy production rate as a quadratic form in some canonical set of driving forces*,

$$\Theta = \sum_{m \leq n} A_m X_{mn} A_n, \tag{8.76}$$

then all the transport coefficients are known:

$$L_{mm} = X_{mm},$$

$$L_{mn} = \frac{1}{2} X_{mn}, \ m \neq n.$$

This result is an immediate consequence of (8.11), (8.67) and Onsager symmetry.

Thus, as we have remarked, the entropy production rate is the central object of linear transport theory. We consider next an efficient way to compute it.

## Variational principle

Here we use the collisional bilinear form $\Psi$ to construct a variational principle for the Spitzer problem. We first multiply both sides of the kinetic equation

$$C^l(\hat{f} f_M) = v_\| f_M \left[ A_1 + \left( x^2 - \frac{5}{2} \right) A_2 \right] \tag{8.77}$$

by $\hat{f}$ to infer that

$$\Psi[\hat{f}, \hat{f}] = P[\hat{f}], \tag{8.78}$$

where we have introduced the *linear* form,

$$P[\hat{g}] \equiv - \int d^3 v v_\| \hat{g} f_M \left[ A_1 + \left( x^2 - \frac{5}{2} \right) A_2 \right]. \tag{8.79}$$

Now suppose we evaluate $\Psi$ on a function $\hat{g}$ that is close to the solution $\hat{f}$:

$$\hat{g} = \hat{f} + \delta f.$$

The first-order correction to $\Psi$, called $\delta\Psi$, is evidently

$$\Psi[\delta f, \hat{f}] + \Psi[\hat{f}, \delta f] = 2\Psi[\delta f, \hat{f}]$$

by self-adjointness. But after multiplying the kinetic equation by $\delta f$ we find that

$$\Psi[\delta f, \hat{f}] = \delta P \equiv P[\delta f],$$

whence

$$\delta\Psi = 2\delta P. \tag{8.80}$$

In other words the function

$$S = 2P - \Psi \tag{8.81}$$

is variational with respect to (8.77):

$$\delta S = 0.$$

Thus the Spitzer function is that function which makes $S$ extremal. Furthermore (8.78) shows that the extremal value of $S$ is

$$S[\hat{f}] = P[\hat{f}] = \Psi[\hat{f}, \hat{f}] = \Theta,$$

the entropy production rate.

These two facts—that $S$ is variational and that its extremal value coincides with $\Theta$—provide a simple and powerful machine for finding transport coefficients. Beginning with a *trial function* representation of the Spitzer problem, one finds the extremal $S$ as a quadratic form in the forces $A_n$. Then variationally accurate transport coefficients can be read off as in (8.76). The key feature of this procedure is that it only requires substituting a known trial function into the Coulomb collision operator, evaluating some integrals and solving simple algebraic equations. It requires nothing nearly as challenging as inverting the collision operator.

The uninstructive details of extremizing $S$ are not presented here. However, for the sake of readers unfamiliar with the variational approach, we outline the procedure. After recalling the general form of the exact solution (8.50), one expresses the trial function as

$$\hat{g} = \frac{v_t}{\nu}\xi\left[A_1 g_1(x) + A_2 g_2(x)\right].$$

Then the $g_n(x)$ are constructed, using *variational parameters*, to allow an approximate representation of the corresponding functions, $h_n(x)$, of the exact solution. We denote the variational parameters by $\alpha_k$ and assume that the $g_n(x)$ depend linearly on them; in practice the $g_n(x)$ are usually chosen as polynomials whose coefficients are the $\alpha_k$.

Now one can evaluate

$$S[\hat{g}] = S(\alpha_1, \alpha_2, \ldots)$$

as an explicit function of the $\alpha_k$. It will evidently include quadratic terms (from $\Psi$) as well as linear terms (from $P$), so that the variational equations

$$\frac{\partial S}{\partial \alpha_k} = 0$$

will be linear and easily solved. After inserting these "best" choices for the $\alpha_k$ into $\hat{g}$, we obtain the variational function denoted by $\hat{g}_*$. Since $\hat{g}_*$ is necessarily linear in the forces, $S[\hat{g}_*]$ will be quadratic: it will have the form of (8.76), from which variationally accurate transport coefficients are directly read off.

There are short cuts in this procedure; for example, one can find the extremal $S$ without explicitly finding the extremizing $\alpha_k$. But only one refinement is worth discussing in any detail: the use of a normalized variational principle. Consider the functional

$$R[\hat{g}] \equiv \frac{P^2[\hat{g}]}{\Psi[\hat{g}, \hat{g}]}. \tag{8.82}$$

The reader can verify that

$$\delta R = 0,$$

and that

$$R[\hat{f}] = \Theta.$$

Thus $R$ has the same variational properties as $S$. It is better than $S$ because it is independent of the size of the trial function: multiplying the latter by a constant has no effect on $R$. Indeed, the normalized variational quantity can be derived from $S$ by choosing an initial trial function

$$\hat{g} = A\hat{g}'$$

and then extremizing $S$ with respect to $A$, for arbitrary $\hat{g}'$. Having been extremized *a priori* with respect to overall size of the trial function, and being hardly more complicated to evaluate than $S$, $R$ provides a more efficient path to the transport matrix.

In what sense does the trial function $\hat{g}_*$ represent the exact solution? The latter is not polynomial in general, nor even expressible in closed form. So the forms of the two functions would not look similar. They also need not agree, or even nearly agree, on any finite set of points: changing $\hat{g}_*$ in a pointwise manner would not effect $S$ or $R$. The trial function resembles the actual solution in a global, integrated sense: one expects the two functions to have similar moments. Most importantly one knows that the two functions yield transport coefficients that closely agree—that a disparity in the functions of order $\delta$ contributes only a much smaller error, of order $\delta^2$, to the transport matrix. This key property, which follows from the variational principle, is what is meant by "variational accuracy."

In practice one finds that trial-function representations closely approximate the Spitzer function in the vicinity of the thermal speed, $x \sim 1$, but are less accurate at the velocity extremes. Therefore the trial functions are not always reliable; for applications outside of transport theory a full numerical solution to the Spitzer problem may be needed. On the other hand, for the specific purpose of computing transport coefficients, even very simple trial functions (for example, two-term polynomials) provide accuracy at the level of a few per cent.

Of course a sufficiently flexible trial function can reproduce the exact coefficients with arbitrary accuracy. These coefficients, for electron parallel transport in a single ion-species plasma with $Z = 1$, are given by

$$L_{mne} = \frac{n\tau_e T_e}{m_e}\lambda_{mn}$$

where $\tau_e$ is the electron collision time of (7.95), and

$$\begin{align}
\lambda_{11} &= 1.98, \tag{8.83}\\
\lambda_{12} &= 1.38,\\
\lambda_{22} &= 4.17.
\end{align}$$

Similarly from (8.60) one finds the ion parallel thermal conductivity to be

$$L_{22i} = 3.52\frac{nT_i\tau_i}{m_i}$$

where the ion collision time $\tau_i$ is defined by (7.82).

From (8.83) we infer the Spitzer conductivity for unit ionic charge,

$$\sigma_\| = 1.98\frac{\tau_e e^2 n}{m_e}. \tag{8.84}$$

The numerical coefficient in (8.84) differs from the large-$Z$ case of (8.55) because the relative importance of energy scattering versus angular scattering changes with effective charge. As we have remarked, this variation is not described by the simple appearance of $Z_{\it eff}$ in $\tau_e$.

## 8.6   Perpendicular transport

### Random walks across the magnetic field

A gyrating particle that suffers a 90° collision begins a new gyro-orbit, displaced
from the original orbit by a distance $\rho$, the gyroradius. The direction of the
jump, depending upon the gyrophase at the time of the scattering, is effectively
random. Figure 8.3 depicts the process in idealized form. Note that the picture
makes sense only if two conditions are satisfied:

1. Collisions must be sufficiently infrequent for gyration to occur; thus we
   must require

$$\nu \ll \Omega, \tag{8.85}$$

   where $\nu$ is the 90° scattering rate and $\Omega$ the gyrofrequency.

2. The electrostatic fluctuations underlying the collisional process must be
   uncorrelated on the spatial scale of the gyro-orbit. Since the correlation
   length is estimated by the Debye length, we require

$$\lambda_D \ll \rho. \tag{8.86}$$

As emphasized in Chapter 7, the abrupt change in direction is also an ide-
alization: most Coulomb collisions involve small angle changes, many of which
must accumulate to create a large-angle scattering. Furthermore the simple
picture leaves out energy scattering, which causes a gradual, random change in
the size of the gyroorbit. But when (8.85) and (8.86) are satisfied, a perpendicular random walk faithfully represents the key collisional effect: every $1/\nu$
seconds, a magnetized particle's guiding center makes a step, perpendicular to
the magnetic field, measured by $\rho$.

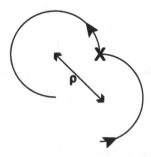

Figure 8.3:  A particle gyrates about a magnetic field perpendicular to the plane of the
figure. At the "x" it suffers a 90° scattering (idealized) and then resumes gyrating with the
new initial velocity. The effect is a random displacement of the guiding center over a distance
measured by ($\sqrt{2}$ times) the gyroradius, in the plane transverse to the magnetic field.

Thus collisions and gyration interact to produce a random walk of guiding
centers across the magnetic field. The corresponding diffusion process is called

"classical" perpendicular diffusion and is the subject of this Section. From the discussion of Section 8.1 we expect classical perpendicular diffusion to be characterized by a diffusion coefficient

$$D_\perp \sim \nu \rho^2; \tag{8.87}$$

recall (8.6).

The two powers of $\rho$ in $D_\perp$ show that classical diffusion occurs in second-order with respect to the gyroradius expansion; it is a correction to the first-order flow,

$$n\mathbf{V}_{s\perp 1} = n\mathbf{V}_E + \frac{1}{e_s B}\mathbf{b} \times \nabla p_s \tag{8.88}$$

computed in Chapter 3. More importantly (8.87) shows that perpendicular diffusion is proportional to collision frequency—opposite to the $1/\nu$-dependence of parallel diffusion. The point is that collisionless motion across the field in limited by gyration, while parallel motion is limited only by the finite mean-free path. Since the collision frequency is proportional to $T^{-3/2}$ while $\rho^2 \propto T$, perpendicular diffusion decreases with temperature:

$$D_\perp \sim T^{-1/2},$$

unlike the $T^{5/2}$-dependence of $D_\parallel$.

Of course Coulomb collisions are binary: there must be another charged particle, not shown in Figure 8.3, to scatter the particle under consideration. Momentum conservation requires the two colliding particles to rebound in opposite directions, so that if both belong to the same species there is no net flow. It follows that classical transport occurs only in the presence of more than one plasma species, and that the different species will diffuse in the same direction only if they gyrate in opposite directions—that is, if they have opposite charge.

## Friction force

The significance of collisions with different species is most easily understood from fluid equations. Hence we recall the exact formula derived in Chapter 3:

$$n\mathbf{V}_\perp = n\mathbf{V}_E + \frac{\mathbf{b}}{m\Omega} \times \left( mn\frac{d\mathbf{V}}{dt} + \nabla \cdot \mathbf{p} - \mathbf{F} \right). \tag{8.89}$$

Only one term here is manifestly proportional to $\nu$: the last term on the right, involving the friction force. Thus, denoting the classical perpendicular transport flow of species $s$ by $n_s \mathbf{V}_{sc}$, we define

$$n_s \mathbf{V}_{sc} \equiv -\frac{\mathbf{b}}{m_s \Omega_s} \times \mathbf{F}_s; \tag{8.90}$$

Our previous comments about the importance of multiple species are immediately clear from (8.90); in particular, there is no friction moment for like-species collisions. We also confirm, since friction is first-order in the gyroradius, that perpendicular diffusion is second order. The question of the relative directions of transport for different species is important enough for a separate discussion which we defer.

The fluid formulation also provides a simple physical interpretation, complementary to that of Figure 8.3. Collisional friction evidently produces a drift analogous to the $\mathbf{E} \times \mathbf{B}$ drift, and for the same reason: it distorts Larmor orbits as discussed in Chapter 2. As a result, the effect of collisions on perpendicular motion is accounted for by simply replacing the $\mathbf{E} \times \mathbf{B}$ drift by an effective drift, $\mathbf{E}_* \times \mathbf{B}$, where

$$\mathbf{E}_* = \mathbf{E} + \frac{\mathbf{F}_s}{e_s n_s},$$

as in (7.97).

Our next task is to compute the friction force,

$$\mathbf{F}_{ss'\perp} = \int d^3 v m_s \mathbf{v}_\perp C_{ss'}. \tag{8.91}$$

Here we have noted that only the perpendicular components of $\mathbf{v}$ can enter $\mathbf{V}_c$. For simplicity we begin by considering a quasineutral plasma with one ion species and compute the friction on electrons, $\mathbf{F}_{ei\perp}$. Since the electrons are magnetized, we can use the distribution computed in Section 8.3. Only the lowest order, Maxwellian form of $F$ is needed to compute $\tilde{f}_e$, so (8.30) provides

$$f_e = -f_{eM} \, \boldsymbol{\rho} \cdot \left[ \frac{\nabla_\perp p_e}{p_e} - \frac{e \mathbf{E}_\perp}{T_e} + \left( x^2 - \frac{5}{2} \right) \frac{\nabla_\perp T_e}{T_e} \right]. \tag{8.92}$$

Here we have noted that the gradient in (8.30) is performed at fixed total energy $\mathcal{E}$; since $f_M \propto \exp[-(\mathcal{E} + e\Phi)]$, its gradient includes a $\nabla \Phi$ term, which combines with the last term in (8.30) to produce the full perpendicular electric field $\mathbf{E}_\perp$.

The collision operator for (8.91) is provided, in terms of the ion flow, by (7.89), for which we use (8.88). After simple manipulation using (7.56) we find

$$\mathbf{F}_{ei\perp} = \frac{m_e n}{\tau_{ei}} \mathbf{V}_{i\perp 1} - \gamma_{ei} n \int d^3 v f_e \frac{\mathbf{v}_\perp}{v^3}$$

whose the second term can be evaluated using (8.92) and straightforward integration. One finds that

$$\mathbf{F}_{ei\perp} = \frac{m_e n}{\tau_{ei}} \left( \mathbf{V}_{i\perp 1} - \mathbf{V}_{e\perp 1} - \frac{3}{2} \frac{1}{eB} \mathbf{b} \times T_e \right). \tag{8.93}$$

Hence (8.89) and (8.90) imply

$$n \mathbf{V}_c = -\frac{1}{m_e \Omega_e^2 \tau_{ei}} \left[ \nabla_\perp (p_i + p_e) - \frac{3}{2} n \nabla_\perp T_e \right]. \tag{8.94}$$

In the special case of equal, constant temperatures, (8.94) reduces to

$$n \mathbf{V}_c = -\frac{2T}{m_e \Omega_e^2 \tau_{ei}} \nabla_\perp n.$$

The reader can verify that this is Fick's law with diffusion coefficient

$$D_{e\perp} = \frac{\rho_e^2}{\tau_{ei}}$$

in agreement with the estimate (8.87): the detailed calculation has confirmed this estimate and revealed that the coefficient is unity.

Other features of (8.94) are also easily understood. First note that the distribution function (8.92) can be written in the form

$$\tilde{f}_e = f_{eM} \frac{2\mathbf{v} \cdot \mathbf{V}_*}{v_{te}^2} \tag{8.95}$$

where

$$\mathbf{V}_* \equiv \mathbf{V}_{e\perp 1} - \frac{T_e}{eB}\left(x^2 - \frac{5}{2}\right)\mathbf{B} \times \frac{\nabla_\perp T_e}{T_e}.$$

Thus, aside from the speed-dependent temperature-gradient term, (8.95) simply describes a Maxwellian displaced by the obvious first-order velocity, and the first term in (8.94) expresses the friction between two species in relative motion; recall (7.79). One can say, in other words, that classical transport occurs because the opposing diamagnetic drifts of ions and electrons generate collisional friction. This explains the appearance of the *total* pressure gradient in (8.94), as well as the absence of the $\mathbf{E} \times \mathbf{B}$ drift.

The temperature gradient term does not fit this simple picture because the response to $\nabla T$ depends on energy; thus the second term in $\mathbf{V}_*$ does not correspond to fluid flow. (The reader can verify that the velocity moment of this second term vanishes.) It should be remarked that most transport calculations are enormously simplified in the absence of temperature gradients.

## Ambipolarity

An interesting feature of the classical flow is that, in the two-species case, it is independent of species:

$$n\mathbf{V}_{ce} = n\mathbf{V}_{ci}$$

as follows immediately from collisional momentum conservation and (8.90). Thus the plasma charge density is unaffected by classical diffusion. Generalizing to the case of any number of species, we compute the current density

$$\begin{aligned}
\mathbf{J}_{c\perp} &= -\sum_s \frac{e_s}{m_s\Omega_s}\mathbf{b} \times \mathbf{F}_s \\
&= -\frac{c}{B}\mathbf{b} \times \sum_s \mathbf{F}_s \tag{8.96}
\end{aligned}$$

which vanishes for the same reason. A flow that yields vanishing current density is called *ambipolar*; (8.96) shows that classical diffusion is necessarily ambipolar.

An example of a flow that need not be ambipolar is parallel transport, studied in the previous section. Notice that the parallel electron flux, which can differ from its ion counterpart, also depends upon the parallel electric field. Hence (if, for example, parallel current implied actual loss of charge) an electrostatic field could in principle arise to modulate the electron flow and maintain quasineutrality. An electrostatic potential that arises in this way is called an *ambipolar potential*; an example is the sheath potential studied in Chapter 4. We see that classical transport cannot produce an ambipolar potential, for two reasons: first, it is intrinsically ambipolar, because of momentum conservation; second, it is independent of the electric field.

Although the calculation leading to (8.94) used specific properties of Coulomb collisions, the result (8.96) does not. Thus any collisional process that conserves momentum will yield ambipolar diffusion across the field of a magnetized plasma. Perhaps the most important consequence of this statement pertains to a single-species plasma: we see that diffusion of such a plasma is ruled out for any momentum-conserving collisional process. Indeed, the confinement of experimental single-species plasmas is found to be limited only by the quality of the collisionless particle orbits.

## Perpendicular conductivity

By "perpendicular conductivity" we mean a coefficient $\sigma_\perp$ such that

$$\mathbf{J}_\perp = \sigma_\perp \mathbf{E}_\perp + \ldots$$

where the ellipsis allows for additional terms, such as thermoelectric contributions. Is there perpendicular conductivity in a magnetized, quasineutral plasma? Certainly not from the single-particle point of view: magnetized particles respond to $\mathbf{E}_\perp$ by $\mathbf{E} \times \mathbf{B}$-drifting across the field, without producing current. Furthermore perpendicular ambipolarity implies that collisions do not yield a $\sigma_\perp$.

In fact, while $\sigma_\perp$ is a more tenuous and less useful concept than $\sigma_\parallel$, it does have occasional application. To make sense of it we use (8.93) to write

$$\mathbf{F}_{ei\perp} = \frac{m_e}{\tau_{ei}} \left( \frac{\mathbf{J}}{e} - \frac{3}{2} \frac{n}{eB} \mathbf{b} \times T_e \right)$$

and then substitute this friction into the electron equation of motion. If we neglect electron inertia and viscosity, the result can be expressed as

$$\mathbf{J} = \sigma_\perp \left( \mathbf{E}_\perp + \frac{\nabla_\perp p_e}{en} + \mathbf{V}_e \times \mathbf{B} \right) + \frac{3}{2} \frac{\mathbf{b}}{B} \times \nabla_\perp T_e, \qquad (8.97)$$

with

$$\sigma_\perp \equiv \frac{e^2 n \tau_{ei}}{m_e}. \qquad (8.98)$$

Thus $\sigma_\perp$ is roughly half the Spitzer conductivity (for $Z = 1$).

The fragility of this result should be clear: in a magnetized plasma the three parenthesized terms in (8.97) combine to vanish! In particular, the electric field term, which otherwise would combine with $\nabla_\perp p_e$ to give a perpendicular version of our canonical force $A_1$, is exactly canceled by the $\mathbf{E} \times \mathbf{B}$ drift in $\mathbf{V}_e$. Thus (8.97) displays the anisotropy of plasma transport as well as the limitations of $\sigma_\perp$; that quantity makes sense as a coefficient relating the friction force and the current, but not as a conductivity in the usual sense.

It might be argued that the formula (8.89) for flow is a consequence of the perpendicular Ohm's law, (8.97)—which would then seem fundamentally important. (Indeed, the $\mathbf{E} \times \mathbf{B}$-drift is sometimes inferred from "infinite conductivity.") But in fact (8.89) is a consequence of small gyroradius alone, as shown in Chapter 3, without reference to conductivity. Resulting from particle drifts and gyro-magnetization, it is valid for any collisionality consistent with a magnetized plasma, $\nu \ll \Omega$, including the case of a collisional plasma, $\nu \gg \omega_t$, which is far from a perfect conductor.

## Other perpendicular transport coefficients

We have computed classical diffusion in terms of a moment of the collision operator, the friction force. Since the moment contains a factor of collision frequency, the part of the distribution that is proportional to $\nu$ was never needed—an enormous calculational advantage. The same moment approach can be used to evaluate other perpendicular collisional processes. For example, the energy flux given by (3.90) includes a third-order collisional moment corresponding to classical energy transport. Again because the factor of $\nu$ is in the collision operator, we can use the simple distribution of (8.92); the integral is only slightly harder to evaluate than that for $\mathbf{F}_{ei}$. Thus one finds the classical collisional heat flows

$$\mathbf{q}_{ic} = -\frac{2p_i}{mi\Omega_i^2\tau_i}\nabla T_i, \tag{8.99}$$

$$\mathbf{q}_{ec} = -\frac{4.66p_e}{me\Omega_e^2\tau_e}\left[\nabla T_e - 0.32\frac{\nabla_\perp(p_i + p_e)}{n}\right]. \tag{8.100}$$

Similarly one can use the formulation of (3.96) to compute the viscosity from a moment of the collision operator. Although this calculation is straightforward, requiring no more kinetic theory than (8.92), the resulting lengthy expression is not reproduced here.

It should be noticed that classical heat transport, like classical diffusion, contains off-diagonal terms: $n\mathbf{V}_c$ can be driven by a temperature gradient, while $\mathbf{q}_{ce}$ can be driven by a pressure gradient. It should also be noticed that the coefficients of these two terms are the same: perpendicular transport obeys Onsager symmetry. The off-diagonal terms for perpendicular transport occur with minus signs, unlike their parallel counterparts. Nonetheless one finds that the perpendicular electron entropy production,

$$\Theta_{ec} = -n\mathbf{V}_c \cdot \nabla_\perp \log(p_i + p_e) - \frac{\mathbf{q}_{ec}}{T_e} \cdot \nabla_\perp \log T_e$$

is always positive.

Thermoelectric flows in the perpendicular direction, like their parallel counterparts, result from the velocity-dependence of Coulomb scattering; recall the discussion of (8.56). The perpendicular particle flow is along the temperature gradient, rather than opposite to it, because perpendicular transport is abetted by collisions, rather than impeded as in the parallel case.

# Additional reading

The Chapman-Enskog procedure is systematically developed in the text by Chapman and Cowling[15]. The details of Onsager symmetry are clearly expounded by de Groot and Mazur[20].

Clear and nearly self-contained treatments of classical plasma transport are due to Braginskii[13] and Hinton[35]. A more detailed and also readable discussion may be found in Balescu[5]. The variational derivation of plasma transport coefficients is expounded by Robinson and Bernstein[60].

# Problems

1. Verify explicitly that, under the approximations stated in the text, (8.30) indeed gives the solution to (8.29).

2. Show that Spitzer conductivity depends only logarithmically on the plasma density; explain this circumstance in physical terms.

3. Show in detail the equivalence, under the stated approximations, of (8.62) and (8.64).

4. Prove the relation (8.71), that the collisional bilinear form is self-adjoint.

5. Compute the second-order variation of the functional $S = 2P - \Psi$. Apply a Schwartz inequality to the result to determine whether the extremal value is a maximum or a minimum.

6. Use the parallel transport relations (8.53) — (8.54) to write down a closed fluid description of short mean-free path electron dynamics in the zero-gyroradius limit. (Neglect the perpendicular electric field for simplicity.) Identify the basic variables and the key time-scales of the resulting system.

7. A short mean-free path, isothermal plasma, magnetized by straight, uniform magnetic field, contains a flux tube of circular cross section, with radius $a$ and length $L$. In terms of the thermal speed $v_t$, gyrofrequency $\Omega$ and collision frequency $\nu$, estimate the characteristic time $\tau_\parallel$ for parallel traversal of the flux tube, and the characteristic time $\tau_\perp$ for radial escape from the tube. Show that the two times are comparable only if $a/L \sim \nu/\Omega$.

## Problems

# Chapter 9

# Turbulent Transport

## 9.1   Turbulence and the closure problem

In the last two chapters we considered the role of fluctuations caused by particle discreteness. The spectrum of these fluctuations is called the thermal fluctuation spectrum. As pointed-out in Chapter 7, however, the thermal fluctuation spectrum represents an irreducible minimum level of fluctuation, approached only near thermodynamic equilibrium.

In practice, plasmas are almost never near thermodynamic equilibrium. Their fluctuations are more intense, and the spectra broader than the thermal limit. They are said to be turbulent. The theory of turbulence is at the forefront of current research, and even its foundations continue to evolve. Here, we present an outline of the most basic results.

It is worthwhile to note that plasma turbulence differs qualitatively from its aquatic relative. The first difference is that essentially all plasma disturbances *propagate*, unlike vortices in a liquid. The second difference is that the dynamics, and indeed the very nature of plasma disturbances, varies in the several frequency regimes of interest. Although one could, in principle, use the Maxwell-Vlasov equations to describe simultaneously a Langmuir soliton and an Alfvén wave, it would be pointless to do so. By contrast, the dynamics of liquid vortices is invariant across the entire spectrum: it is adequately and *irreducibly* described by the Navier-Stokes equation at all relevant wavelengths. This leads to the concept, fundamental in the theory of Navier Stokes turbulence, of the universality of the spectrum. Although this concept can be applied to plasma turbulence over comparatively narrow spectral regions, its practical importance is far from compelling.

The theory of plasma turbulence may be viewed as the natural extension of the Balescu-Lenard theory to cases where the reference state either has unstable modes, or is driven away from equilibrium by external forces. As for collisional transport, the aim is to devise closures for the hierarchy of equations describing the average values of measurable quantities.

For the most frequently encountered case of quadratic nonlinearities, the evolution equations can be cast in the form

$$\left(\frac{d}{dt} + i\omega_\alpha\right) \Psi_\alpha(t) = \sum_{\beta\gamma} N_{\alpha\beta\gamma} \Psi_\beta(t)\Psi_\gamma(t) \tag{9.1}$$

where the $N_{\alpha\beta\gamma}$ are the coupling coefficients and the $\Psi_\alpha$ are the fluctuation amplitudes. The basic assumption underlying (9.1) is that the dynamical state can always be represented in terms of a complete basis of eigenmodes of the reference state. The subscripts $\alpha, \beta$, and $\gamma$ label the eigencomponents of $\Psi$ in this basis. In a homogeneous reference state, $\alpha = (\mathbf{k}, \ell)$ where $\mathbf{k}$ is the wavevector and $\ell$ is an index describing the roots of the dispersion relation. More generally we will denote the set of numbers describing the spatial mode structure with $a, b, c$, and write $\alpha = (a, \ell)$. Note that we may assume without loss of generality that

$$N_{\alpha\beta\gamma} = N_{\alpha\gamma\beta}.$$

Consider the average of the dynamical equations over an ensemble of realizations of the experiment (or, equivalently, over a suitably defined period of time):

$$\left(\frac{d}{dt} + i\omega_\alpha\right) \langle\Psi_\alpha(t)\rangle = \sum_{\beta\gamma} N_{\alpha\beta\gamma} \langle\Psi_\beta(t)\Psi_\gamma(t)\rangle. \tag{9.2}$$

The evolution equation for the first-order expectation $\langle\Psi_\alpha(t)\rangle$ depends on the second-order quantity $\langle\Psi_\beta(t)\Psi_\gamma(t)\rangle$. An equation for the latter can be obtained by multiplying the equation for $\Psi_\beta(t)$ by $\Psi_\gamma(t)$ and taking the average. This leads, of course, to the appearance of an unknown third-order moment: we are faced with the usual closure problem.

Useful results can be obtained in the special case when one of the dynamical quantities has negligible effect on the evolution of the others. These results apply to problems such as the transport of a diffuse impurity entrained in a turbulent plasma (passive advection), or the propagation of a small-amplitude wave through a medium with a randomly varying index of refraction. More generally, the passive evolution problem serves to illuminate some of the properties of the general problem.

We begin this chapter by describing the passive evolution problem in some detail, using it to make two main points. First, we show that the random fluctuation of the frequency of an oscillator interacting with a turbulent bath causes damping of the oscillation, and an associated broadening of the oscillator resonance. Second, we describe the role of the ratio between the correlation time and the "eddy-turnover time," or the time required for the turbulent field to affect a test field. We show how approximate solutions can be found in the limit of short autocorrelation time, thereby motivating the more detailed development of quasilinear theory presented in Sec. 9.5. We conclude the chapter by describing the quasilinear theory for the relaxation of a gentle bump on the tail of a distribution function.

## 9.2 Passive evolution

Consider a passive field or small-amplitude wave corresponding to the $j$-th degree of freedom,

$$\psi_a = \Psi_{a,j}.$$

This may represent, for example, the helium density in a plasma or the amplitude of a compressional Alfvén wave in the magnetosphere. More generally it describes any quantity that is subjected to turbulent fluctuations it has no effect on. The evolution of $\psi_a$ is described by

$$\left(\frac{d}{dt} + i\omega_a\right)\psi_a(t) = \sum_b A_{ab}(t)\psi_b(t) + \xi_a(t), \tag{9.3}$$

where

$$A_{ab}(t) = \sum_\gamma N_{(a,j);(b,j);\gamma}\Psi_\gamma(t)$$

is a stochastic "spring-constant" and

$$\xi_a(t) = \sum_{\beta,\gamma \neq j} N_{(a,j);\beta;\gamma}\Psi_\beta(t)\Psi_\gamma(t)$$

is a stochastic force. By stochastic, we mean that we may view each experiment as equivalent to picking a set of functions $A_{a,b}$ and $\xi_a$ at random out of a bag full of functions. For example the bag might contain 5 Maxwellians, 2 Bessel functions, and 3 Coulomb wave functions. Kolmogorov has shown that specifying all the probabilities for drawing a particular function in the bag (in the example above P(Maxwellian)=0.5, P(Bessel)=0.2, etc...) is equivalent to specifying the complete sequence of probability distributions $\{P_1, P_2, \ldots\}$ for the *values* taken by the functions. Here

$$P_N = P_N(x_1, t_1; x_2, t_2; \ldots; x_N, t_N)$$

is the probability that the function takes the value $x_1$ at time $t_1$, $x_2$ at time $t_2$, and so on.

The problem posed by the passive evolution equation (9.3) is to determine the probability distribution for the solution $\psi_a$ knowing the probability distribution for the $\xi_a$ and $A_{ab}$ functions. The key simplification contained in (9.3) is that the dynamic nonlinearity of the original system (9.1) is replaced by the stochastic nonlinearity $A_{ab}\psi_b$. This is a drastic simplification, but one that plays a fundamental role in the development of closures for the dynamically nonlinear system. Indeed, it can be shown that the lowest-order effect of the interaction between a given triplet of waves is described by an equation essentially equivalent to (9.3).

For $A_{ab}(t) = 0$, (9.3) has the same form as the Langevin equation used to describe Brownian motion. The Langevin equation has been extensively studied. Its most basic property is that the oscillators must be damped, $\text{Im}(\omega) <$

0, in order for the system to have well-behaved solutions. In a wide class of applications of interest in plasma physics, however, Im$(\omega)$ vanishes. We might expect, for these applications, that the amplitude and spectral width of the passive fields will grow without bounds!

That this does not happen is perhaps the most important result of the theory of stochastic differential equations. The reason is that the random variation of the spring-constant $A_{ab}(t)$ provides an effective damping. We will demonstrate this damping in the next section by solving the evolution equations for a simple scalar model. Before doing this, however, we note that the random force $\xi_a(t)$ may be assimilated into the random spring-constant matrix $A$ by allowing the index $a$ to extend over an additional variable $\psi_{N+1}(t) = 1$. Extending the matrix $A$ by $A_{a,N+1} = \xi_a$, $A_{N+1\,a} = 0$, the evolution equations take the form

$$\left(\frac{d}{dt} + i\omega_a\right)\psi_a(t) = \sum_b^{N+1} A_{ab}(t)\psi_b(t), \tag{9.4}$$

We next demonstrate stochastic damping in the simplest possible context: a scalar model describing the gyration of a particle in a tangled magnetic field.

## 9.3   The random gyrator

The evolution of the velocity of a magnetized particle in a stochastic magnetic field—a magnetic field that varies in an unpredictable way—is governed by

$$\left(\frac{d}{dt} + i\Omega_0\right)u_\perp(t) = i\delta\Omega(t)u_\perp(t), \tag{9.5}$$

where $\mathrm{Re}(u_\perp) = u_{\perp 1}$ and $\mathrm{Im}(u_\perp) = u_{\perp 2}$ are the two components of the perpendicular velocity, and $\Omega = \Omega_0 + \delta\Omega$ is the cyclotron frequency. Here,

$$\langle \delta\Omega \rangle = 0$$

and $\Omega$ is assumed to be real.

Equation (9.5) is clearly a special case of (9.4) with only two degrees of freedom. It is easily integrated,

$$u_\perp(t) = u_\perp(0)\exp\left[i\Omega_0 t + i\int_0^t dt'\,\delta\Omega(t')\right]. \tag{9.6}$$

The difficulty is to evaluate the average, or expectation value, of $u_\perp$. The first method that comes to mind is to expand the exponential in a Taylor series,

$$\langle u_\perp(t)\rangle = u_\perp(0)e^{i\Omega_0 t}\left[1 - \frac{1}{2!}\int_0^t\int_0^t dt'\,dt''\,\langle\delta\Omega(t')\delta\Omega(t'')\rangle\right.$$
$$\left. -\frac{i}{3!}\int_0^t\int_0^t\int_0^t dt'\,dt''\,dt'''\,\langle\delta\Omega(t')\delta\Omega(t'')\delta\Omega(t''')\rangle + \ldots\right].\tag{9.7}$$

Unfortunately, the successive moments of $\delta\Omega$ do not decay as their order increases. The Taylor series is thus only accurate for small times. We will use it here to introduce some of the key features of the passive evolution problem.

The first significant feature of (9.7) is the appearance of the function

$$C(t, t') = \langle \delta\Omega(t)\delta\Omega(t')\rangle \tag{9.8}$$

called the auto-correlation function. In a statistically stationary system,

$$C(t, t') = C(t - t').$$

The auto-correlation function plays a fundamental role in all statistical theories of turbulence. To understand its meaning, recall that random variables are statistically independent when their probability distribution factors,

$$P(1, 2, \ldots, n) = P(1, 2, \ldots, j)P(j + 1, \ldots, n).$$

If $\delta\Omega(t + \tau)$ becomes statistically independent of $\delta\Omega(t)$ after a time $T$, for example, the correlation function $C(\tau)$ will vanish for all $\tau > T$. The correlation function describes thus the rate at which stochasticity causes a system to lose memory.

When $\lim_{\tau\to\infty} C(\tau) = 0$, $\Omega(t + \tau)$ is said to be asymptotically independent of $\Omega(t)$. The width of $C(\tau)$,

$$\tau_{ac} = \int_0^\infty d\tau\, \frac{C(\tau)}{C(0)}. \tag{9.9}$$

is called the *autocorrelation time*. $\tau_{ac}$ measures the time after which successive observations of $\delta\Omega$ become independent.

Having identified the auto-correlation time, it is natural to ask to what extent the particle gyration is affected by the frequency perturbation during an auto-correlation time. This is indicated by the parameter

$$K = \delta\Omega_\sigma \tau_{ac}, \tag{9.10}$$

called the *Kubo number K*. Here

$$\delta\Omega_\sigma = \langle \delta\Omega(t)^2 \rangle^{1/2}$$

is the characteristic amplitude of the frequency fluctuation. For a stationary distribution,

$$\delta\Omega_\sigma = \langle \delta\Omega(0)^2 \rangle^{1/2} = [C(0)]^{1/2}$$

whence

$$K = \int_0^\infty d\tau \frac{C(\tau)}{[C(0)]^{1/2}}$$

in the stationary case. Under various guises, the Kubo number plays a central role in essentially all theories of turbulence. In hydrodynamic turbulence, for example, the Kubo number is given by the ratio of the correlation time to the eddy turn-over time. We will discuss the significance of this number in more detail after we have introduced an expansion with better convergence properties, called the *cumulant expansion*.

## Cumulant expansion

Before presenting the cumulant expansion it is useful to review a few elementary statistical concepts. First, recall that the expectation value for a function $f$ of a random variable $x$ is

$$\langle f(x) \rangle = \int_{-\infty}^{\infty} dx \, f(x) P(x),$$

where $P(x)$ is the probability distribution for $x$. The moments of $P$,

$$\langle x^n \rangle = \int_{-\infty}^{\infty} dx \, x^n P(x),$$

can be conveniently calculated in terms of the characteristic function

$$G(k) = \langle \exp(ikx) \rangle. \tag{9.11}$$

One finds

$$\langle x^n \rangle = (i)^{-n} \left. \frac{d^n G(k)}{dk^n} \right|_{k=0}. \tag{9.12}$$

Unfortunately, even a well-behaved probability distribution $P$ cannot be described by a few of its lowest-order moments. This is because the magnitude of the moments increases rapidly with order. For a Gaussian distribution for example,

$$P(x) = (2\pi\sigma^2)^{-1/2} e^{-\frac{x^2}{2\sigma^2}},$$

the moments are

$$\langle x^{2n} \rangle = (2n-1)!! \, \sigma^{2n}.$$

The need for an expansion that can be truncated at low order motivates the introduction of the cumulants $\langle\langle x^n \rangle\rangle$. These are defined as the coefficients of the Taylor expansion of the *logarithm* of the characteristic function, or by

$$F(k) = \exp\left[ \sum_{n=1}^{\infty} \frac{(ik)^n}{n!} \langle\langle x^n \rangle\rangle \right]. \tag{9.13}$$

The second cumulant is equal to the variance $\sigma^2$. It is easy to show that for a Gaussian distribution, all higher-order cumulants vanish.

We next apply the cumulant expansion to evaluate the average of the solution (9.6). For fixed $t$, we consider

$$x = \int_0^t dt' \, \delta\Omega(t')$$

as the stochastic variable (not function!) in terms of which we carry out the cumulant expansion. The expansion is

$$\langle u_\perp(t) \rangle = u_\perp(0) e^{i\Omega_0 t} \exp\left[ -\frac{1}{2} \int_0^t \int_0^t dt_1 \, dt_2 \, \langle\langle \delta\Omega(t_1)\delta\Omega(t_2) \rangle\rangle \right.$$
$$\left. -\frac{i}{3!} \int_0^t \int_0^t \int_0^t dt_1 \, dt_2 \, dt_3 \, \langle\langle \delta\Omega(t_1)\delta\Omega(t_2)\delta\Omega(t_3) \rangle\rangle + \dots \right] \tag{9.14}$$

This differs from (9.7) in that the expansion is here inside the argument of the exponential. The first few averages are related to the cumulants by

$$\langle 1 \rangle = \langle\langle 1 \rangle\rangle;$$
$$\langle 12 \rangle = \langle\langle 1 \rangle\rangle\langle\langle 2 \rangle\rangle + \langle\langle 12 \rangle\rangle;$$
$$\langle 123 \rangle = \langle\langle 1 \rangle\rangle\langle\langle 2 \rangle\rangle\langle\langle 3 \rangle\rangle + \langle\langle 12 \rangle\rangle\langle\langle 3 \rangle\rangle + \langle\langle 23 \rangle\rangle\langle\langle 1 \rangle\rangle + \langle\langle 31 \rangle\rangle\langle\langle 2 \rangle\rangle + \langle\langle 123 \rangle\rangle,$$

where we have used the shorthand notation $"n" \equiv \delta\Omega(t_n)$.

The advantage of the cumulant expansion resides in the following property: *the cumulant of a set of variables vanishes whenever this set can be divided into two statistically independent subsets.*

In order to apply this property to the expansion (9.14), we note that we expect two frequencies $\delta\Omega(t_j)$ and $\delta\Omega(t_k)$ to be independent if the delay between the times of observation $t_j$ and $t_k$ is much greater than the correlation time. That is,

$$\langle\langle 1 \ldots j \ldots k \ldots n \rangle\rangle = 0 \qquad \text{for} \qquad |t_j - t_k| \gg \tau_{\text{ac}}.$$

The argument of the integrals is thus negligible when *any two* of the times are separated by more than $\tau_{\text{ac}}$. It follows that the integrals in (9.14) scale as $t(\tau_{\text{ac}})^{n-1}\delta\Omega_\sigma^n$. The expansion is thus a power series in $K = \delta\Omega_\sigma \tau_{\text{ac}}$, and we expect its truncations to be accurate for all time whenever the Kubo number is small. The Taylor series, by contrast, requires $\delta\Omega_\sigma t \ll 1$.

The asymptotic behavior predicted by the cumulant expansion for $t \gg \tau_{\text{ac}}$ is

$$\langle u_\perp(t) \rangle = u_\perp(0)e^{[i(\Omega_0 + \Delta\Omega) - \gamma]t}. \tag{9.15}$$

That is, the fluctuation of the frequency gives rise to a *damping* given, to order $\tau_{\text{ac}}\delta\Omega_\sigma^2$, by

$$2\gamma = \int_0^\infty dt\, C(t) = K\,\delta\Omega_\sigma \tag{9.16}$$

and to a frequency shift $\Delta\Omega = O(\tau_{\text{ac}}^2\delta\Omega_\sigma^3)$. A simple and widely used method for estimating the fluctuation amplitude in an unstable system is to equate this damping coefficient to the growth rate of the instability.

An attractive property of the cumulant expansion is that it applies regardless of whether the frequency perturbation is a Gaussian process. The importance of this property follows from the fact that turbulent systems almost never obey Gaussian statistics. For a centered, Gaussian process, however, all cumulants other than the auto-correlation vanish. The truncation of the cumulant expansion is then exact, independently of the Kubo number. We may thus use the Gaussian case to investigate the role of the Kubo number.

## Case of finite correlation time

In order to explore the effect of a finite Kubo number, we consider the case where the frequency fluctuation $\delta\Omega$ is a stationary, Gaussian stochastic process with

$$\langle \delta\Omega(t)\delta\Omega(t') \rangle = F(t - t'). \tag{9.17}$$

In this case (9.5) has the exact solution

$$\langle u_\perp(t)\rangle = u_\perp(0)\exp\left[-\int_0^t dt'\,(t-t')F(t')\right].\tag{9.18}$$

We may choose

$$F(t) = \delta\Omega_\sigma e^{-t/\tau_{ac}}.$$

Substituting this choice of $F$ in (9.18), we obtain the exact, explicit solution

$$\langle u_\perp(t)\rangle = u_\perp(0)\exp\left[-K^2(\tau - 1 + e^{-\tau})\right].\tag{9.19}$$

where $\tau = t/\tau_{ac}$. The above solution is shown in Fig. 9.1 for values of $K$ ranging from $K = 0.5$ to $K = 2.0$.

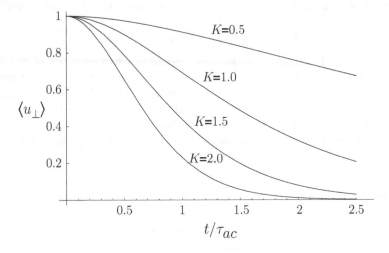

Figure 9.1: Response of a mode to stochastic variation of the frequency for a gaussian frequency process. $K$ denotes the Kubo number.

The small-time limit of the solution, (9.19), has the form

$$\langle u_\perp(t)\rangle = u_\perp(0)\exp\left[-K^2\tau^2/2\right].\tag{9.20}$$

For large $K$, this small-time limit remains valid until $\langle u_\perp(t)\rangle$ is negligible. We conclude that the decay of the oscillation is more rapid for large $K$, both functionally and through the faster decay rate $\gamma = \delta\Omega_\sigma \gg K\delta\Omega_\sigma$. The large-$K$ limit may be understood by noting that for a stationary process with vanishing mean, $\langle \delta\Omega\rangle = 0$, large $K$ implies nearly constant $\delta\Omega(t)$. It follows that in this limit the average $\langle\psi\rangle$ is simply the characteristic function of $\delta\Omega(0)$ (that is, the Fourier transform of the probability distribution for $\delta\Omega(0)$).

It can be shown that the absorption spectrum for a wave traversing a medium consisting of such random gyrators is proportional to the Fourier transform of the average amplitude $\langle u_\perp(t)\rangle$. In the small Kubo number limit, this leads to the well-known Lorentz line-shape. In the large Kubo number limit, by contrast, the absorption line is broadened by the frequency fluctuations.

We may summarize the role of the Kubo number as follows. In the limit of small $K$, the random frequency variation is rapid and of small amplitude. In this limit the gyration amplitude does not vary much during an auto-correlation time of the perturbation, so that during this time the gyrator approximately follows its unperturbed orbit. The gyration amplitude, however, decays with a decay rate given by the product of the auto-correlation time with the variance of the frequency fluctuation.

In the limit of large $K$, by contrast, the perturbation is nearly constant in time, and must be included even in the lowest-order description of the gyration. Special methods exist to treat this limit. Of particular interest is the case of the passive transport of heat in a two-dimensional turbulent fluid. In this case the diffusion coefficient is related to the probability of finding a stream line that connects the hot plate to the cold plate. this probability can be calculated using methods of percolation theory.

Heuristically, we may view $K$ as indicating whether stochasticity or nonlinearity dominates the character of the passive evolution. For the dynamically nonlinear problem, however, stochasticity is a direct result of nonlinearity. It follows that the Kubo number is almost always of order unity in dynamically nonlinear problems. The case $K \approx 1$ is somewhat confusingly referred to as the case of "strong turbulence."

## 9.4 Transport in a known spectrum

In the previous section we considered the evolution of an oscillator with a single degree of freedom under the effect of random variations of the spring constant. We now return to the more general problem (9.3), describing the passive evolution of a field—an infinite degree of freedom dynamical system. To compensate for the added difficulty associated with the infinite degree of freedom, we restrict consideration to the case of small Kubo number.

The generalization of the cumulant expansion method to the case of infinite degrees of freedom is complicated by the fact that the frequencies $\Omega_0$ and $\delta\Omega$ become noncommuting time-dependent operators. To avoid issues related to the time-ordering of these operators in the cumulant expansion, we present a simpler expansion scheme that is equivalent to the cumulant expansion to lowest significant order in the Kubo number.

First, note that the evolution equation is equivalent to

$$\psi_a(t) = \psi_a(0) + \int_0^t d\tau \, e^{-i\omega_a \tau} \sum_b A_{ab}(t - \tau)\psi_b(t - \tau), \qquad (9.21)$$

This may be written symbolically

$$\psi(t) = \int_0^t d\tau \, e^{-i\Omega \tau} \mathbf{A}(t - \tau)\psi(t - \tau) \qquad (9.22)$$

where $\psi$ is the vector with components $\psi_a$, $\mathbf{\Omega}$ is the diagonal matrix (operator) consisting of the eigenfrequencies $\omega_a$ of the unperturbed system, and $\mathbf{A}$ is the matrix with elements $A_{ab}$. Note that since $\mathbf{\Omega}$ is diagonal, $\exp(i\mathbf{\Omega}t)$ is the diagonal matrix with elements $e^{i\omega_a t}$.

The above result may be iterated. Substituting (9.22) into (9.3) and averaging yields

$$\frac{d\langle\psi(t)\rangle}{dt} + i\mathbf{\Omega}\langle\psi(t)\rangle = \int_0^t d\tau \,\langle \mathbf{A}(t)e^{-i\mathbf{\Omega}\tau}\mathbf{A}(t-\tau)\psi(t-\tau)\rangle \tag{9.23}$$

For small Kubo number, we know from our study of the scalar case that the effect of the perturbation on the evolution of $\psi$, over the duration of a few correlation times, is of order $K^2$. It follows that the correlation between $\psi(t-\tau)$ and the $\mathbf{A}$ matrices is of order $K^2$. We may thus write

$$\frac{d\langle\psi(t)\rangle}{dt} + i\mathbf{\Omega}\langle\psi(t)\rangle = \int_0^t d\tau \,\langle \mathbf{A}(t)e^{-i\mathbf{\Omega}\tau}\mathbf{A}(t-\tau)\rangle\langle\psi(t-\tau)\rangle \tag{9.24}$$

This is called the Bourret approximation. For $t \ll \tau_{ac}$, the Bourret approximation is equivalent to the Taylor series solution. In the limit $t \gg \tau_{ac}$, by contrast, the correlation between $\mathbf{A}(t)$ and $\mathbf{A}(t-\tau)$ vanishes. In this limit, we may estimate $\langle\psi(t-\tau)\rangle$ by assuming that it evolves according to the unperturbed equation (the justification is the same as that used to neglect the correlation between $\psi$ and $\mathbf{A}$). There follows

$$\frac{d\langle\psi\rangle}{dt} + i\mathbf{\Omega}\langle\psi\rangle = -\Gamma\langle\psi(t)\rangle, \quad t \gg \tau_{\text{ac}} \tag{9.25}$$

where

$$\Gamma = -\int_0^\infty d\tau \,\langle \mathbf{A}(t)e^{-i\mathbf{\Omega}\tau}\mathbf{A}(t-\tau)\rangle e^{i\mathbf{\Omega}\tau} \tag{9.26}$$

This may easily be seen to be equivalent, for the case of a scalar $\psi$, to the lowest order approximation of the cumulant expansion.

## Particle transport in Langmuir turbulence

We now apply the above analysis to the problem of a tracer element carried in a turbulent spectrum of Langmuir waves (described in Section 6.3). For simplicity, we consider the one-dimensional case. The evolution of the distribution function for the tracer is described by

$$\frac{\partial f}{\partial t} + v\frac{\partial f}{\partial x} = -\frac{e_s}{m_s}E\frac{\partial f}{\partial v}. \tag{9.27}$$

We assume that the statistics of the electric field are known and are uninfluenced by the distribution $f$ of the tracer element.

Equation (9.27) may be solved by iteration. The equivalent of (9.21) is

$$f_1(x, v, t) = f_1(x - vt, v, 0) - \frac{e_s}{m_s}\int_0^t d\tau E(x - v\tau, t - \tau)\frac{\partial}{\partial v}f_0(v, t - \tau), \tag{9.28}$$

where $f_0$ represents the slowly varying and $f_1$ the rapidly varying parts of the distribution function. Substituting this in (9.27) and taking the average, we find

$$\frac{\partial \langle f_0 \rangle}{\partial t} = \frac{e_s^2}{m_s^2} \int_0^t d\tau \left\langle E(x,t) \frac{\partial}{\partial v} \left( E(x - v\tau, t - \tau) \frac{\partial}{\partial v} f_0(v, t - \tau) \right) \right\rangle, \quad (9.29)$$

where we have dropped the initial-value term on the grounds that phase-mixing will rapidly reduce it to irrelevance. (Note, from Chapter 7, that the initial value term is plays a crucial role in the corresponding derivation of the collision operator.) We next use the independence approximation. There follows

$$\frac{\partial \langle f_0 \rangle}{\partial t} = \frac{e_s^2}{m_s^2} \frac{\partial}{\partial v} \int_0^t d\tau \, C(x, t; x - v\tau, t - \tau) \frac{\partial}{\partial v} \langle f_0(v, t - \tau) \rangle, \quad (9.30)$$

where

$$C(x, t; x - v\tau, t - \tau) = \langle E(x, t) E(x - v\tau, t - \tau) \rangle \quad (9.31)$$

is the auto-correlation function of the electric field between two successive points on the unperturbed particle trajectory. We will refer to this quantity as the Lagrangian auto-correlation. In (9.29), we assumed that the turbulence is homogeneous and discarded the gradient of $\langle f \rangle$. For small Kubo numbers, the approximations made to obtain (9.25) yield, for $t \gg \tau_{ac}$,

$$\frac{\partial \langle f_0 \rangle}{\partial t} = \frac{\partial}{\partial v} \left[ D(v) \frac{\partial \langle f_0 \rangle}{\partial v} \right]. \quad (9.32)$$

We find that that the stochastic electric fields cause diffusion of the average tracer distribution in velocity space. The diffusion coefficient,

$$D(v) = \frac{e^2}{m^2} \int_0^\infty d\tau C(x, t; x - v\tau, t - \tau) \rangle, \quad (9.33)$$

is given in terms of the Lagrangian auto-correlation. This is a standard feature of passive advection problems.

We next evaluate the diffusion coefficient $D_{sh}$ for a stationary and homogenous electric field distribution, expressing the result in terms of the auto-correlation of the *Fourier coefficients* of the electric field.

For stationary and homogenous electric field distributions, we may replace the statistical average in (9.33) by an average over time and space

$$D_{sh}(v) = \lim_{L,T \to \infty} \frac{e^2}{m^2} \int_0^\infty d\tau \int_{-L}^L \frac{dx}{2L} \int_{-T}^T \frac{dt}{2T} E(x, t) E(x - v\tau, t - \tau)$$

Expressing $E(x, t)$ by its Fourier transform and evaluating the averages yields:

$$D_{sh}(v) = \frac{e^2}{2m^2} \int_{-\infty}^\infty d\tau \int_{-\infty}^\infty \frac{dk}{2\pi} \int_{-\infty}^\infty \frac{d\omega}{2\pi} E_{k,\omega} E_{-k,-\omega} \exp[i(\omega - kv)\tau] \quad (9.34)$$

where we have used the symmetry of the integrand under $\tau \to -\tau$ to extend
the $\tau$ integral from $-\infty$ to $\infty$. Note that the Fourier components must obey

$$E_{k,\omega} = E_{-k,-\omega}$$

in order for the electric field to be real. Evaluating the $\tau$ integral yields

$$D_{sh}(v) = \frac{e^2}{2m^2} \int_{-\infty}^{\infty} \frac{dk}{2\pi} |E_{k,kv}|^2, \qquad (9.35)$$

The result (9.35) shows that diffusion is due to the resonant interaction of
particles with waves such that $\omega = kv$.

When the field, rather than being prescribed, evolves self-consistently, its
correlation function reflects the propagation properties of the waves. In particular in the small amplitude, weak turbulence limit, the wave propagation must
obey the dispersion relation. It follows that in this limit

$$|E_{k,\omega}|^2 = |E(\mathbf{k})|^2 \delta(\omega - \omega_{\mathbf{k}}) \qquad (9.36)$$

where $\omega_{\mathbf{k}}$ is a root of the dispersion relation. We will evaluate the diffusion
coeficient for self-consistently evolving fields in Sec. 9.5.

## Resonance broadening

The resonant nature of the diffusion coefficient in the limit of short correlation
time is due to the fact that our approximation method treats the particle motion
as unperturbed. Specifically, the Dirac functions representing the resonances in
the diffusion coefficient originate from our perturbative solution (9.28) of the
passive advection equation (9.27). If, instead of integrating the effect of the
perturbation along the unperturbed trajectory $x = x(0) + v\tau$, we integrate the
perturbation along the actual trajectory, the term $kv\tau$ in (9.34) is replaced by

$$k\Delta x = kv\tau + k \int_0^\tau dt\, \Delta v(t).$$

The stochastic velocity $\Delta v$ is evidently correlated with the electric field perturbation. An interesting result is obtained, however, if we make the independence
ansatz

$$\langle E_{k,\omega} E_{k,\omega}^* e^{-ik\Delta x} \rangle = \langle E_{k,\omega} E_{k,\omega}^* \rangle \langle e^{-ik\Delta x} \rangle.$$

This ansatz makes it possible to calculate the effect of the velocity fluctuations
by the method of cumulant expansion. We find

$$\left\langle e^{-ik \int_0^\tau dt\, \Delta v(t)} \right\rangle = e^{-k^2 \int_0^\tau dt\, t \langle\!\langle \Delta v(0)\Delta v(t)\rangle\!\rangle} = e^{-k^2 D(v)t^3/3}. \qquad (9.37)$$

Substituting this in place of the term $e^{ikv\tau}$ in (9.34) yields a nonlinear equation
for the diffusion coefficient in which the singular terms $\delta(\omega - kv)$ are replaced
by

$$R(kv - \omega, D) = \mathrm{Re} \int_o^\infty dt\, e^{i(kv-\omega)t - k^2 Dt^3/3}.$$

The width of this function,

$$\Delta\omega_{\text{RB}} = (k^2 D/3)^{1/3},\tag{9.38}$$

is called the resonance-broadening width.

The resonance-broadening idea may be applied to the case of diffusion in configuration ($\mathbf{x}$) space under the influence of a turbulent convective velocity. In that case $\Delta x$ itself undergoes a random walk and the resonance-broadening term is simply $e^{-k^2 Dt}$. This result may be used to estimate the saturation level of fully developed turbulence by assuming that saturation occurs when the linear growth rate is balanced by the resonance-broadening term. There follows the estimate

$$D \equiv \frac{\gamma}{k^2}\tag{9.39}$$

This result and the reasoning behind it are essentially equivalent to Prandtl's mixing-length estimate of the turbulent viscosity in hydrodynamics.

In the next section, we consider the quasilinear evolution of the Langmuir-wave turbulence responsible for the relaxation of a particle distribution with a gentle bump.

## 9.5   Quasilinear theory

We consider the evolution of a spectrum of Langmuir waves excited by the weak gentle-bump instability described in Sec. 6.2. We assume that the wavenumber spectrum is restricted by the finite length $L$ of the system to the discrete values $k = 2\pi n/L$. A key question is the number and density of unstable modes. This is determined by the number of values of $v_{ph} = \omega_{pe}/k$ lying in the range of velocities where $f_0'(v_{ph}) > 0$. In the case of a single unstable mode, we saw in Sec. 6.7 that the mode will saturate once its amplitude is such that the bounce frequency is comparable to the linear growth rate.

When there is more than one mode present, the rapid oscillation of the electric field can no longer be eliminated by studying the motion in the frame moving at the phase velocity of the wave: the Hamiltonian for the particle motion is irreducibly non-stationary. A fundamental result of classical mechanics is that particle motion becomes chaotic when the potential wells for the single-wave motion overlap (Fig. 9.2). Consider a box of length $L$ and a wave field given by its Fourier series,

$$\phi = \sum_l \phi_l \exp(ik_l x),$$

where $k_l = 2\pi l/L$. The chaotic threshold is then

$$2\left(\frac{2e\phi}{m}\right)^{1/2} > \left|\frac{\omega_1}{k_1} - \frac{\omega_2}{k_2}\right|.$$

where the $\omega_i$ and $k_i$ correspond to neighboring resonances in velocity space. When this condition is satisfied the electrons near the separatrix of one wave

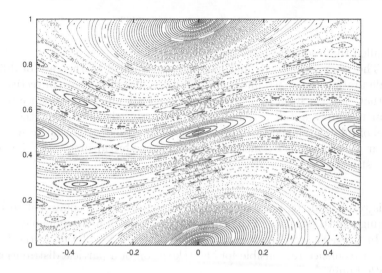

Figure 9.2: Puncture plot of the trajectories of particles in a system near the onset of mode overlap. Courtesy of A. Wurm.

can be pulled away by the other wave. The result is a Brownian motion in velocity space. This greatly simplifies the problem by allowing a statistical analysis.

## Self-consistent diffusion

In Sec. 9.4 we showed that a stochastic electric field causes the particle distribution to diffuse in velocity-space according to (9.32). We found that the diffusion coefficient (9.33) is proportional to the integral of the Lagrangian correlation function for the electric field. Here we consider the effect of the particle diffusion on the electric field, supposing a weakly unstable Vlasov plasma. We know from Chapter 6 that the growth rate is given by

$$\gamma = \frac{\pi}{2} \frac{\omega_{pe}^3}{k^2} f_0' \left( \frac{\omega_{pe}}{k} \right).$$

(9.40)

Of course this rate must be positive: our distribution has positive slope at some resonant velocity $\omega_p/k$, as shown in Figure 9.3. When the growth rate is small enough to neglect wave-wave nonlinearities, the saturation of the instability results from the effect on the distribution of the particles' random walk.

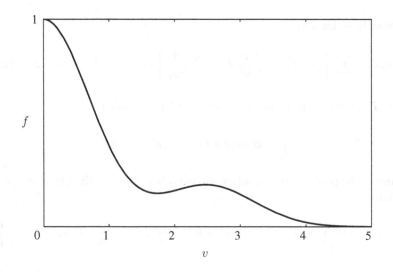

Figure 9.3: Weakly unstable distribution function with a "gentle bump.'

We look for solutions of the form

$$E(x,t) = \sum_k E_k(t) \exp(ikx - i\omega_k t + i\phi_k), \qquad (9.41)$$

where $\omega_k$ and $\phi_k$ are taken to be *real*. Any growth or damping of the electric field is assumed to relatively slow and taken into account by the coefficients $E_k(t)$; in particular we assume

$$\frac{d}{dt}|E_k|^2 = 2\gamma(k,t)|E_k|^2$$

where $\gamma(k,t)$ is given by (9.40).

Next consider the correlation function of (9.31). To evaluate $C(x,t; x - v\tau, t - \tau)$ we express the field in terms of its Fourier components, remembering that in order for the field to be real we must have $E_{-k} = E_k^*$:

$$C(x,t; x - v\tau, t - \tau) = \sum_k E_k^*(t) E_k(t - \tau) \exp[i(\omega_k - kv)\tau].$$

The sum over $k$ leads to the decay of $C$ with the decay time

$$\tau_C \sim |\Delta\omega - v\Delta k|^{-1} \sim |v - d\omega/dk|^{-1}(\Delta k)^{-1}$$

where $\Delta k$ is the width of the spectrum of excited waves. We assume that this decay time is shorter than the time for growth or decay of $f_0$ and $E_k(t)$. It follows that the field amplitude $E_k(t)$ may be expanded in Taylor series about $t = \tau$ in the equation for the correlation function. Thus

$$C(v; t, \tau) = \sum_k \left\{ |E_k(t)|^2 - \frac{1}{2}\tau \frac{d}{dt}|E_k(t)|^2 \right\} \exp[i(\omega_k - kv)\tau].$$

We now have for $D(v,t)$

$$D((v,t) = \sum_k \left\{ |E_k(t)|^2 + \frac{1}{2}\frac{d}{dt}|E_k(t)|^2 i\frac{\partial}{\partial\omega_k} \right\} (\pi\delta(\omega_k - kv) + iP(\omega_k - kv)^{-1}),$$

where $P$ denotes a principal value and we have recalled

$$\int_0^\infty d\tau \, \exp(iy\tau) = \pi\delta(y) + iP\,y^{-1}.$$

Summing the positive and negative wavenumbers and recalling that $\omega_{-k} = -\omega_k$, we find

$$D(v,t) = \sum_{k>0} \left\{ |E_k(t)|^2 2\pi\delta(\omega_k - kv) - \frac{d}{dt}|E_k(t)|^2 \frac{\partial}{\partial\omega_k}P(\omega_k - kv)^{-1} \right\}$$

Notice that substitution of this result into (9.32) leads to a pathological diffusion equation, because of the delta-function and the discreteness of the sum over $k$. In particular, the only diffusional equilibrium would have $f'(v) = 0$ for all $v = \omega_k/k$ corresponding to admissible values of the wave-number. Noting that replacing the sum by an integral over $k$ removes this pathology, we at this point assume that unspecified nonlinear effects broaden the velocity-space resonance sufficiently to allow approximating $D(v,t)$ by an integral. (Recall that any sum is well approximated by an integral when the integrand varies little from one summand to the next. It is sometimes said that the replacement of the sum by an integral constitutes a renormalization of the linear propagator.) If the width of the broadened resonance is small compared to the $k$–scale for variation of $|E_k(t)|^2 \equiv \mathcal{E}(k;t)$, then we can make the replacement

$$\sum_k |E_k(t)|^2 \to \int_0^\infty \frac{dk}{2\pi}\mathcal{E}(k;t),$$

and are finally left with

$$D(v,t) = \int_0^\infty dk\mathcal{E}(k;t)\delta(\omega_k - kv) \\ - \frac{\partial}{\partial t}\int_0^\infty \frac{dk}{2\pi}\mathcal{E}(k;t)\frac{\partial}{\partial\omega_k}P(\omega_k - kv)^{-1}. \tag{9.42}$$

We next consider the interpretation of the two terms in (9.42).

## Quasilinear diffusion

Quasilinear theory is a closed description of plasma behavior based on diffusion of the distribution function, described by (9.32) and (9.42), together with evolution of the field amplitude at the evolving linear growth rate $\gamma$. Thus it is summarized by the equations

$$\frac{\partial f_0}{\partial t} = \left(\frac{e}{m}\right)^2 \frac{\partial}{\partial v}\left(D(v,t)\frac{\partial f_0}{\partial v}\right)$$

$$\frac{d\mathcal{E}(k,t)}{dt} = 2\gamma(k,t)\mathcal{E}(k,t).$$

Note that these equations are self-consistent in the sense that the effect of the changing distribution function is taken into account in the evolution of the electric field spectral energy density; the growth rate at any time is evaluated from (9.40) using the distribution function at that time. The evolution of the particle distribution function, likewise, accounts for the evolution of the electric field.

The spatial average of the distribution function is usefully separated into two components, corresponding to the two terms in (9.42):

$$f_0(v,t) = \langle f(x,v,t)\rangle = f_B(v,t) + f_W(v,t).$$

The second component, $f_W$, reflects the prompt action of the wave on the $k = 0$ component of the distribution function. It is small, being *in magnitude* directly proportional to the electric field energy:

$$f_W(v,t) = -\left(\frac{e}{m}\right)^2 \frac{\partial}{\partial v}\left[f_0'(v,t)\int_0^\infty \frac{dk}{2\pi}\mathcal{E}(k;t)\frac{\partial}{\partial\omega_k}P(\omega_k - kv)^{-1}\right]$$

$$= -\left(\frac{e}{m}\right)^2 \frac{\partial}{\partial v}\left[f_B'(v,t)\int_0^\infty \frac{dk}{2\pi}\mathcal{E}(k;t)\frac{\partial}{\partial\omega_k}P(\omega_k - kv)^{-1}\right]$$

We have replaced the distribution function by the background distribution function in the second line, since the two differ only by $f_W$ which is of order $\mathcal{E}$. This second component ensures the conservation of energy and momentum, as will be shown below.

The first component, $f_B$, can be thought of as the background distribution. It evolves diffusively:

$$\frac{\partial f_B}{\partial t} = \frac{\partial}{\partial v}\left(D_W(v,t)\frac{\partial f_B}{\partial v}\right), \tag{9.43}$$

with a diffusion coefficient given by

$$D_W(v,t) = \left(\frac{e}{m}\right)^2 \int_0^\infty \frac{dk}{2\pi}\mathcal{E}(k;t)2\pi\delta(\omega_k - kv). \tag{9.44}$$

The characteristic of $f_B$ is that its *rate of evolution* is proportional to the electric field energy. We next consider the conservation of momentum and energy.

## Conservation of momentum and energy

The momentum associated with $f_W$ is

$$P_W = \int dv\, mv f_W(v,t).$$

We can evaluate this momentum by integrating by parts:

$$P_W = \frac{e^2}{m} \int_0^\infty \frac{dk}{2\pi} \mathcal{E}(k;t) \frac{\partial}{\partial \omega_k} \int dv\, f_B'(v,t) P(\omega_k - kv)^{-1}.$$

Recall that the real part of the plasma dielectric function is given for real $\omega_k$ by

$$\epsilon_R(\omega_k, k) = \epsilon_0 + \frac{e^2}{mk} \int dv\, f_B'(v,t) P(\omega_k - kv)^{-1},$$

so that the wave momentum density is

$$P_W = \int_0^\infty \frac{dk}{2\pi} \mathcal{E}(k;t) k \frac{\partial \epsilon_R(\omega_k, k)}{\partial \omega_k}.$$

The rate of change of the wave momentum is

$$\frac{\partial P_W}{\partial t} = \int_0^\infty \frac{dk}{2\pi} 2\gamma_k \mathcal{E}(k;t) k \frac{\partial \epsilon_R(\omega_k, k)}{\partial \omega_k}.$$

Using the expression for the growth rate,

$$\gamma_k \frac{\partial}{\partial \omega_k} \int_0^\infty dv\, f_B'(v,t) P(\omega_k - kv)^{-1} = \frac{\pi}{k} f_B'(\omega_k/k),$$

this is

$$\frac{\partial P_W}{\partial t} = \frac{e^2}{m} \int_0^\infty dk\, \mathcal{E}(k;t) \int_{-\infty}^\infty dv\, \delta(\omega_k - kv) f_B'(v,t).$$

The momentum change due to diffusive evolution is

$$\frac{dP_D}{dt} = \int dv\, mv \frac{\partial f_B}{\partial t}.$$

$$= -\int dv\, D(v,t) \frac{\partial f_B}{\partial v}$$

We have shown that the total momentum is conserved.

To show that energy is also conserved, the particle contribution to the energy density is evaluated by taking the second moment of $f_W$,

$$K_W(t) = \int dv\, \frac{1}{2} mv^2 f_W(v,t)$$

$$= \int_0^\infty \frac{dk}{2\pi} \mathcal{E}(k;t) \frac{\partial}{\partial \omega_k} \{\omega_k [\epsilon_R(\omega_k, k) - 1]\}$$

adding this to the electric field energy yields the familiar wave energy,

$$K_E(t) + K_W(t) = \int_0^\infty \frac{dk}{2\pi} \mathcal{E}(k;t) \frac{\partial}{\partial \omega_k} [\omega_k \epsilon_R(\omega_k, k)]$$

The full self-consistent quasi linear theory unfortunately has a quite limited domain of application. Away from marginal stability, wave-wave nonlinearities rapidly become important: in practice plasmas are always in the strong turbulence regime. It has been found, however, that the transport coefficients can nevertheless be predicted surprisingly reliably by rescaling the quasilinear results. That is, the dependence of the transport coefficient on the macroscopic parameters seems to be described well by quasilinear theory. The rescaling coefficient can be determined either through experiment or nonlinear simulations. Many linear stability codes routinely provide quasilinear estimates of transport coefficients.

# Additional reading

The most enjoyable presentation of the theory of passive evolution in a turbulent medium is the review by Van Kampen.[73] The derivation of the quasilinear diffusion equation given here is due to Kaufman.[41] A less methodical but very popular derivation is given by Sagdeev and Galeev in their book on nonlinear plasma theory[64] and in their review of weak turbulence theory.[28] These authors also discuss the nature of quasilinear diffusion in more than one dimension, and derive the "wave-kinetic" equation. The problem of how to determine the saturation amplitude and spectrum of the turbulence is discussed in a very readable form in two papers by Dupree.[22, 23] Lastly, an enlightening presentation of the direct interaction approximation and of its application to problems of plasma physics is given by Krommes.[44]

# Problems

1. Calculate the relationship between the cumulants and the moments for a simple stochastic variable and for a bivariate distribution $P(x_1, x_2)$.

2. Show that for a stationary process, the second order term in the cumulant expansion of the solution of the random gyrator model is

$$\int_0^t dt' \, (t - t')C(t')$$

   Evaluate this integral in the regime $t \ll 1/\delta\Omega_\sigma \ll \tau_{ac}$ relevant to the early-time evolution of a large Kubo-number system and compare your result to the early-time expansion of the exact result for the Gaussian process, (9.19). Evaluate the same integral in the limit $\tau_{ac} \ll t$ to verify (9.16

3. An infinitely correlated stochastic process is specified by functions that take random but constant values. Let $P(\delta\Omega)$ be the probability distribution that such a constant function take the value $\delta\Omega$. Write an equation for the average amplitude of the random gyrator, $\langle u_\perp(t) \rangle$. Evaluate this average when $P$ is gaussian, and compare it to the large Kubo-number limit of the exact solution (9.19).

4. Show that the quasilinear equations conserve energy.

# Chapter 10

# Description of a Relativistic Plasma

## 10.1   Relativistic plasma

A plasma in which the thermal energy of a typical particle approaches the particle's rest energy $mc^2$, or a plasma in which flow speeds approach the speed of light, is called relativistic. Relativistic plasmas occur in several astrophysical environments, such as galactic jets; they are also created in laboratory experiments, usually by means of high intensity lasers.

The basic notions describing a relativistic plasma are not very different from the non-relativistic case emphasized in this book. The essential change is the replacement of the non-relativistic kinetic equation (3.22) by an equation that is consistent with special relativity. This chapter derives the relativistic kinetic equation, and then proceeds to develop its fluid moments, the relativistic fluid equations. Thus we present a relativistic parallel to the discussion of the Chapter 3. We also consider the moments of the distribution that replaces the Maxwellian of non-relativistic theory. Without studying the applications of the relativistic plasma theory, we try to establish the tools underlying such applications.

Electromagnetic radiation plays an important role in many relativistic plasmas. Furthermore, in the ultrarelativistic case, particle creation must be taken into account. The present introductory treatment omits such effects.

For the sake of a self-contained treatment, and also to introduce our notation, we begin with a review of relativistic mechanics. Note that in this chapter exclusively we use units, nearly universal in the relativistic literature, in which the light speed is unity: $c \rightarrow 1$

## 10.2    Relativistic kinematics

### Lorentz transformation

We use the Minkowski tensor $\eta_{\alpha\beta}$ with $\eta_{00} = -1$, $\eta_{ii} = 1$; Greek indices vary from 0 to 3, Roman indices vary from 1 to 3. We will always use boldface to indicate a 3-vector, so that

$$A^\mu = (A^0, \mathbf{A})$$

and

$$A^\mu A_\mu = \mathbf{A}^2 - (A^0)^2$$

A general Lorentz transformation combines a rotation in coordinate space with a "boost" in velocity. (The rotation group is a subgroup of the Lorentz group.) A boost is interpreted in the passive sense: we transform coordinate frames, thus changing the coordinates of tensors, but consider the physical system to be unchanged. The boosted system is observed to move, relative to the original system, with velocity $\mathbf{v}$; obviously the corresponding boosted frame has the opposite velocity.

It is sometimes convenient to consider the original frame to be the rest frame of some system, $R$, and to call the boosted frame the "lab frame," $S$. Thus the system is observed to move with velocity $\mathbf{v}$ in the lab frame. If some quantity has the value $Q$ in the lab frame, then its value measured in the rest frame is denoted by $Q_R$.

$Q$ is related by $Q_R$ by a Lorentz boost $\mathbf{v}$, whose matrix is denoted by $\Lambda(\mathbf{v})$. Thus a 4-vector transforms according to

$$V^\mu = \Lambda^\mu{}_\nu(\mathbf{v}) V_R^\nu$$

We recall that

$$\Lambda^\mu{}_\nu(\mathbf{v}) = \begin{bmatrix} \gamma & \gamma v_1 & \gamma v_2 & \gamma v_3 \\ \gamma v_1 & 1 + \hat{v}_1^2(\gamma-1) & \hat{v}_1\hat{v}_2(\gamma-1) & \hat{v}_1\hat{v}_3(\gamma-1) \\ \gamma v_2 & \hat{v}_1\hat{v}_2(\gamma-1) & 1 + \hat{v}_2^2(\gamma-1) & \hat{v}_2\hat{v}_3(\gamma-1) \\ \gamma v_3 & \hat{v}_1\hat{v}_3(\gamma-1) & \hat{v}_2\hat{v}_3(\gamma-1) & 1 + \hat{v}_3^2(\gamma-1) \end{bmatrix} \tag{10.1}$$

where $\hat{\mathbf{v}} \equiv \mathbf{v}/v$. An alternative expression for $\Lambda$ is

$$\Lambda^0{}_0 = \gamma \tag{10.2}$$
$$\Lambda^i{}_j = \delta^i{}_j + \hat{v}_i\hat{v}_j(\gamma-1) \tag{10.3}$$
$$\Lambda^0{}_j = \gamma v_j \tag{10.4}$$

To show the Lorentz boost more explicitly, we consider an arbitrary four–vector $K^\mu = (K^0, \mathbf{K})$. Then we find that

$$K'^\mu = \Lambda^\mu{}_\nu K^\nu$$

is given by

$$K'^0 = \gamma(K^0 + \mathbf{v} \cdot \mathbf{K}), \tag{10.5}$$
$$\mathbf{K}' = \mathbf{K} + \gamma \mathbf{v} K^0 + (\gamma-1)\frac{\mathbf{v}\mathbf{v} \cdot \mathbf{K}}{v^2} \tag{10.6}$$

It is not hard to verify the expected invariance:

$$\boldsymbol{K}^2 - (K^0)^2 = \boldsymbol{K}'^2 - (K'^0)^2$$

## Instantaneous rest frame

We observe a particle (or a fluid element—any object having a single velocity) from a frame called the "lab frame," and find that it has varying velocity $\boldsymbol{v}(t)$. At the time $t^*$ we can imagine a frame with velocity $\boldsymbol{v}^* \equiv \boldsymbol{v}(t^*)$, in which the particle will be instantaneously at rest. This frame is called the instantaneous rest frame (IRF) of the moving particle. Since the IRF has fixed velocity, it does not continue to move with the particle; the particle is at rest in the IRF for only an infinitesimal time interval. But the IRF is an inertial frame, to which we can apply Lorentz transformation rules.

We denote the coordinates in the lab frame by $(t, \boldsymbol{x})$, and those in the IRF by $(t', \boldsymbol{x}')$. In the lab frame, the particle is boosted relative by $\boldsymbol{v}^*$, so that, for example,

$$x^\mu = \Lambda^\mu{}_\nu(\boldsymbol{v}^*)x'^\nu$$

With this in mind, consider the interval $\Delta x^\mu$ between two neighboring space–time locations of the particle. It is clear that this interval is a four–vector, and that its spatial components are arbitrarily small in the IRF:

$$\Delta x'^\mu = (\Delta t', 0)$$

This interval—the interval between two events measured in the IRF—is a Lorentz scalar called the "proper time" and denoted by $\Delta \tau = \Delta t'$. According to (10.6)

$$\Delta x^\mu = \gamma(\boldsymbol{v}^*)\Delta\tau(1, \boldsymbol{v}^*) = \Delta\tau u^\mu$$

where we have introduced the "four–velocity"

$$u^\mu = \gamma(\boldsymbol{v})(1, \boldsymbol{v})$$

It is clear that the $u^\mu$ is a Lorentz four–vector.

Notice that the time intervals in the two frames are related by

$$\Delta t = \gamma \Delta \tau,$$

We can suppose that the $\Delta \tau$ measures the time between two clicks of a watch carried by the particle; the laboratory observer sees a longer time interval between the two clicks and concludes that the moving watch is slow.

## Phase space

The four-velocity of a particle is denoted by

$$u^\mu = (\gamma, \gamma\mathbf{v})$$

and the four-momentum by $p^\mu = mu^\mu$, where $m$ is the rest–mass. The speed of light is equated to unity, $c \to 1$, so that the 4-momentum of a particle of mass $m$ and 3-momentum $\mathbf{p}$ is given by

$$p^\mu = (p^0, \mathbf{p})$$

It must be kept in mind that these four components are not independent. The (squared) total energy $p^0$ is a sum of the kinetic energy and the rest energy, so that the four-vector $p^\mu$ satisfies the "mass-shell" restriction

$$p^\mu p_\mu = -m^2 \tag{10.7}$$

or

$$(p^0)^2 - \mathbf{p}^2 = m^2$$

When $p^0$ is evaluated on the mass–shell, we denote it by

$$E \equiv p^0(\mathbf{p}) = \sqrt{m^2 + \mathbf{p}^2} \tag{10.8}$$

The positive square–root is assumed.

Phase–space can be considered as the product of coordinate space and momentum space, restricted to the mass–shell. However, in some contexts it is convenient to relax the mass–shell restriction, re–imposing it explicitly through a $\delta$–function.

## Invariance

Suppose that $(x, p)$ are the coordinates, measured in a frame $S$, of some point in phase space ($x$ and $p$ are 4-vectors). The same point as measured in a different Lorentz frame, $S'$, has coodinates $(x' = \Lambda x, p' = \Lambda p)$ and a function $F(x, p)$ is a *Lorentz scalar* if its value as measured by an observer in $S'$ is given by

$$F'(x', p') = F(x, p). \tag{10.9}$$

That is, $F$ has the same value to both observers when measured at same phase–space point. Similarly a 4–component object $V$ is a Lorentz vector or 4–vector if its components in the two frames are related by

$$V'(x', p') = \Lambda V(x, p),$$

and so on. Tensors (of whatever rank) that transform according to these Lorentz rules, with the number of $\Lambda$-factors corresponding to the rank, are Lorentz tensors.

Notice that being a Lorentz tensor puts no constraint on the function form of the tensor components; it only specifies the transformed components. There is a distinct property, however, that does constrain the functional dependence. We say that a function $G(x, p)$ is *invariant* under Lorentz transformation if

$$G'(x', p') = G(x', p')$$

In other words the functional form is preserved. Notice that this definition is meaningful only when a rule for determining $G'$ from $G$ is given. For example, if $G$ is a Lorentz scalar, then the invariance property becomes $G(x', p') = G(x, p)$ or

$$G(\Lambda x, \Lambda p) = G(x, p),$$

which does indeed constrain the form of $G$. An example of an invariant (scalar) function is

$$p \cdot x = p^\mu x_\mu$$

In this sense the 4-dimensional volume element is invariant,

$$d^4 x = d^4 x'$$

because volume elements transform with the Jacobian, and the Jacobian of a (proper) Lorentz matrix is unity:

$$|\Lambda| = 1$$

For the same reason,

1. $d^4 p$, where $p$ denotes 4-momentum, is invariant.

2. The 4-dimensional Dirac delta functions,

$$\delta^4(x - \bar{x}) = \delta(x^1 - \bar{x}^1)\delta(x^2 - \bar{x}^2)\delta(x^3 - \bar{x}^3)\delta(x^4 - \bar{x}^4)$$

   and $\delta^4(p - \bar{p})$ are invariant. Here $\bar{x}^\mu$ and $\bar{p}^\mu$ are specified four vectors.

3. Since the right-hand side of (10.7) is manifestly scalar, the $\delta$−function

$$\delta((p^0)^2 - \mathbf{p}^2 - m^2) = \frac{1}{E(\mathbf{p})}\delta(p^0 - E(\mathbf{p})) \qquad (10.10)$$

   is scalar. (Here we have discarded the negative energy root.)

## 10.3 Relativistic electrodynamics

### Equations of motion

The relativistic electromagnetic Hamiltonian for a single particle is expressed in terms of the 4-vector potential

$$A^\mu = (\Phi, \mathbf{A})$$

where $\mathbf{A}$ is the vector potential and $\Phi$ the electrostatic potential. We introduce the canonical momentum

$$P^\mu = p^\mu + qA^\mu$$

which is obviously a 4-vector, in order to write the Hamiltonian as

$$H(\mathbf{x}, \mathbf{P}) = P^0 = \sqrt{m^2 + (\mathbf{P} - q\mathbf{A})^2} + q\Phi \qquad (10.11)$$

Thus the Hamiltonian is the temporal component of a four–vector. Notice that

$$\dot{\mathbf{x}} = \frac{\partial H}{\partial \mathbf{P}} = \frac{\mathbf{P} - q\mathbf{A}}{\sqrt{m^2 + (\mathbf{P} - q\mathbf{A})^2}}$$

so that the relation between velocity $\dot{\mathbf{x}}$ and momentum $\mathbf{p}$ is independent of the magnetic field:

$$\dot{\mathbf{x}}\sqrt{m^2 + \mathbf{p}^2} = \mathbf{p}$$

After solving this relation for $\mathbf{p}$, one finds that

$$\mathbf{p} = m\gamma\dot{\mathbf{x}} \tag{10.12}$$

where

$$\gamma \equiv (1 - \dot{\mathbf{x}}^2)^{-1/2}$$

Notice that $\gamma$ has the equivalent expression

$$\gamma = \sqrt{1 + \mathbf{p}^2/m^2};$$

which implies

$$p^0 = m\gamma. \tag{10.13}$$

The other set of Hamilton's equations

$$\dot{\mathbf{P}} = -\frac{\partial H}{\partial \mathbf{x}} \tag{10.14}$$

reproduce the familiar electromagnetic equation of motion. To show how the Lorentz force law is contained in (10.14), we first compute the gradient,

$$\nabla_i(\sqrt{m^2 + (\mathbf{P} - q\mathbf{A})^2} + q\Phi) = \frac{q}{p^0}(P_j - qA_j)\nabla_i A_j - q\nabla_i\Phi.$$

to write (10.14) as

$$\dot{p}^i + q\dot{A}_i = \frac{q}{p^0}\nabla_i(P_j A_j) - \frac{q^2}{p^0}A_j\nabla_i A_j - q\nabla_i\Phi$$

We next use a standard vector identity for $\nabla(\mathbf{P}\cdot\mathbf{A})$, and then eliminate $\mathbf{P} = \mathbf{p} + q\mathbf{A}$ in favor of $\mathbf{p}$. The resulting quadratic terms in $\mathbf{A}$ precisely cancel,

$$\mathbf{A}\times\nabla\times\mathbf{A} + \mathbf{A}\cdot\nabla\mathbf{A} - \nabla(\mathbf{A}^2/2) = 0,$$

leaving

$$\dot{\mathbf{p}} + q\dot{\mathbf{A}} = \frac{q}{p^0}(\mathbf{p}\cdot\nabla\mathbf{A} + \mathbf{p}\times\mathbf{E}) - q\nabla\Phi$$

or, in view of (10.13) and (10.12),

$$\dot{\mathbf{p}} = q(\dot{\mathbf{x}}\times\mathbf{E} - \nabla\Phi) - q(\dot{\mathbf{A}} - \dot{\mathbf{x}}\cdot\nabla\mathbf{A})$$

In the last term here, $\dot{\mathbf{A}}$ is the total time derivative

$$\dot{\mathbf{A}} = \frac{\partial \mathbf{A}}{\partial t} + \dot{\mathbf{x}} \cdot \nabla \mathbf{A}$$

Hence we have

$$\dot{\mathbf{p}} = q \left( \dot{\mathbf{x}} \times \mathbf{B} - \nabla \Phi - \frac{\partial \mathbf{A}}{\partial t} \right) \equiv \mathbf{F}_L \tag{10.15}$$

where $\mathbf{F}_L = q(\mathbf{E} + \mathbf{v} \times \mathbf{B})$ is the usual Lorentz force.

Next we write the equations of motion in covariant form. This requires the "4-force" $F^\mu$ satisfying

$$\frac{dp^\mu}{d\tau} = F^\mu \tag{10.16}$$

where $\tau$ is the proper time. Since

$$dt = \gamma d\tau$$

we see that the spatial components of the 4-force are

$$F^i = \gamma F_L^i \tag{10.17}$$

Thus the ordinary 3-vector force $\mathbf{F}_L$ is not part of a 4-vector. For the remaining component, we equate $dp^0$, the energy change, to the work done by $\mathbf{F}_L$:

$$dp^0 = \mathbf{F}_L \cdot \mathbf{v} dt.$$

But (10.13) implies that $\mathbf{v}dt = m^{-1}\mathbf{p}d\tau$, so

$$F^0 = \frac{dp^0}{d\tau} = \frac{\mathbf{F}_L \cdot \mathbf{p}}{m} = \frac{\mathbf{F} \cdot \mathbf{p}}{E}, \tag{10.18}$$

or $F^0 = \mathbf{F} \cdot \mathbf{v}$.

It is helpful to recall here that

$$\left( \frac{\partial}{\partial \mathbf{p}} \right) \cdot \mathbf{F}_L = 0$$

It follows from (10.17) that the spatial components of the 4-force satisfy a slightly different condition:

$$\left( \frac{\partial}{\partial \mathbf{p}} \right) \cdot \left( \frac{\mathbf{F}}{E} \right) = 0 \tag{10.19}$$

since $E = m\gamma$.

## Faraday tensor

It is useful to express the electromagnetic force in terms of the field strength tensor, or Faraday tensor,

$$F^{\mu\nu} = \nabla^\mu A^\nu - \nabla^\nu A^\mu$$

Here

$$\nabla^\mu = \eta^{\mu\nu} \frac{\partial}{\partial x^\nu} = \left(-\frac{\partial}{\partial t}, \nabla\right)$$

Explicitly,

$$F^{\mu\nu} = \begin{bmatrix} 0 & E_x & E_y & E_z \\ -E_x & 0 & B_z & -B_y \\ -E_y & -B_z & 0 & B_x \\ -E_z & B_y & -B_x & 0 \end{bmatrix} \tag{10.20}$$

Since lowering the first index reverses the sign of the $0^{th}$ row, and lowering the second reverses the sign of the $0^{th}$ column, we have

$$F_{\mu\nu} = \begin{bmatrix} 0 & -E_x & -E_y & -E_z \\ E_x & 0 & B_z & -B_y \\ E_y & -B_z & 0 & B_x \\ E_z & B_y & -B_x & 0 \end{bmatrix} \tag{10.21}$$

The action of the Faraday tensor on a four-vector $K$ can be seen from direct multiplication:

$$F^{\mu\kappa} K_\kappa = (\mathbf{E} \cdot \mathbf{K}, -EK_0 + \mathbf{K} \times \mathbf{B}) \tag{10.22}$$

In particular the product $F^{\mu\nu} p_\nu$ is easily computed from (10.22):

$$\begin{aligned} F^{\mu\nu} p_\nu &= (\mathbf{E} \cdot \mathbf{p}, -Ep_0 + \mathbf{p} \times \mathbf{B}) \\ &= m\gamma(\mathbf{E} \cdot \mathbf{v}, \mathbf{E} + \mathbf{v} \times \mathbf{B}) \end{aligned}$$

According to (10.17), the vector components here coincide with those of the 4-force, while the temporal component agrees with (10.18). Thus we have

$$F^{\mu\nu} p_\nu = \frac{m}{q} F^\mu \tag{10.23}$$

and the covariant equation of motion (10.16) becomes

$$\frac{dp^\mu}{dt} = \frac{q}{m} F^{\mu\nu} p_\nu \tag{10.24}$$

One other second–rank tensor is linear in the field components: the tensor dual to $F_{\mu\nu}$, which we denote by

$$\mathcal{F}^{\mu\nu} \equiv \frac{1}{2} \epsilon^{\mu\nu\kappa\lambda} F_{\kappa\lambda} \tag{10.25}$$

Note that $\mathcal{F}$ is dual to $F$ in two senses: because it satisfies (10.25) and also because it satisfies

$$\mathcal{F}^{\mu\nu}(\mathbf{E}, \mathbf{B}) = F^{\mu\nu}(\mathbf{E} \to \mathbf{B}, \mathbf{B} \to -\mathbf{E}) \tag{10.26}$$

It is straightforward to show that

$$\mathcal{F}^{\mu\kappa}F_{\kappa\nu} = \mathcal{F}_{\mu\kappa}F^{\kappa\nu} = \eta^{\mu}_{\nu}\mathbf{E} \cdot \mathbf{B} \tag{10.27}$$

We infer that $\mathbf{E} \cdot \mathbf{B}$ is a Lorentz scalar. A second Lorentz scalar can be computed from the field strength tensor:

$$F_{\kappa\lambda}F^{\kappa\lambda} = 2(E^2 - B^2) \tag{10.28}$$

We label the two scalars using the notations

$$W_1 \equiv B^2 - E^2 \tag{10.29}$$
$$W_2 \equiv \mathbf{E} \cdot \mathbf{B} \tag{10.30}$$

## Gyration

Suppose the electric force is negligible (in some Lorentz frame). Then the equation of motion (10.16) has vector components given by

$$m\frac{d\gamma\mathbf{v}}{d\tau} = q\gamma\mathbf{v} \times \mathbf{B} \tag{10.31}$$

or, equivalently,

$$m\frac{d\mathbf{u}}{d\tau} = q\mathbf{u} \times \mathbf{B}$$

Hence the four-velocity $\mathbf{u}(\tau)$ gyrates identically to $\mathbf{v}(t)$ in non-relativistic theory.

It is generally more convenient to use the "lab–frame" measure of time, $t$, instead of $\tau$. The transformation $\tau \to t$ is easy in this case because $v = |\mathbf{v}|$ is constant, implying $\gamma = $ constant. Hence the $\gamma$–factors on the two sides of (10.31) cancel, and there remains only a factor from $dt = \gamma d\tau$:

$$m\gamma\frac{d\mathbf{v}}{dt} = q\mathbf{v} \times \mathbf{B}$$

This can be solved in the same way as its non-relativistic version (because $\gamma$ is constant), leading to the conclusion that the particle gyrates with frequency

$$\omega_B = \frac{qB}{\gamma m}$$

Note here that the field $B$ is, like $t$, that measured in the lab frame. In other words, relativistic gyration is the same as the non–relativistic case, except that the gyrofrequency is modified by the relativistic mass increase.

## Energy-momentum tensor

Symmetry considerations show that the energy-momentum tensor of an ideal fluid, when measured in the rest-frame of the fluid, will have the form [49]

$$
T_R^{\mu\nu} =
\begin{bmatrix}
u & 0 & 0 & 0 \\
0 & p & 0 & 0 \\
0 & 0 & p & 0 \\
0 & 0 & 0 & p
\end{bmatrix}
\tag{10.32}
$$

where $u$ is the energy density, $p$ is the pressure and the $R$–subscript refers to the state of rest. Notice that this definition makes $p$ and $u$ Lorentz scalars. The fact that the symbol 'p' has more than one meaning will rarely cause confusion.

Let us denote the fluid velocity measured in the lab frame, by $\mathbf{V}$. When the energy-momentum tensor of the moving fluid is measured in the lab frame, its components will be boosted by $\mathbf{V}$. Since $\mathbf{V}$ is arbitrary, we denote the tensor observed in the lab frame simply by $T$:

$$
T = \Lambda(\mathbf{V}) \cdot \Lambda(\mathbf{V}) \cdot T_R
$$

where $\Lambda(\mathbf{V})$ is the Lorentz boost of (10.1). Straightforward calculation yields

$$
T^{00} = \Lambda^0{}_\alpha \Lambda^0{}_\beta T_R^{\alpha\beta} = \gamma^2(u + V^2 p);
$$

$$
T^{0i} = \Lambda^0{}_\alpha \Lambda^i{}_\beta T_R^{\alpha\beta} = \gamma^2 V_i(u + p);
$$

and

$$
T^{ij} = \Lambda^i{}_\alpha \Lambda^j{}_\beta T_R^{\alpha\beta} = \delta_{ij}p + \gamma^2(u + p)V_i V_j.
$$

Recall that the combination $U + p\mathcal{V}$, where $U$ is the internal energy and $\mathcal{V}$ is the volume, is a thermodynamic potential called enthalpy. Therefore the scalar $u+p$ is enthalpy per unit volume, or enthalpy density. It is conventionally denoted by

$$
u + p \equiv h
$$

A simpler expression for $T$ is

$$
T^{\alpha\beta} = \eta^{\alpha\beta}p + hU^\alpha U^\beta
\tag{10.33}
$$

where

$$
U^\alpha = \gamma(\mathbf{V})(1, \mathbf{V})
$$

is the 4–vector for fluid flow. It is not hard to verify that the two expressions for the energy-momentum tensor agree.

## 10.4 Kinetic theory

### Scalar distribution function

The key fact of relativistic kinetic theory is that the distribution function $f(\mathbf{x}, \mathbf{p}, t)$ is a Lorentz scalar. Here we demonstrate the Lorentz invariance of $f$.

The distribution function $f(\mathbf{x}, \mathbf{p}, t)$ is defined such that

$$n(\mathbf{x}, t) = \int d^3p f \tag{10.34}$$

is the density of particles at $\mathbf{x}$ at time $t$. We suppose that the particle trajectories, $\mathbf{x}_i(t)$ and $\mathbf{p}_i(t)$, are known. Then the distribution function can be expressed as the ensemble average of the microscopic distribution

$$\mathcal{F} \equiv \sum_i \delta^3(\mathbf{x} - \mathbf{x}_i(t))\delta^3(\mathbf{p} - \mathbf{p}_i(t))$$

We next introduce a separate time variable $t_i$ for each $i$ in order to write

$$\mathcal{F} = \sum_i \int dt_i \delta(t - t_i)\delta^3(\mathbf{x} - \mathbf{x}_i(t_i))\delta^3(\mathbf{p} - \mathbf{p}_i(t_i))$$

or, choosing $x_i^0 = t_i$,

$$\mathcal{F} = \sum_i \int dt_i \delta^4(\mathbf{x} - \mathbf{x}_i(t_i))\delta^3(\mathbf{p} - \mathbf{p}_i(t_i))$$

Now the proper time interval for the $i$th particle is

$$d\tau_i = \frac{dt_i}{\gamma_i}$$

where $\gamma_i$ is the relativistic factor for the $i$th particle. We recall (10.13) in order to write

$$dt_i = \frac{p_i^0}{m}d\tau_i \tag{10.35}$$

whence

$$\mathcal{F} = \sum_i \int d\tau_i \frac{p_i^0}{m}\delta^4(x - x_i(\tau_i))\delta^3(\mathbf{p} - \mathbf{p}_i(\tau_i))$$

We multiply this function by the Lorentz scalar appearing in (10.10):

$$F(\mathbf{x}, \mathbf{p}, t) \equiv \frac{1}{p^0}\delta(p^0 - \sqrt{\mathbf{p}^2 + m^2})\mathcal{F}(\mathbf{x}, \mathbf{p}, t)$$

and note that the mass-shell factor can be put inside the sum and then evaluated at each $\mathbf{p}_i$:

$$F = \sum_i \int d\tau_i \frac{p_i^0}{mp^0}\delta^4(x - x_i)\delta^3(\mathbf{p} - \mathbf{p}_i)\delta(p^0 - p_i^0)$$

or

$$F = \frac{1}{m} \sum_i \int d\tau_i \delta^4(x - x_i) \delta^4(p - p_i)$$

This quantity is manifestly a scalar. Since it differs from $\mathcal{F}$ by a scalar factor, and since the physical distribution $f(\mathbf{x}, \mathbf{p}, t)$ is the ensemble average of $\mathcal{F}$, we conclude that $f(\mathbf{x}, \mathbf{p}, t)$ is a Lorentz scalar.

It is worthwhile to recall the significance of this fact. Suppose that the distribution function of some fluid is known in the fluid rest frame, $R$: $f'(x', p') = f_R(x', p')$. Then, if the local fluid velocity (as measured in the "lab frame" $S$) is $\mathbf{V}$, we have $x' = \Lambda(\mathbf{V})x$, $p' = \Lambda(\mathbf{V})p$ and the distribution observed in the lab frame is, according to (10.9),

$$f(x, p) = f_R(\Lambda x, \Lambda p). \tag{10.36}$$

In the presence of an electromagnetic field, $f_R$ can depend on position both directly (through its density, for example), and also through the (rest-frame) field variables $A'^\mu(x')$. It is natural to express the lab-frame distribution in terms of the lab-frame fields $A^\mu(x)$, and therefore to write the transformation law in the form

$$f(x, p, A(x)) = f_R(\Lambda x, \Lambda p, \Lambda A(\Lambda x)) \tag{10.37}$$

The fact that the distribution function in $(x, p)$–space is a scalar fixes the transformation properties of key moments of $f$. More importantly, it makes transparent the form and derivation of the relativistic kinetic equation. We examine both issues after the following brief digression.

## Momentum-space volume

It follows from (10.10) and the scalar nature of $d^4p$ that the quantity $d^3p/E$ is a Lorentz scalar. This fact is sufficiently important to deserve an alternative derivation. Therefore we next compute the metric tensor $g^s_{ij}$ measuring distance on the mass shell. The volume element on the mass shell is given by the Jacobian $\sqrt{|g^s|}$, and it will turn out that

$$\sqrt{|g^s|} = m/E. \tag{10.38}$$

We denote the infinitesimal infinitesimal "length" in momentum space, constrained by the mass–shell condition (10.7), by $(dp^\mu dp_\mu)^s$. After making the substitution

$$dp^0 = \frac{\mathbf{p} \cdot d\mathbf{p}}{E}$$

we find, after some manipulation,

$$(dp^\mu dp_\mu)^s = g^s_{ij} dp^i dp^j$$

where

$$g^s_{ij} = \begin{vmatrix} 1 - \hat{p}_x^2 & -\hat{p}_x\hat{p}_y & -\hat{p}_x\hat{p}_z \\ -\hat{p}_x\hat{p}_y & 1 - \hat{p}_y^2 & -\hat{p}_y\hat{p}_z \\ -\hat{p}_x\hat{p}_z & -\hat{p}_y\hat{p}_z & 1 - \hat{p}_z^2 \end{vmatrix} \tag{10.39}$$

and $\hat{p}_i = p_i/E$. Now a straightforward calculation gives

$$|g^s| = \det(g) = 1 - \frac{p^2}{E^2} = \frac{m^2}{E^2}$$

as was to be demonstrated.

## Kinetic equation

Once the scalar nature of the distribution is established, it is straightforward to write an invariant kinetic equation with the correct nonrelativistic limit:

$$\frac{p^\mu}{m} \frac{\partial f}{\partial x^\mu} + F^\mu \frac{\partial f}{\partial p^\mu} = C \tag{10.40}$$

where $C$ is a collision operator, and $F^\mu$ is the 4-vector force constructed in Sec. 10.2. Notice that the covariance of (10.40) depends upon the scalar nature of $f$. The form of the relativistic collision operator is not considered here [24].

The form of (10.40) becomes obvious when one notices that it can be written as

$$\frac{dx^\mu}{d\tau} \frac{\partial f}{\partial x^\mu} + \frac{dp^\mu}{d\tau} \frac{\partial f}{\partial p^\mu} = C$$

To see how this equation reduces to the familiar version, we write it more explicitly as

$$\frac{p^0}{m} \frac{\partial f}{\partial t} + \gamma \dot{x}^i \frac{\partial f}{\partial x^i} + \frac{\mathbf{p} \cdot \mathbf{F}}{p^0} \frac{\partial f}{\partial p^0} + \mathbf{F} \cdot \frac{\partial f}{\partial \mathbf{p}} = C$$

and then observe that the mass-shell restriction

$$p^0 = \sqrt{m^2 + p^2}$$

allows us to write

$$\frac{\partial f(p^0(\mathbf{p}), \mathbf{p})}{\partial \mathbf{p}} = \frac{dp^0}{d\mathbf{p}} \frac{\partial f}{\partial p^0} + \frac{\partial f}{\partial \mathbf{p}}$$

by the chain rule. Hence, since

$$\frac{dp^0}{d\mathbf{p}} = \frac{\mathbf{p}}{p^0}$$

the $p^0$-derivative in the kinetic equation can be considered part of the $\mathbf{p}$-derivative and suppressed, giving an obvious relativistic version of the usual kinetic equation,

$$\frac{p^\mu}{m} \frac{\partial f}{\partial x^\mu} + F^i \frac{\partial f}{\partial p^i} = C \tag{10.41}$$

Notice that in (10.41) the derivative is performed on the mass–shell, while in (10.40) it is performed as if $p^0$ were an independent variable. The two agree essentially because of the form of the temporal component of the Lorentz force.

## 10.5   Moments of kinetic equation

### Tensor moments

It is clear that the density $n(\mathbf{x}, t)$ is not a scalar, since $d^3p$ is not. To construct moments of $f$ that are Lorentz tensors, we use the scalar momentum-space volume element $d^3p/E$, where, as before,

$$E(\mathbf{p}) \equiv \sqrt{\mathbf{p}^2 + m^2} = p^0(\mathbf{p}) = m\gamma$$

Thus the moment

$$M^{\alpha\beta\dots\nu} \equiv \int \frac{d^3p}{E} p^\alpha p^\beta \dots p^\nu f(\mathbf{x}, \mathbf{p}, t) \tag{10.42}$$

is a Lorentz tensor—the general tensor moment of the distribution $f$.

Next consider the quantity (not a 4-vector)

$$\frac{p^\alpha}{E} = \left(1, \frac{\mathbf{p}}{m\gamma}\right).$$

Recalling (10.35) we see that

$$\frac{p^\alpha}{E} = \frac{dx^\alpha}{dt} \tag{10.43}$$

Hence our tensor can be written as

$$M^{\alpha\beta\dots\nu} \equiv \int d^3p \frac{dx^\alpha}{dt} p^\beta \dots p^\nu f(\mathbf{x}, \mathbf{p}, t)$$

showing that it measures the flow of $p^\beta \dots p^\nu$.

When necessary we indicate the rank of our moment tensor with a subscript: $M_r$ is the tensor with $r$-factors of 4-momentum in the integrand. Examples are

1. $M_0 = \rho_m$ is the scalar mass density,

$$\rho_m = m^2 \int \frac{d^3p}{E} f$$

2. $M_1^\alpha = \Gamma^\alpha$ is the flow 4-vector

$$\Gamma^\alpha = \int \frac{d^3p}{E} p^\alpha f$$

whose time-component is the density,

$$\Gamma^0 = \int d^3p f = n, \tag{10.44}$$

and whose spatial components give the fluid mean-flow vector

$$\Gamma^k = nV^k = \int d^3p v^k f \tag{10.45}$$

3. $M_2^{\alpha\beta} = T^{\alpha\beta}$ is the energy-momentum tensor,

$$T^{\alpha\beta} = \int \frac{d^3p}{E} p^\alpha p^\beta f,$$

which we have already considered.

The flow-vector provides a simple definition of the fluid rest frame, as the frame in which $\Gamma$ has only a temporal component:

$$\Gamma_R^\mu = (n_R, 0) \tag{10.46}$$

Here the $R$-subscript stands for "rest-frame." Notice that unlike $n$, $n_R$ is a Lorentz scalar. From the general rule (10.2) we see that

$$n = \gamma(\mathbf{V}) n_R \tag{10.47}$$

It is often convenient to refer to the four-vector fluid velocity,

$$U^\mu \equiv \Gamma^\mu / n_R = (\gamma(\mathbf{V}), \gamma(\mathbf{V})\mathbf{V}) \tag{10.48}$$

Notice that contraction, together with the mass–shell condition, provides identities relating moments of different rank:

$$M^{\alpha\beta\cdots\gamma\lambda}{}_\lambda = -m^2 M^{\alpha\beta\cdots\gamma} \tag{10.49}$$

For example,

$$T^\alpha{}_\alpha = -\rho_m$$

## General moment

We return to the kinetic equation (10.40)

$$\frac{p^\mu}{m} \frac{\partial f}{\partial x^\mu} + F^\mu \frac{\partial f}{\partial p^\mu} = C \tag{10.50}$$

and recall the tensor moment

$$M^{\alpha\cdots\gamma} \equiv \int \frac{d^3p}{E} f p^\alpha \cdots p^\gamma$$

Here we operate on (10.50) with

$$\int \frac{d^3p}{E} p^{\alpha\cdots}p^\gamma$$

to obtain a sequence of equations for the moments $M$.

The moment of the first, convective term in (10.50) is simply

$$\int \frac{d^3p}{E} p^\alpha \cdots p^\gamma \frac{p^\mu}{m} \frac{\partial f}{\partial x^\mu} = \frac{1}{m} \frac{\partial M^{\mu\alpha\cdots\gamma}}{\partial x^\mu}$$

In the second term

$$\mathcal{F}^{\alpha\cdots\gamma} \equiv \int \frac{d^3p}{E} p^\alpha \cdots p^\gamma F^\mu \frac{\partial f}{\partial p^\mu}$$

we note that the denominator $E(\mathbf{p})$ is a function of $\mathbf{p}$ and independent of $p^0$. Since $F^0 \propto \mathbf{F} \cdot \mathbf{p}$ is also independent of $p^0$, we conclude from (10.19) that

$$\frac{\partial}{\partial p^\mu} \frac{F^\mu}{E} = 0$$

Therefore an integration by parts yields

$$\mathcal{F}^{\alpha\cdots\gamma} = -\int \frac{d^3p}{E} f F^\mu \frac{\partial}{\partial p^\mu}(p^\alpha \cdots p^\gamma)$$

or, in view of (10.23),

$$\mathcal{F}^{\alpha\cdots\gamma} = -\frac{q}{m} F^{\mu\nu} \int \frac{d^3p}{E} f p_\nu \frac{\partial}{\partial p^\mu}(p^\alpha \cdots p^\gamma)$$

After performing the derivative we find that

$$\mathcal{F}^{\alpha\cdots\gamma} = -\frac{q}{m} F^{\{\alpha\nu} M_\nu{}^{\beta\cdots\gamma\}} \tag{10.51}$$

Here

$$M_\nu{}^{\beta\cdots\gamma} \equiv \int \frac{d^3p}{E} f p_\nu p^\beta \cdots p^\gamma$$

and the curly brackets instruct us to symmetrize by exchanging the superscript $\alpha$ with each of the superscript indices on $M$:

$$F^{\{\alpha\nu} M_\nu{}^{\beta\cdots\gamma\}} = F^{\alpha\nu} M_\nu{}^{\beta\cdots\gamma} + F^{\beta\nu} M_\nu{}^{\alpha\cdots\gamma} + \cdots$$

It is clear that $M$ transforms like a mixed tensor.

We denote the corresponding moment of the collision operator by

$$\mathcal{C}^{\alpha\cdots\gamma} = m \int \frac{d^3p}{E} p^\alpha \cdots p^\gamma C$$

Then the general moment of the kinetic equation can be expressed as

$$\frac{\partial M^{\mu\alpha\cdots\gamma}}{\partial x^\mu} - qF^{\{\alpha\nu} M_\nu{}^{\beta\cdots\gamma\}} = \mathcal{C}^{\alpha\cdots\gamma} \tag{10.52}$$

## Examples

1. The zeroth moment describes particle conservation:

$$\frac{\partial \Gamma^\mu}{\partial x^\mu} = 0 \tag{10.53}$$

Here we have noted that the collision operator vanishes for any operator that conserves particles.

2. The first moment describes conversion of particle momentum and energy into electromagnetic field energy-momentum and collisional dissipation (friction),

$$\frac{\partial T^{\mu\alpha}}{\partial x^\mu} - qF^{\alpha\nu}\Gamma_\nu = \mathcal{C}^\alpha \tag{10.54}$$

## 10.6 Maxwellian distribution

### Relativistic Maxwellian

Maxwell showed that the equilibrium distribution depend exponentially on the quantities that are conserved by the collisional process: the so-called additive invariants. If the distribution is observed in its rest–frame, than the only relevant collisional invariant is the energy. Hence the Lorentz scalar distribution that corresponds to thermal equilibrium is expressed in terms of the rest–frame coordinates by

$$f_{MR}(x,p) \equiv N_M e^{-H(\mathbf{x},\mathbf{p})/T} \tag{10.55}$$

where

$$H(\mathbf{x},\mathbf{p}) = \sqrt{\mathbf{p}^2 + m^2} + q\Phi(\mathbf{x})$$

is the Hamiltonian and $N_M$ and $T$ are Lorentz–scalar constants. It is evident that the normalization $N_M$ measures density, while $T$ is related to temperature.

It is significant that the Hamiltonian is the sum of the time-components of two 4-vectors, $p^\mu = (\sqrt{\mathbf{p}^2 + m^2}, \mathbf{p})$ and $A^\mu = (\Phi, \mathbf{A})$. Thus, in terms of the canonical momentum $P^\mu = p^\mu + qA^\mu$ we can write

$$f_{MR}(x,p) = N_M e^{-P^0(x,p)/T} \tag{10.56}$$

Because $H$ is a constant of the motion, $f_M$ obviously satisfies the kinetic equation.

In terms of the coordinates of an arbitrary frame, the Maxwellian distribution has the form [recall (10.37)]

$$f_M(x,p,A(x)) = f_{MR}(\Lambda x, \Lambda p, \Lambda A(\Lambda x)) \tag{10.57}$$

We can also express $f_M$ in manifestly invariant form:

$$f_M(x,p) = N_M e^{U_\mu P^\mu(x,p)/T} \tag{10.58}$$

where $V_\mu$ is the (covariant) fluid four-velocity defined in (10.48). This version has the advantage of displaying immediately the form of $f_M$ when it is observed from a moving system:

$$f_M(x,p) = N_M e^{-\gamma q(\Phi - \mathbf{V}\cdot\mathbf{A})/T} e^{-\gamma(E - \mathbf{V}\cdot\mathbf{p})/T} \tag{10.59}$$

Notice that the exponent here is consistent with the Lorentz rule

$$P'^0 = \Lambda^0{}_\mu P^\mu = \gamma(P^0 - \mathbf{P}\cdot\mathbf{V}).$$

We remark that this rule coincides with the non-relativistic version

$$\frac{1}{2}mv^2 \to \frac{1}{2}m(\mathbf{v} - \mathbf{V})^2$$

through terms quadratic in $V$, but not beyond the quadratic terms.

## Maxwellian moments in the rest frame

Using (10.42) we can compute the Maxwellian tensors

$$
\begin{aligned}
M_{MR}^{(r)\alpha\beta\dots\nu}(\mathbf{x}) &\equiv \int \frac{d^3p}{E} p^\alpha p^\beta \dots p^\nu f_{MR}(\mathbf{x}, \mathbf{p}) \\
&= N_M m^{r+2} e^{-\Phi/T} \int \frac{d^3s}{\sqrt{1+s^2}} s^\alpha \dots s^\nu e^{-z\sqrt{1+s^2}}
\end{aligned}
$$

where $s^\alpha = p^\alpha/m$,

$$z \equiv m/T \ (= mc^2/T)$$

and $r$ is the rank of the tensor. The isotropy of $f_{MR}$ implies that odd-rank tensors vanish and that the second-rank tensor is diagonal. Of primary interest are the three lowest-rank moments.

1. The scalar mass density $\rho_m(x) = m^2 M_{MR}^{(0)}$ is given by

$$\rho_m = 4\pi N_M m^4 e^{-\Phi/T} \int \frac{ds\, s^2}{\sqrt{1+s^2}} e^{-z\sqrt{1+s^2}}$$

2. The flow vector $\Gamma^\mu = M_{MR}^{(1)\mu}$ has only a temporal component in the rest frame,

$$\Gamma_R^0 = n_R = N_M m^3 e^{-\Phi/T} \int d^3s\, e^{-z\sqrt{1+s^2}} \tag{10.60}$$

Finally the stress tensor, or energy-momentum tensor ($r = 2$), which has only diagonal components in the rest frame; recall (10.32). Thus we need the mean energy $u = T^{00}$ and the pressure $p = T^{ii}/3$; in the Maxwellian case we have

$$u = 4\pi N_M m^4 e^{-\Phi/T} \int \frac{ds}{\sqrt{1+s^2}} (s^2 + s^4) e^{-z\sqrt{1+s^2}}$$

and

$$p = \frac{4\pi}{3} N_M e^{-\Phi/T} m^4 \int_0^\infty \frac{ds}{\sqrt{1+s^2}} s^4 e^{-z\sqrt{1+s^2}}$$

Notice that, since the Maxwellian contains only two independent parameters, the quantities $p, u, \rho_m$, and $n_R$ are not independent.

All the integrals are performed using the formula

$$\int_0^\infty \frac{ds}{\sqrt{1+s^2}} s^{2n} e^{-z\sqrt{1+s^2}} = \frac{1 \cdot 3 \cdots (2n-1) K_n(z)}{z^n} \tag{10.61}$$

where $K_n$ is the MacDonald function. These functions satisfy

$$\frac{d}{dz}(z^n K_n(z)) = -z^n K_{n-1}(z) \tag{10.62}$$

$$\frac{d}{dz}(z^{-n} K_n(z)) = -z^{-n} K_{n+1}(z) \tag{10.63}$$

and have limiting forms given by

$$K_\nu(z) \sim \frac{\Gamma(\nu)}{2}\left(\frac{2}{z}\right)^\nu \tag{10.64}$$

for $z \to 0$, and

$$K_\nu(z) \sim \sqrt{\frac{\pi}{2z}} e^{-z}\left[1 + \frac{4\nu^2 - 1}{8z} + O(z^{-2})\right] \tag{10.65}$$

for $z \to \infty$.

Our formulae have become

$$\rho_m = 4\pi N_M m^3 e^{-\Phi/T}\frac{K_1(z)}{z} \tag{10.66}$$

$$n_R = 4\pi N_M m^2 e^{-\Phi/T}\frac{K_2(z)}{z} \tag{10.67}$$

$$u = 4\pi N_M m^4 e^{-\Phi/T}\left(\frac{K_1}{z} + \frac{3K_2}{z^2}\right) \tag{10.68}$$

$$p = 4\pi N_M e^{-\Phi/T} m^4 \frac{K_2}{z^2} \tag{10.69}$$

There is one subtlety here: the integral for $n_R$ involves

$$I_0 \equiv \int ds s^2 e^{-z\sqrt{1+s^2}}$$

which does not fit the mold of (10.61). However it is not hard to show that

$$I_0 = -\frac{d}{dz}\left(\frac{K_1}{z}\right)$$

so (10.63) implies $I_0 = K_2/z$ and (10.67) follows.

It is convenient to express the normalization in terms of $n_R$:

$$N_M = \frac{n_R e^{\Phi/T}}{4\pi m^2 T K_2(z)} \tag{10.70}$$

Then the Maxwellian can be expressed in terms of its rest–frame density as

$$f_M = \frac{n_R e^{-E/T}}{4\pi m^2 T K_2(z)} \tag{10.71}$$

with $E = p^0 = \sqrt{p^2 + m^2}$.

After substituting (10.70) into our expression for the moments we find

$$\rho_m = mn_R \frac{K_1(z)}{K_2(z)} \tag{10.72}$$

$$u = mn_R \left( \frac{K_1(z)}{K_2(z)} + \frac{3}{z} \right) \tag{10.73}$$

$$p = n_R T \tag{10.74}$$

All quantitites appearing in these relations are scalars. Notice in particular that

$$u = \rho_m + 3p \tag{10.75}$$

$$= 3p + mn_R \frac{K_1(z)}{K_2(z)} \tag{10.76}$$

In fact (10.75) merely states the identity (10.49), since

$$T^\alpha{}_\alpha = -u + 3p = -\rho_m$$

## 10.7　Magnetized plasma

In Section 3.5 of Chapter 3 we found that a strong magnetic field allowed partial closure of the non-relativistic fluid equations: the dynamics perpendicular to the magnetic field could be expresssed in terms of the electric field and a few scalar moments. The same manipulation of moment equations is effective in the relativistic case, where it leads to closely similar predictions [4, 33]. The relativistic treatment differs in essentially two ways: first, some care is needed to define a *magnetized* plasma in the relativistic case; and second, the unit vector **b** needs to be replaced by a covariant quantity, in order to give invariant meaning to the terms "parallel" and "perpendicular". We address these two differences in the following two subsections.

### Definition

The electric and magnetic field components transform as parts of the second–rank Faraday tensor, (10.21). If **E** and **B** are the fields measured in some laboratory frame, then those measured in the IRF are given by

$$\mathbf{E}_R = \gamma(\mathbf{E} + \mathbf{v} \times \mathbf{B}) - \frac{\gamma^2}{\gamma + 1} \mathbf{v}(\mathbf{v} \cdot \mathbf{E}) \tag{10.77}$$

and

$$\mathbf{B}_R = \gamma(\mathbf{B} - \mathbf{v} \times \mathbf{E}) - \frac{\gamma^2}{\gamma + 1} \mathbf{v}(\mathbf{v} \cdot \mathbf{B}) \tag{10.78}$$

In the case of a plasma fluid, we can identify the frame velocity **v** with the local fluid velocity **V**. It is then natural to ask whether a rapidly moving plasma fluid that is magnetized in the lab frame will be magnetized in other frames, such as its rest frame.

A magnetized plasma must satisfy two conditions: the gradient scale lengths of the fluid variables must exceed the gyroradius, and, as noted in (3.69), the parallel electric field must be relatively small. An invariant version of the first requirement is easily constructed: one simply defines $\delta$ as the ratio of gyroradius to gradient scale length, when both lengths are measured in the fluid rest frame.

To bound the parallel electric field in an invariant way, we use the two Lorentz scalars defined by (10.29) and (10.30). Thus we obtain the relativistic generalization of (3.69):

$$W_2/W_1 \sim \delta \tag{10.79}$$

## Projection operators

It is clear from (10.78) that the unit vector **b**, as well as such definitions as (1.13), behave awkwardly under Lorentz transformation. Yet the notion of alignment along the magnetic field does have a Lorentz covariant expression. Consider the second rank tensor

$$b_\mu{}^\nu \equiv \eta_\mu{}^\nu + F_{\mu\kappa}F^{\kappa\nu}/W_1 \tag{10.80}$$

where $W_1 = B^2 - E^2$ is the Lorentz invariant noted in (10.29). Alternatively $b_\mu{}^\nu$ can be expressed in terms of the dual Faraday tensor as

$$b_\mu{}^\nu = \mathcal{F}_{\mu\kappa}\mathcal{F}^{\kappa\nu}/W_1 \tag{10.81}$$

In a magnetized plasma this tensor selects, approximately, the parallel components of arbitrary four–vector $K_\mu = (K_0, \mathbf{K})$. Explicitly,

$$
\begin{aligned}
b_\mu{}^\kappa K_\kappa \;=\; & W_1^{-1}[B^2 K_0 + \mathbf{E} \times \mathbf{B} \cdot \mathbf{K}, \\
& - \mathbf{E} \times \mathbf{B} K_0 + \mathbf{B}\mathbf{B} \cdot \mathbf{K} - \mathbf{E} \times (\mathbf{K} \times \mathbf{E})]
\end{aligned}
\tag{10.82}
$$

To see the relation between this expression and the ordinary three–vector $\mathbf{b}K_\parallel$, one notes from (10.77) that, for any electromagnetic field, there is a Lorentz frame in which

$$E_\perp \ll B \tag{10.83}$$

In a magnetized plasma, the rest–frame satisfies this ordering, and furthermore has $E_\parallel \ll B$, in view of (10.79). But (10.82) shows that, in such a frame, the vector components of $b_\mu{}^\kappa K_\kappa$ approximately coincide with $K_\parallel$.

We conclude that the tensor $b_\mu{}^\kappa$ acts, in the rest frame of a magnetized plasma, as an approximate projector in the direction of the magnetic field [33]. An exact projection operator, which does not require small $E_\parallel$, can also be written down [25], but is much more complicated.

## MHD flow

The relativistic treatment of a magnetized plasma provides special insight into the $E \times B$ drift. Consider the energy-momentum conservation law, (10.54), in the limit in which the electromagnetic interaction dominates. It is clear that this regime, which includes the magnetized plasma case, is described by the limit $q \to \infty$; thus we have, in lowest order,

$$qF^{\mu\nu}\Gamma_\nu = 0 \tag{10.84}$$

It can be seen from (10.22) that this relation has no non-trivial solution for a general magnetic field. However, in the magnetized case, we can assume that the parallel electric field also vanishes in lowest order. Then (10.84) implies that that the plasma flow is given by the $E \times B$ drift, plus an arbitrary flow along the magnetic field.

To understand this conclusion, we first write $\Gamma_\nu = n_R U_\nu$ and consider the fluid four–velocity $U_\nu$. For vanishing $E_\parallel$, we see from (10.22) that

$$U^0\mathbf{E} \;=\; \mathbf{B} \times \mathbf{U}, \tag{10.85}$$
$$\mathbf{U} \cdot \mathbf{E} \;=\; 0 \tag{10.86}$$

whence

$$\mathbf{U} = \mathbf{B}U_\parallel + \alpha\mathbf{E} \times \mathbf{B} \tag{10.87}$$

where $U_\parallel$ and $\alpha$ are free parameters. Then (10.85) implies

$$U^0 = \alpha B^2 \tag{10.88}$$

To proceed further we need more information than is provided by (10.84): we need to specify that $U^\nu$ designates the components of a four–velocity. This fact implies that

$$U^0 = \gamma(\mathbf{V}) \tag{10.89}$$

where $\mathbf{V} = \mathbf{U}/\gamma$ is the associated three–velocity. Hence we have

$$\alpha = \gamma/B^2$$

and (10.87) becomes

$$\mathbf{U} = \gamma(\mathbf{b}V_\parallel + \mathbf{V}_E) \tag{10.90}$$

where

$$\mathbf{V}_E \equiv \frac{\mathbf{E} \times \mathbf{B}}{B^2}$$

is the familiar $E \times B$ velocity.

## Additional reading

A concise summary of relativistic electrodynamics can be found in Weinberg [76]. The book on relativistic kinetic theory by de Groot et al.[21] is both thorough and readable. It has strongly influenced the approach of this chapter. Plasma and astrophysical applications of relativistic fluid theory are studied by Anile [4]. The phenomenology and theory of astrophysical jets are surveyed in a famous review by Begelman et al.[7]. An extensive treatment of laser plasma interactions is due to Kruer[45].

# Problems

1. Show that the scalar property of the distribution $f$ is consistent with (10.34) and Lorentz contraction.

2. At what temperature do you expect a stationary, nearly Maxwellian plasma to display relativistic behavior? State your answer in terms of electron Volts.

3. Write the $\alpha = 0$ component of (10.54) explicitly, using the ideal tensor of (10.33). Find the non-relativistic limit of this equation and compare it to the conventional energy conservation law.

4. Show that the spatial components of (10.33) reproduce (3.20) in the non-relativistic limit.

5. Since charge density is not a Lorentz scalar, the usual definition of quasineutrality is not Lorentz invariant. Find an invariant statement of quasineutrality.

6. Derive the relation between the scalar mass density $\rho_m$ and the rest–frame density $n_R$.

7. Verify directly that the general moment $M^{\alpha\cdot\gamma}$ is a Lorentz tensor, as follows. First express the moment in terms of an integral of the rest–frame distribution, $f_R$. Then transform the integration variable $p \to p' = \Lambda p$ to express $M^{\alpha\cdot\gamma}$ in terms of its rest–frame components.

8. Show that the Faraday tensor, considered as a matrix operator, has a two-dimensional null space in the magnetized limit, $W_2 \to 0$. It follows that there are two linearly independent solutions to (10.84). One of these corresponds to MHD flow, as show in the text. Find the other solution.

9. By deriving the nonrelativistic limit of the internal energy $u$ and mass density $\rho_m$, verify the limits

$$u \rightarrow mn \left( \frac{K_1}{K_2} + \frac{3T}{m} \right) = mn \left( 1 + \frac{3T}{2m} \right)$$

$$\rho_m \rightarrow mn \left( 1 - \frac{3T}{2m} \right)$$

# Appendix A

# Derivation of Balescu-Lenard operator

This Appendix provides a detailed derivation of the Balescu-Lenard equation [27], thus filling in the details of the analysis outlined in Chapter 7. For the sake of readability, the present treatment is mostly self-contained.

## A.1 Preliminaries

### Statistics

We consider an $N$-particle system with $N \gg 1$. (To reduce notational clutter we consider a single particle species; the generalization to multi-species is straightforward.) The corresponding $6N$ dimensional phase-space has coordinates $(\mathbf{z}_1, \mathbf{z}_2, \ldots \mathbf{z}_N)$, where $\mathbf{z}_i = (\mathbf{x}_i, \mathbf{v}_i)$, and distribution function

$$\mathcal{D}(\mathbf{z}_1, \mathbf{z}_2, \ldots \mathbf{z}_N)$$

This function describes an ensemble of identical systems and can be assumed to be smooth. Because the particles are also identical, $\mathcal{D}$ is symmetric under any permutation of its variables.

Any physical quantity $A$ describing the system is in principle a function of the $\mathbf{z}_i$ and therefore can be ensemble-averaged according to

$$\langle A \rangle = \int d^{6N} z \mathcal{D} A$$

Note the normalization

$$\int d^{6N} z \mathcal{D} = 1$$

We often denote the average by an overbar and make the standard decomposition

$$A = \bar{A} + \tilde{A}$$

307

Next consider the exact one-particle distribution

$$\mathcal{F}(\mathbf{z}, t; \mathbf{z}_1, \dots \mathbf{z}_N) = \sum_i \delta(\mathbf{z} - \mathbf{z}_i(t))$$

where $\mathbf{z}_i(t)$ denotes the exact trajectory of the $ith$ particle. Notice that the normalization is

$$\int d^6 z \mathcal{F} = N$$

The smooth distribution that describes the system generically is

$$\langle \mathcal{F} \rangle = \int d^6 z_1 \cdots d^6 z_N \mathcal{D} \sum_i \delta(\mathbf{z} - \mathbf{z}_i(t))$$

Here the integral immediately decomposes into $N$ identical integrals (since the particles are identical) and we have

$$\langle \mathcal{F} \rangle = N \int d^6 z_2 \cdots d^6 z_N \mathcal{D}(\mathbf{z}, \mathbf{z}_2, \dots, \mathbf{z}_N)$$

This agrees with more conventional definitions of the one-particle distribution. From here on we use the simplified notation

$$\langle \mathcal{F} \rangle \equiv f$$

whence

$$\mathcal{F} = f + \tilde{f}.$$

Also useful is the two-particle function

$$\langle \mathcal{F}(\mathbf{z})\mathcal{F}(\mathbf{z}') \rangle = \int d^6 z_1 \cdots d^6 z_N \left[ \delta(\mathbf{z} - \mathbf{z}_1) + \cdots + \delta(\mathbf{z} - \mathbf{z}_N) \right]$$
$$\times \left[ \delta(\mathbf{z}' - \mathbf{z}_1) + \cdots + \delta(\mathbf{z}' - \mathbf{z}_N) \right] \mathcal{D}(\mathbf{z}_1, \dots, \mathbf{z}_N)$$

Because $\mathcal{D}$ is symmetrical under exchanges of the $\mathbf{z}_i$, this sum consists of $N$ identical diagonal terms, and $N(N-1)$ identical non-diagonal terms. Thus we have

$$\langle \mathcal{F}(\mathbf{z})\mathcal{F}(\mathbf{z}') \rangle = N\delta(\mathbf{z} - \mathbf{z}') \int d^6 z_2 \cdots d^6 z_N \mathcal{D}(\mathbf{z}, \mathbf{z}_2, \dots, \mathbf{z}_N)$$
$$+ N(N-1) \int d^6 z_1 \cdots d^6 z_N \delta(\mathbf{z} - \mathbf{z}_1)\delta(\mathbf{z}' - \mathbf{z}_2)\mathcal{D}(\mathbf{z}_1, \dots, \mathbf{z}_N)$$

That is,

$$\langle \mathcal{F}(\mathbf{z})\mathcal{F}(\mathbf{z}') \rangle = \delta(\mathbf{z} - \mathbf{z}')f(\mathbf{z}) + N(N-1) \int d^6 z_3 \cdots d^6 z_N \mathcal{D}(\mathbf{z}, \mathbf{z}', \dots, \mathbf{z}_N) \quad (A.1)$$

**Uncorrelated particles** As an example we consider the case when the N particles move independently, so that

$$\mathcal{D}(\mathbf{z}_1, \mathbf{z}_2, \dots \mathbf{z}_N) = \prod_{i=1}^{N} d(\mathbf{z}_i)$$

We then compute

$$f = \langle \mathcal{F} \rangle = N d(\mathbf{z}) \int d(\mathbf{z}_2) d^6 z_2 \cdots \int d(\mathbf{z}_N) d^6 z_N$$

Here the integral is $N$-factors of unity, so we have

$$d(\mathbf{z}) = \frac{1}{N} f$$

That is,

$$\mathcal{D}(\mathbf{z}_1, \dots, \mathbf{z}_N) = N^{-N} f(\mathbf{z}_1) \cdots f(\mathbf{z}_N)$$

in the uncorrelated case.

The form of the two-particle distribution in the uncorrelated case follows immediately from the general version (A.1):

$$\langle \mathcal{F}(\mathbf{z}) \mathcal{F}(\mathbf{z}') \rangle = \delta(\mathbf{z} - \mathbf{z}') f(\mathbf{z}) + N(N-1) d(\mathbf{z}) d(\mathbf{z}')$$

or, neglecting a term of order $1/N$,

$$\langle \mathcal{F}(\mathbf{z}) \mathcal{F}(\mathbf{z}') \rangle = \delta(\mathbf{z} - \mathbf{z}') f(\mathbf{z}) + f(\mathbf{z}) f(\mathbf{z}')$$

It is easily seen that this relation implies the important result

$$\langle \tilde{f}(\mathbf{z}) \tilde{f}(\mathbf{z}') \rangle = \delta(\mathbf{z} - \mathbf{z}') f(\mathbf{z}) \tag{A.2}$$

## Linear kinetic theory

We assume that fluctuations from the mean distribution are small: $\tilde{f} \ll f$. We also consider a plasma with neither magnetic field nor equilibrium electric field, $\mathbf{E} = \tilde{\mathbf{E}}$, and with at most slow spatial variation. Then the fluctuating distribution can be computed from the linearized Vlasov equation,

$$\frac{\partial \tilde{f}}{\partial t} + \mathbf{v} \cdot \nabla \tilde{f} = -\frac{e}{m} \mathbf{E} \cdot \frac{\partial f}{\partial \mathbf{v}}$$

Solution by characteristics gives

$$\tilde{f}(\mathbf{x}, \mathbf{v}, t) = \tilde{f}(\mathbf{x} - \mathbf{v}t, \mathbf{v}, 0) - \frac{e}{m} \int_0^\infty d\tau \mathbf{E}(\mathbf{x} - \mathbf{v}\tau, t - \tau) \cdot \frac{\partial f}{\partial \mathbf{v}} \tag{A.3}$$

Of course this result is incomplete without specification of the electric field. In this regard we recall the Fourier-transformed results,

$$f_k = \frac{if_k(\mathbf{v}, 0)}{\omega - ku} - \frac{ie}{m} \frac{f_\mathbf{v}}{\omega - ku} \cdot \mathbf{E}_k \tag{A.4}$$

$$\phi_k = \frac{ie}{\epsilon_0 k^2 K(\omega)} \int d^3v \frac{f_k(\mathbf{v}, 0)}{\omega - ku} \tag{A.5}$$

Here the transform is denoted by

$$f_k \equiv \int_0^\infty dt \int d^3x e^{i(\omega t - \mathbf{k} \cdot \mathbf{x})} f(\mathbf{x}, \mathbf{v}, t)$$

Notice that it is a function of $\mathbf{k}$ and $\omega$; on the other hand, the initial value $f_k(\mathbf{v}, 0)$ depends only on $\mathbf{k}$. We also abbreviate $f_\mathbf{v} = \partial f / \partial \mathbf{v}$ and

$$ku \equiv \mathbf{k} \cdot \mathbf{v}$$

Finally we use the electrostatic potential, so that $\mathbf{E}_k = -i\mathbf{k}\phi_k$, and we have recalled the dispersion function,

$$K(\omega) = K(\omega, \mathbf{k}) = 1 + \frac{e^2}{\epsilon_0 mk^2} \int d^3v \frac{\mathbf{k} \cdot f_\mathbf{v}}{\omega - ku} \tag{A.6}$$

## Dispersion function

Here we decompose the function $K$ of (A.6) into its real and imaginery parts, in the limiting case of real frequency:

$$\omega_i \to 0$$

with $\omega_i \equiv Re(\omega) > 0$. This limit is significant when there are no unstable normal modes—when $K$ has no zeroes in the upper half $\omega$-plane. Denoting the component of $\mathbf{v}$ that is parallel to $\mathbf{k}$ by $u$ as usual, and the other two components of $\mathbf{v}$ by $\mathbf{w}$, we have

$$K(\omega) = 1 + \frac{e^2}{\epsilon_0 mk^2} \int d^2w \int du \frac{f_u}{\omega - ku}$$

As $\omega$ approaches the real axis from above, we have

$$\frac{1}{\omega - ku} \to \frac{1}{\omega - ku + i0} = P\left(\frac{1}{\omega - ku}\right) - i\pi\delta(\omega - ku)$$

Hence in this limit we have

$$K = K_r + iK_i$$

with

$$K_r(\omega) = 1 + \frac{e^2}{\epsilon_0 mk^2} \int d^3v (\mathbf{k} \cdot f_\mathbf{v}) P\left(\frac{1}{\omega - \mathbf{k} \cdot \mathbf{v}}\right) \tag{A.7}$$

and

$$K_i(\omega) = -\pi \frac{e^2}{\epsilon_0 mk^2} \int d^3v (\mathbf{k} \cdot f_\mathbf{v}) \delta(\omega - \mathbf{k} \cdot \mathbf{v}) \tag{A.8}$$

## Collision operator

The collision operator represents the correlation between the fluctuating electric force and the fluctuating distribution. It enters the ensemble–averaged kinetic equation through the term

$$C = -\langle \frac{e}{m}\mathbf{E} \cdot \frac{\partial \tilde{f}}{\partial \mathbf{v}} \rangle \qquad (A.9)$$

Our objective is to express this correlation in terms of the averaged distribution, $f$.

From Fokker-Planck theory we know that the collision operator consists of diffusion in velocity space, with diffusion tensor $D_{\alpha\beta}$, and dynamical friction. These contributions come from the second and first terms of (A.3) respectively:

$$
\begin{aligned}
C &= -\frac{e}{m}\frac{\partial}{\partial v_\alpha}\langle E_\alpha(\mathbf{x},t)\tilde{f}(\mathbf{x}-\mathbf{v}t,\mathbf{v},0)\rangle \\
&+ \frac{e^2}{m^2}\frac{\partial}{\partial v_\alpha}\int_0^\infty d\tau \langle E_\alpha(\mathbf{x},t)E_\beta(\mathbf{x}-\mathbf{v}\tau,t-\tau)\rangle \frac{\partial f}{\partial v_\beta}
\end{aligned}
\qquad (A.10)
$$

It is noteworthy that the two terms in (A.10) have equal importance. Note that the quasilinear theory discussed in Chapter 9 omits the initial value term. This is because quasilinear theory follows the growth and saturation of an unstable mode, while the present discussion assumes a stable plasma.

## A.2 Diffusion tensor

Next we study the diffusion tensor,

$$D_{\alpha\beta} = \frac{e^2}{m^2}\int_0^\infty d\tau \langle E_\alpha(\mathbf{x},t)E_\beta(\mathbf{x}-\mathbf{v}\tau,t-\tau)\rangle$$

We assume that the fluctuation statistics are stationary, so that

$$\langle E_\alpha(\mathbf{x},t)E_\beta(\mathbf{x}-\mathbf{v}\tau,t-\tau)\rangle = \langle E_\alpha(\mathbf{v}\tau,\tau)E_\beta(0,0)\rangle \equiv C_{\alpha\beta}(\tau) \qquad (A.11)$$

Here $C_{\alpha\beta}(\tau)$ is the autocorrelation of the electric field; the width of its support in $\tau$ measures the autocorrelation time $\tau_c$. Since the $\tau-$integral can be taken to start at a time slightly greater than zero, we see that the diffusion tensor,

$$D_{\alpha\beta} = \frac{e^2}{m^2}\int_0^\infty d\tau C_{\alpha\beta}(\tau) \qquad (A.12)$$

vanishes in the limit of vanishing correlation time. Indeed we can estimate

$$D_{\alpha\beta} \sim \frac{e^2}{m^2}\langle E^2\rangle \tau_c$$

We next express the electric field in terms of the Fourier-transformed potential to find that

$$C_{\alpha\beta}(\tau) = (2\pi)^{-8} \int d^3k d^3k' d\omega d\omega' e^{i(\omega-ku)\tau} k_\alpha k'_\beta \langle \phi_k \phi^*_{k'} \rangle \qquad (A.13)$$

Here we use the complex conjugate for convenience as usual. Notice that the $\omega$-integral is performed along the contour $C_+$, in the upper–half plane, but because of complex conjugation the $\omega'$-integral is performed below the real-$\omega$ axis.

It is straightforward, using (A.5), to express the potential correlation in (A.13) in terms of the correlation of the initial fluctuation of the distribution:

$$\begin{aligned}
C_{\alpha\beta}(\tau) &= \frac{e^2}{\epsilon_0^2 (2\pi)^8} \int d^3k d^3k' d\omega d\omega' d^3v' d^3v'' \\
&\times \frac{k_\alpha k'_\beta e^{i(\omega-ku)\tau} \langle f_k(\mathbf{v}',0) f_{k'}(\mathbf{v}'',0) \rangle}{(\omega-ku)(\omega-ku')K(\omega)K(\omega')k^2 k'^2}
\end{aligned}$$

At this point we introduce the key statistical *ansatz*: that the initial statistics are uncorrelated. That is, in order to calculate the correlations that arise from the fluctuating field and its interaction with particles, we assume that no correlations exist at $t = 0$, before the field has acted. Recalling (A.2), we write our *ansatz* explicitly as

$$\langle \tilde{f}(\mathbf{x},\mathbf{v},0) \tilde{f}(\mathbf{x}',\mathbf{v}',0) \rangle = \delta(\mathbf{x}-\mathbf{x}')\delta(\mathbf{v}-\mathbf{v}')f(\mathbf{v})$$

This relation is equivalent to

$$\langle f_k(\mathbf{v},0) f^*_{k'}(\mathbf{v}',0) \rangle = (2\pi)^3 \delta(\mathbf{k}-\mathbf{k}')\delta(\mathbf{v}-\mathbf{v}')f(\mathbf{v}) \qquad (A.14)$$

After substituting (A.14) into our expression for $C_{\alpha\beta}$ we can integrate over $\mathbf{k}'$ and $\mathbf{v}''$ to obtain the form

$$C_{\alpha\beta}(\tau) = \frac{e^2}{\epsilon_0^2 (2\pi)^5} \int d^3k d\omega d\omega' d^3v' \frac{f(\mathbf{v}')k_\alpha k_\beta e^{i(\omega-ku)\tau}}{(\omega-ku')(\omega'-ku')K(\omega)K^*(\omega')k^4} \qquad (A.15)$$

Next we perform the integral over $\omega$. Here of course we analytically continue the dispersion function $K(\omega)$ through its cut on the real axis; for simplicity the continuation is also denoted by $K$. Since the plasma is assumed stable, $K(\omega)$ has no zeroes in the upper half $\omega$–plane; its zeroes in the lower–half plane will contribute exponentially damped terms that we ignore. Thus, when the $\omega$-contour is closed by a large semi-circle in the lower half–plane, the dominant contribution at long times comes from the (ballistic) pole at $\omega = ku'$:

$$\int d\omega \frac{e^{i\omega\tau}}{(\omega-ku')K(\omega)} = -2\pi i \frac{e^{iku'\tau}}{K(ku')}$$

The $\omega'$ integral is the complex conjugate of the $\omega$ integral, evaluated in the limit $\tau \to 0$ (with $\tau > 0$ for the initial value problem). Thus

$$\int d\omega' \frac{1}{(\omega-ku')K^*(\omega')} = 2\pi i \frac{1}{K^*(ku')}$$

Substituting these results into (A.15) we obtain

$$C_{\alpha\beta}(\tau) = \frac{e^2}{\epsilon_0^2 (2\pi)^3} \int d^3k d^3v' f(\mathbf{v}') \frac{k_\alpha k_\beta e^{ik(u'-u)\tau}}{|K(ku')|^2 k^4} \tag{A.16}$$

Now the diffusion tensor of (A.12) has become

$$D_{\alpha\beta} = \frac{e^4}{(2\pi)^3 m^2 \epsilon_0^2} \int d^3k d^3v' \frac{k_\alpha k_\beta f(\mathbf{v}')}{|K(ku')|^2 k^4} \int_0^\infty d\tau e^{ik(u'-u)\tau}$$

To evaluate the $\tau$–integral we replace

$$k(u' - u) \to k(u' - u) + i0$$

and find

$$D_{\alpha\beta} = \frac{ie^4}{(2\pi)^3 m^2 \epsilon_0^2} \int d^3k d^3v' \frac{k_\alpha k_\beta f(\mathbf{v}')}{|K(ku')|^2 k^4} \left[ P\left(\frac{1}{k(u'-u)}\right) - i\pi\delta(k(u'-u)) \right]$$

Here the principal value term cannot contribute because it is odd in $\mathbf{k}$; therefore

$$D_{\alpha\beta} = \frac{e^4}{8\pi^2 m^2 \epsilon_0^2} \int d^3k d^3v' \frac{k_\alpha k_\beta f(\mathbf{v}')}{|K(ku')|^2 k^4} \delta(\mathbf{k} \cdot (\mathbf{v}' - \mathbf{v})) \tag{A.17}$$

# A.3  Dynamical friction

The dynamical friction is given by the first term in (A.10):

$$-\frac{\partial}{\partial v_\alpha} R_\alpha f(\mathbf{v})$$

where

$$R_\alpha f(\mathbf{v}) \equiv \frac{e}{m} \langle E_\alpha(\mathbf{x}, t) \tilde{f}(\mathbf{x} - \mathbf{v}t, \mathbf{v}, 0) \rangle$$

It is easy to express $R_\alpha$ in terms of the Fourier transforms:

$$R_\alpha f(\mathbf{v}) = -i(2\pi)^{-7} \frac{e}{m} \int d^3k d^3k' d\omega e^{-i(\omega t - \mathbf{k}\cdot\mathbf{x} + \mathbf{k}'\cdot\mathbf{x} - \mathbf{k}'\cdot\mathbf{v}t)} \mathbf{k} \langle \phi_k f_{k'}^*(\mathbf{v}, 0) \rangle$$

Then we can recall (A.5) to write

$$\langle \phi_k f_{k'}^*(\mathbf{v}, 0) \rangle = \frac{ie}{\epsilon_0 k^2} \int d^3v' \frac{\langle f_k(\mathbf{v}', 0) f_{k'}^*(\mathbf{v}, 0) \rangle}{K(\omega)(\omega - \mathbf{k}\cdot\mathbf{v}')}$$

Here the correlation in the integrand is given by (A.14), which makes the integral trivial and yields

$$\langle \phi_k f_{k'}^*(\mathbf{v}, 0) \rangle = \frac{ie}{\epsilon_0 k^2} \frac{(2\pi)^3 \delta(\mathbf{k} - \mathbf{k}') f(\mathbf{v})}{K(\omega)(\omega - \mathbf{k}\cdot\mathbf{v})}$$

After substitution we have

$$R_\alpha = \frac{e^2}{(2\pi)^4\epsilon_0 m} \int \frac{d^3k\, d\omega\, e^{-i(\omega - ku)t} k_\alpha}{k^2 K(\omega)(\omega - ku)}$$

or, after performing the $\omega$–integral,

$$R_\alpha = \frac{-ie^2}{(2\pi)^3\epsilon_0 m} \int \frac{d^3k\, k_\alpha}{k^2 K(ku)} \tag{A.18}$$

It is remarkable that this expression can be put in a form closely parallel to the diffusion tensor. To obtain the parallel form, we write

$$\frac{1}{K} = \frac{K_r - iK_i}{|K|^2}$$

and recall (A.7) and (A.8). We note in particular that $K_r(ku)$ is even in $\mathbf{k}$, so it does not contribute to $R_\alpha$. There remains

$$R_\alpha = \frac{e^4 f(\mathbf{v})}{8\pi^2\epsilon_0^2 m^2} \int \frac{d^3k\, d^3v'\, k_\alpha k_\beta \delta(\mathbf{k}\cdot(\mathbf{v} - \mathbf{v}'))}{k^4 |K(\mathbf{k}\cdot\mathbf{v})|^2} \frac{\partial f}{\partial v'_\beta} \tag{A.19}$$

## A.4   Conclusion

We insert (A.17) and (A.19) into (A.10), or, equivalently, into

$$C(f, f) = \frac{\partial}{\partial v_\alpha} D_{\alpha\beta} \frac{\partial f}{\partial v_\beta} - \frac{\partial(R_\alpha f)}{\partial v_\alpha}$$

and obtain the Balescu-Lenard collision operator. We notice that both terms contain the same over–all constant

$$c_0 \equiv \frac{e^4}{8\pi^2\epsilon_0^2 m^2},$$

as well as the same $k$–integral:

$$K_{\alpha\beta}(\mathbf{v}, \mathbf{v}') \equiv \int d^3k \frac{k_\alpha k_\beta \delta(\mathbf{k}\cdot(\mathbf{v} - \mathbf{v}'))}{k^4 |K(\mathbf{k}\cdot\mathbf{v})|^2}$$

Hence the combined expression is relatively simple:

$$C(f, f) = c_0 \frac{\partial}{\partial v_\alpha} \int d^3v'\, K_{\alpha\beta}(\mathbf{v}, \mathbf{v}') \left[ f(\mathbf{v}') \frac{\partial f(\mathbf{v})}{\partial v_\beta} - f(\mathbf{v}) \frac{\partial f(\mathbf{v}')}{\partial v'_\beta} \right] \tag{A.20}$$

# Bibliography

[1] J. Ahearne and C. Surko. Plasma science: from fundamental research to technological applications. Technical report, National Research Council, 1995.

[2] H. Alfvén. *Cosmical Electrodynamics*. Clarendon Press, Oxford, 1962.

[3] W. P. Allis, S. J. Buchsbaum, and A. Bers. *Waves in anisotropic plasma*. MIT Press, Cambridge MA., 1963.

[4] A. M. Anile. *Relativistic fluids and magneto-fluids: with applications in astrophysics and plasma physics*. Cambridge University Press, Cambridge, 1989.

[5] R. Balescu. *Transport Processes in Plasmas*, volume 1. North-Holland, Amsterdam, 1988.

[6] G. Bateman. *MHD Instabilities*. MIT Press, Cambridge, 1980.

[7] M. C. Begelman, R. D. Blandford, and M. J. Rees. Theory of extragalactic radio sources. *Reviews of Modern Physics*, 56:255–351, 1984.

[8] I. B. Bernstein. Geometric optics in space- and time-varying plasma. *Physics of Fluids*, 18:320–324, 1975.

[9] I. B. Bernstein, J. M. Greene, and M. D. Kruskal. Exact nonlinear plasma oscillations. *Physical Review*, 108:546, 1957.

[10] Dieter Biskamp. *Nonlinear Magnetohydrodynamics*. Cambridge University Press, 1993.

[11] D. Bohm. *The Characterisitics of Electrical Discharges in Magnetic Fields*. McGraw-Hill, New York, 1949.

[12] A. H. Boozer. Time-dependent drift Hamiltonian. *The Physics of Fluids*, 27:2441–2445, 1984.

[13] S. I. Braginskii. Transport processes in a plasma. In M. A. Leontovich, editor, *Reviews of Plasma Physics*, volume 1. Consultants Bureau, New York, 1965.

315

[14] E. Buckingham. On physically similar systems. *Physical Review*, 4:345, 1914.

[15] S. Chapman and T. G. Cowling. *The Mathematical Theory of Non-uniform Gases*. Cambridge University Press, London, 2nd edition, 1953.

[16] F. F. Chen. *Introduction to Plasma Physics and Controlled Fusion*, volume 1. Plenum Press, New York, 1974.

[17] G. L. Chew, M. L. Goldberger, and F. E. Low. The Boltzmann equation and the one-fluid hydromagnetic equations in the absence of particle collisions. *Proceedings of the Royal Society of London A*, 236:112–118, 1956.

[18] P. C. Clemmow and J. P. Dougherty. *Electrodynamics of Particles and Plasmas*. Addison Wesley, Reading, Mass., 1969.

[19] J. Dawson. On Landau damping. *Physics of Fluids*, 4:869, 1961.

[20] S. R. de Groot and P. Mazur. *Non-equilibrium Thermodynamics*. North-Holland Publishing C., Amsterdam, 1962.

[21] S. R. de Groot, W. A. van Leeuwen, and Ch. G. van Weert. *Relativistic kinetic theory: principles and applications*. North Holland, Amsterdam, 1980.

[22] T. H. Dupree. Kinetic theory of plasma and the electromagnetic field. *Physics of Fluids*, 6:1714, 1963.

[23] T. H. Dupree. A perturbation theory for strong plasma turbulence. *Physics of Fluids*, 9:1733, 1966.

[24] D. I. Dzhavakhishvili and N. L. Tsintsadze. Transport phenomena in a completely ionized ultrarelativistic plasma. *Zh. Eksp. Teor. Fiz.*, 64:1314–1325, 1973.

[25] D. M. Fradkin. Covariant electromagnetic projection operators and a covariant description of charged particle guidiing centre motion projection operators and a covariant dexcription of charged particle guiding centre motion. *Journal of Physics A*, 11:1069, 1978.

[26] J. P. Freidberg. *Ideal Magnetohyrodynamics*. Plenum Press, New York, 1987.

[27] B. D. Fried. Statistical mechanical foundations. In W. B. Kunkel, editor, *Plasma Physics in Theory and Application*. McGraw-Hill, New York, 1966.

[28] A. Galeev and R. Z. Sagdeev. Theory of weakly turbulent plasma. In *Handbook of Plasma Physics, Volume 1: Basic Plasma Physics I*. edited by A. A. Galeev and R. N. Sudan, North Holland, Amsterdam, 1983.

[29] C. S. Gardner. Bound on the energy available from a plasma. *Physics of Fluids*, 6:839–840, 1963.

[30] R. J. Goldston and P. H. Rutherford. *Introduction to Plasma Physics*. Institute of Physics Publishing, Bristol and Philadelphia, 1995.

[31] Harold Grad. Plasmas. *Physics Today*, 19:34–44, December 1969.

[32] G. W. Hammett, W. Dorland, M. A. Beer, and F. W. Perkins. The gyrofluid approach to simulating tokamak turbulence. In *New Ideas in Tokamak Confinement*, pages 199–216, 1993.

[33] R. D. Hazeltine and S. M. M. Mahajan. Fluid description of relativistic, magnetized plasma. *The Astrophysics Journal*, 567:1262–1271, March 2002.

[34] R. D. Hazeltine and J. D. Meiss. Shear-Alfvén dynamics of toroidally confined plasmas. *Physics Reports*, 121:1–164, 1985.

[35] F. L. Hinton. Collisional transport in plasmas. In A. A. Galeev and R. N. Sudan, editors, *Basic Plasma Physics*, volume 1. North Holland, Amsterdam, 1983.

[36] W. Horton. Nonlinear drift waves and transport in magnetized plasma. *Physics Reports*, 192:1–177, 1989.

[37] Fred Hoyle. *The Black Cloud*. Harper, New York, 1957.

[38] S. Ichimaru. *Statistical Plasma Physics*, volume 1. Addison-Wesley, Reading, Massachussetts, 1992.

[39] S. Ichimaru. *Statistical Plasma Physics*, volume 2. Addison-Wesley, Reading, Massachussetts, 1994.

[40] B. B. Kadomtsev. Hydromagnetic stability of a plasma. In *Reviews of Plasma Physics*, volume 2. Consultants Bureau, New York, 1965.

[41] Allan N. Kaufman. Reformulation of quasi-linear theory. *Journal of Plasma Physics*, 8:1–5, 1972.

[42] N. A. Krall. Drift waves. In *Advances in Plasma Physics*. Interscience, New York, 1968.

[43] N. A. Krall and A. W. Trivelpiece. *Principles of Plasma Physics*. McGraw Hill, New York, 1973.

[44] John A. Krommes. Statistical descriptions and plasma physics. In *Handbook of Plasma Physics, Volume 2: Basic Plasma Physics II*. edited by A. A. Galeev and R. N. Sudan, North Holland, Amsterdam, 1983.

[45] William L. Kruer. *The Physics of Laser Plasma Ineractions*. Addison Wesley, Reading, Mass., 1988.

[46] M. Kruskal. Asymptotic theory of Hamiltonian and other systems with all solutions nearly periodic. *Journal of Mathematical Physics*, 3:806, 1962.

[47] M. D. Kruskal and C. R. Oberman. On the stability of plasma in static equilibrium. *The Physics of Fluids*, 1:275–280, 1958.

[48] L. D. Landau. On the vibrations of the electronic plasma. *Journal of Physics (U.S.S.R.)*, 10:25, 1946.

[49] L. D. Landau and E. M. Lifschitz. *Fluid Mechanics*. Addison-Wesley, Reading, Mass., 1959.

[50] R. G. Littlejohn. Hamiltonian formulation of guiding center motion. *Journal of Plasma Physics*, 24:1730–1749, 1981.

[51] R. G. Littlejohn. Variational principles of guiding center motion. *Journal of Plasma Physics*, 29:111–125, 1983.

[52] A. I. Morozov and L. S. Solov'ev. Motion of charged particles in electromagnetic fields. In *Reviews of Plasma Physics*, volume 2. Consultants Bureau, New York, 1966.

[53] Dwight R. Nicholson. *Introduction to Plasma Theory*. Krieger Publishing Company, Malabar, Florida, 1983.

[54] T. G. Northrop. *The Adiabatic Motion of Charged Particles*. Wiley, New York, 1963.

[55] T. M. O'Neil, J. H. Winfrey, and J. H. Malmberg. Nonlinear interaction of a small cold beam and a plasma. *Physics of Fluids*, 14:1204, 1971.

[56] E. N. Parker. The solar flare phenomenom and the theory of reconnection and annihilation of magnetic fields. *Astrophysics Journal supplement series*, 8:177–211, 1963.

[57] Eugene N. Parker. Dynamics of the interplanetary gas and magnetic fields. *Astrophysics Journal*, 128:664, 1958.

[58] O. Penrose. Electrostatic instabilities of a uniform non-Maxwellian plasma. *Physics of Fluids*, 3:258, 1960.

[59] K. V. Roberts and J. B. Taylor. Magnetohydrodynamic equations for finite Larmor radius. *Physical Review Letters*, 8:197–198, 1962.

[60] B. B. Robinson and I. B. Bernstein. A variational description of transport phenomena in a plasma. *Annals of Physics*, 18:110–169, 1962.

[61] M. N. Rosenbluth, William M. MacDonald, and David L. Judd. Fokker-Planck equation for an inverse-square force. *The Physical Review*, 107:1–6, 1957.

[62] M. N. Rosenbluth and N. Rostoker. Theoretical structure of plasma equations. *The Physics of Fluids*, 2:23–30, 1959.

[63] M. N. Rosenbluth and A. Simon. Finite Larmor radius equations with nonuniform electric fields and velocities. *The Physics of Fluids*, 8:1300–1325, 1965.

[64] R. Z. Sagdeev and A. Galeev. *Nonlinear Plasma Theory*. Benjamin, New York, 1969.

[65] G. Schmidt. *Physics of High Temperature Plasmas*. Academic Press, New York, 1979.

[66] V. D. Shafranov. Plasma equilibrium in a magnetic field. In *Reviews of Plasma Physics*, volume 2. Consultants Bureau, New York, 1966.

[67] Lyman Spitzer, Jr. *Physics of Fully Ionized Gases*. John Wiley and Sons, New York, 1962.

[68] T. H. Stix. *Waves in Plasmas*. American Institute of Physics, New York, 1992.

[69] L. R. O. Storey. An investigation of whistling atmospherics. *Philosophical Transactions of the Royal Society London, Ser. A*, 246:113, 1953.

[70] P. A. Sturrock. *Plasma Physics*. Cambridge University Press, Cambridge, 1994.

[71] S. I. Syrovatskii. Formation of current sheets in a plasma with a frozen-in strong magnetic field. *Soviet Physics JETP*, 33:933–940, 1971.

[72] Michael Tendler and Daniel Heifetz. Neutral particle kinetics in fusion devices. *Fusion Technology*, 11:289–310, 1987.

[73] N. G. Van Kampen. Stochastic differential equations. *Physics Reports*, 24:172–228, 1976.

[74] N. G. Van Kampen and B. U. Felderhof. *Theoretical Methods in Plasma Physics*. North Holland, Amsterdam, 1967.

[75] Steven Weinberg. Eikonal method in magnetohydrodynamics. *Physical Review*, 126:1899–1909, 1962.

[76] Steven Weinberg. *Gravitation and Cosmology: Principles and Applications of the General Theory of Relativity*. John Wiley & Sons, 1972.

[77] G. B. Whitham. Linear and nonlinear waves. *Journal of Fluid Mechanics*, 22:273–283, 1974.

[78] G. B. Whitham. *Linear and Nonlinear Waves*. John Wiley & Sons, 1974.

# Index